Springer Tracts in Modern Physics 94

Springer Tracts in Modern Physics

73 **Excitons at High Density** Edited by H. Haken, S. Nikitine. With contributions by V. S. Bagaev, J. Biellmann, A. Bivas, J. Goll, M. Grosmann, J. B. Grun, H. Haken, E. Hanamura, R. Levy, H. Mahr, S. Nikitine, B. V. Novikov, E. I. Rashba, T. M. Rice, A. A. Rogachev, A. Schenzle, K. L. Shaklee

74 **Solid-State Physics** With contributions by G. Bauer, G. Borstel, H. J. Falge, A. Otto

75 **Light Scattering by Phonon-Polaritons** By R. Claus, L. Merten, J. Brandmüller

76 **Irreversible Properties of Type II Superconductors** By H. Ullmaier

77 **Surface Physics** With contributions by K. Müller, P. Wißmann

78 **Solid-State Physics** With contributions by R. Dornhaus, G. Nimtz, W. Richter

79 **Elementary Particle Physics** With contributions by E. Paul, H. Rollnick, P. Stichel

80* **Neutron Physics** With contributions by L. Koester, A. Steyerl

81 **Point Defects in Metals I:** Introductions to the Theory (2nd Printing) By G. Leibfried, N. Breuer

82 **Electronic Structure of Noble Metals, and Polariton-Mediated Light Scattering** With contributions by B. Bendow, B. Lengeler

83 **Electroproduction at Low Energy and Hadron Form Factors** By E. Amaldi, S. P. Fubini, G. Furlan

84 **Collective Ion Acceleration** With contributions by C. L. Olson, U. Schumacher

85 **Solid Surface Physics** With contributions by J. Hölzl, F. K. Schulte, H. Wagner

86 **Electron-Positron Interactions** By B. H. Wiik, G. Wolf

87 **Point Defects in Metals II:** Dynamical Properties and Diffusion Controlled Reactions With contributions by P. H. Dederichs, K. Schroeder, R. Zeller

88 **Excitation of Plasmons and Interband Transitions by Electrons** By H. Raether

89 Giant Resonance Phenomena in **Intermediate-Energy Nuclear Reactions** By F. Cannata, H. Überall

90* **Jets of Hadrons** By W. Hofmann

91 **Structural Studies of Surfaces** With contributions by K. Heinz, K. Müller, T. Engel, and K. H. Rieder

92 **Single-Particle Rotations in Molecular Crystals** By W. Press

93 **Coherent Inelastic Neutron Scattering in Lattice Dynamics** By B. Dorner

94 **Exciton Dynamics in Molecular Crystals and Aggregates** With contributions by V. M. Kenkre and P. Reineker

95 **Projection Operator Techniques** in Nonequilibrium Statistical Mechanics By H. Grabert

96 **Hyperfine Structure in 4d- and 5d-Shell Atoms** By S. Büttgenbach

97 **Elements of Flow and Diffusion Processes in Separation Nozzles** By W. Ehrfeld

98 **Narrow-Gap Semiconductors** With contributions by G. Nimtz and B. Schlicht

99 **Property of IV—VI Compounds** With contributions by H. Bilz, A. Bussmann-Holder and W. Jantsch

100* **Quarks and Nuclear Forces** Edited by D.C. Fries and B. Zeitnitz

---

* denotes a volume which contains a Classified Index starting from Volume 36.

Exciton Dynamics in Molecular Crystals and Aggregates

Contributions by
V. M. Kenkre   P. Reineker

With 37 Figures

Springer-Verlag
Berlin Heidelberg GmbH 1982

Professor Dr. Vasudev M. Kenkre
Department of Physics and Astronomy, The University of Rochester
Rochester, N.Y. 14627, USA

Professor Dr. Peter Reineker
Universität Ulm, Abteilung für Theoretische Physik I,
D-7900 Ulm, Fed. Rep. of Germany

Manuscripts for publication should be addressed to:

Gerhard Höhler
Institut für Theoretische Kernphysik der Universität Karlsruhe
Postfach 6380, D-7500 Karlsruhe 1, Fed. Rep. of Germany

Proofs and all correspondence concerning papers in the process of publication should be addressed to:

Ernst A. Niekisch
Haubourdinstrasse 6, D-5170 Jülich 1, Fed. Rep. of Germany

ISBN 978-3-662-15774-9 ISBN 978-3-540-39068-8 (eBook)
DOI 10.1007/978-3-540-39068-8

Library of Congress Cataloging in Publication Data. Kenkre, V. M. (Vasudev M.), 1946–. Exciton dynamics in molecular crystals and aggregates. (Springer tracts in modern physics; 94). Bibliography: p. Includes index. 1. Exciton theory. 2. Molecular crystals. I. Reineker, P. (Peter), 1940– . II. Title. III. Series. [QCI.S797] vol.94 [QC176.8.E9]
539s [530.4'1] 81-23246 AACR2

2153/3130 – 5 4 3 2 1 0

Contents

The Master Equation Approach: Coherence, Energy Transfer, Annihilation, and Relaxation

By *V.M. Kenkre*. With 9 Figures

Stochastic Liouville Equation Approach: Coupled Coherent and Incoherent Motion, Optical Line Shapes, Magnetic Resonance Phenomena

By *P. Reineker*. With 28 Figures

The Master Equation Approach: Coherence, Energy Transfer, Annihilation, and Relaxation

By V. M. Kenkre

1. Background

1.1 Introduction

Exciton motion in molecular systems is both an important and an incompletely understood field of physics and has been undergoing active theoretical and experimental investigation. Many reviews /1.1-5/ on various aspects of the subject have appeared recently. The field derives its importance from its being a part of the general area of energy transfer and its consequent connections with a variety of disciplines, even those outside physics, such as the study of photosynthesis in biology /1.6/. From the theorist's point of view the field is a unique meeting place of non-equilibrium statistical mechanics, specifically transport theory, and of the subject of the interaction of light with matter: the creation as well as the radiative decay of excitons are part of the latter, whereas the motion belongs to the former. This review describes the theoretical work done recently in this field within the framework of the master equation approach. Master equations have found wide use in various contexts and several detailed expositions of them exist in the literature /1.7-11/.

This review is divided into three parts. The first part, the background, introduces the subject and states the questions to be addressed. The second part is concerned with the specific problem of describing, in a unified manner, the coherent and incoherent motion of excitons and employs the method of generalized master equations (GME). This is the largest part of the review and covers Chaps. 2 and 3. Chapter 2 describes the motivation for the GME method and its general formalism, including the relations which it bears to other approaches such as the stochastic Liouville equation (SLE) approach reviewed in the following article in this book. The third part deals with questions other than that of coherence and employs not the GME method, but techniques based on ordinary Master equations.

1.2 The System and the Questions

The physical system under investigation is an aggregate of molécules, in most cases a crystal (thus possessing translational invariance), but in some cases a dimer (a molecule pair), an n-mer, a solution, a mixed crystal, or even a spatially complex structure such as a photosynthetic unit. The species being studied is the Frenkel exciton. The reader is referred to well-known reviews /1.12-14/ for the general subject of excitons but is reminded that if the mth molecule in the aggregate is electronically excited and the others are unexcited, a Frenkel exciton may be said to occupy the mth site. Viewed in this way, the exciton may move from site to site, carrying energy from one location to another. This constitutes the subject of Frenkel exciton dynamics.

The interactions responsible for exciton motion from one molecule to another have been known since the work of FÖRSTER and DEXTER /1.15,16/. Their nature is not the subject of this review. With them as inputs in the exciton Hamiltonian, and given that the excitons interact with phonons, or generally with a bath, we address the macroscopic behavior of the system. Our interest is therefore the usual one of the practitioner of statistical mechanics whose purpose is to arrive at an understanding of macroscopic phenomena by considering microscopic interactions as given entities.

Of the many fundamental questions that arise in this field, the coherence question is perhaps the most interesting. It is posed in many forms. Is the exciton motion coherent or incoherent? Does an exciton move in a wave-like fashion as an electron in a metal might, or does it possess the characteristics of a random walker? Is the Schrödinger equation appropriate for its description, or must one resort to a Master equation or a diffusion equation? Are exciton transfer rates proportional to the matrix element of the site-to-site interaction or to its square? Do oscillations and reversibility characterize the motion, or are irreversibility and lack of ringing its dominant traits? Can these two types be considered as well-defined limits of a unified kind of motion? If so, how may the degree of coherence be ascertained theoretically and measured experimentally? These and related questions have been raised and discussed by many authors, one of the first being FÖRSTER /1.17/, who presented an incomplete broken plot of the general exciton transfer rate as a function of the intermolecular distance and hoped that the plot would be completed through a unified theory. These various issues concerning coherence will occupy us through Chaps. 2 and 3.

An unrelated question, often incorrectly identified with the coherence question, concerns the interplay of vibrational relaxation with exciton motion. Does the exciton move from one molecule to another before, while or after vibrational relaxation occurs? In Chap. 4 we shall discuss this issue of hot versus cold motion of the exciton.

There are numerous other questions that are being continually raised in this field. Is the band picture of Frenkel excitons in molecular crystals valid? What is the nature of trapping interactions that lead to capture of excitons by guest molecules introduced into a host crystal? Can sensitized luminescence be described through a time-independent host-guest "energy transfer rate"? If time-dependent, is it correctly given by the CHANDRASEKHAR - SMOLUCHOWSKI /1.18-20/ expression? Does exciton annihilation obey the standard bilinear rate equation? Why is there a large discrepancy of several orders of magnitude in reported values of the singlet-exciton diffusion constant /1.20/? Some of these and similar questions will also be addressed in this review.

There is, regrettably but unavoidably, a large number of noteworthy omissions in this review. Thus, the two modern treatments of exciton motion with intermediate degree of coherence, which predate the generalized master equation approach, have been mentioned only briefly. These are the work of HAKEN, REINEKER, and others, based on stochastic considerations and the work of SILBEY and others, based on a polaron-like treatment of microscopic Hamiltonians. However, these topics belong more properly to the second review in this book. Furthermore, space considerations and the fact that this review attempts to provide not an overview of the field but only a description of work within one particular theoretical viewpoint, that of master equations, should explain many other omissions of important theoretical work. Experimental work is also referred to only in the context of direct applications of the particular approach discussed.

A note on terminology before we begin: Practitioners of statistical mechanics usually reserve the term "master equations" to denote equations for probabilities and use other expressions for equations describing the evolution of nondiagonal elements of the density matrix. This is the terminology followed in this book. The quantum optics usage /1.9,10/ refers to all reduced density matrix equations, even the stochastic Liouville equation described in the second article in the book, as master equations.

2. Coherence and the Generalized Master Equation Approach: Formalism

2.1 Introduction

To sharpen the coherence question, consider an exciton moving between only two lattice sites 1 and 2. The exciton energies on the two sites are equal in the absence of the intersite interaction. The latter is represented by its matrix element V, taken real for simplicity, between kets $|1\rangle$ and $|2\rangle$. The Schrödinger equation for the exciton evolution is then

$$i\hbar\frac{dc_1}{dt} = V c_2 \tag{2.1}$$

with a similar equation for c_2. Here $2\pi\hbar$ is Planck's constant and $c_{1,2}(t) = \langle 1,2 \mid \psi(t)\rangle$ are the amplitudes. The probabilities of occupation of sites 1 and 2 oscillate in time, and the exciton, if placed on site 1 at $t=0$, rings back and forth between the sites:

$$P_1(t) = \cos \frac{2V}{\hbar} t \quad . \tag{2.2}$$

For the two-site system this motion is termed coherent.

If the two sites are considered to provide, not sharp levels, but "smeared" states because of bath interactions, one often describes the motion through a Master equation

$$\frac{dP_1(t)}{dt} = F\,[\,P_2(t) - P_1(t)\,] \tag{2.3}$$

with a similar equation for P_2. The transition (or hopping) rate F is given usually by the Fermi Golden rule and is proportional to V^2. The probabilities exhibit no oscillations and approach monotonically the value 1/2. The exciton, if placed initially on site 1, finds itself eventually on both sites equally:

$$P_1(t) = \frac{1}{2}\,(1 + e^{-2Ft}) \quad . \tag{2.4}$$

For the two-site system this motion is termed incoherent.

Clearly, the two kinds of motion are profoundly different. Reversibility, oscillations, wave-like behaviour typify one, and irreversibility, approach to equilibrium, diffusive behaviour typify the other. A characteristic time is proportional to $1/V$ in the coherent case but to $1/V^2$ in the incoherent one. The inverse of this time, which is a frequency in the former and a real rate in the latter case, has been often used in the literature /2.1-4/ as a "transfer rate". The coherence question has therefore often been worded as follows. What is the dependence of the singlet exciton transfer rate on R, the intermolecular separation, in general, and is it $1/R^3$ (coherent) or $1/R^6$ (incoherent) in particular? The powers of R correspond to the fact that for singlet excitons the interaction V, being dipole-dipole, is proportional to $1/R^3$.

To emphasize the basic differences in coherent and incoherent motion, let us examine two further examples. If exciton motion occurs through nearest-neighbour interactions V on an infinite chain (one-dimensional crystal) of equivalent sites m, n, etc., coherent motion is represented by

$$i\hbar \frac{dc_m(t)}{dt} = V\,[\,c_{m+1}(t) + c_{m-1}(t)\,] \quad . \tag{2.5}$$

Although the infinite size of the crystal destroys Poincaré recurrences and true ringing does not result, wave-like characteristics are certainly present in the solution. Initial placement of the exciton at $m = 0$ gives

$$P_m(t) = J_m^2\left(\frac{2V}{\hbar} t\right) \quad , \tag{2.6}$$

where the J's are regular Bessel functions. If, on the other hand, exciton motion is incoherent and occurs through nearest-neighbour transition rates F,

$$\frac{dP_m(t)}{dt} = F\,[P_{m+1}(t) + P_{m-1}(t) - 2P_m(t)]\ , \tag{2.7}$$

one sees diffusive or irreversible behaviour with no oscillations. An exciton initially placed at m = 0 travels according to

$$P_m(t) = e^{-2Ft}\ I_m(2Ft)\ , \tag{2.8}$$

where the I's are modified Bessel functions. Straightforward Fourier techniques /1.11/ yield (2.6) and (2.8) from (2.5) and (2.7).

Yet another example is given by the solutions of the classical wave equation on one hand and of the diffusion equation on the other. It is well known that

$$\frac{\partial^2 P(x,t)}{\partial t^2} = c^2\,\frac{\partial^2 P(x,t)}{\partial x^2} \tag{2.9}$$

represents waves moving at speed c, whereas

$$\frac{\partial P(x,t)}{\partial t} = D\,\frac{\partial^2 P(x,t)}{\partial x^2} \tag{2.10}$$

represents diffusive motion with diffusion constant D. The solutions of (2.9) are what we term coherent and differ drastically from the incoherent consequences of (2.10). Although this last example is classical, and therefore not entirely representative of exciton motion, it serves to focus on the coherence problem.

What then are our specific questions in the context of transport coherence? Given a system, such as a particular molecular crystal, we wish to know whether exciton motion in it proceeds in the extreme limits illustrated above. If the motion is intermediate, we require a unified theory of the motion. And we need practical prescriptions to ascertain the degree of coherence from experimental observations.

2.2 Motivation for the GME Approach

If the moving exciton is subject to no bath interactions, i.e., if no phonons or similar interacting species exist, Schrödinger equations such as (2.1) and (2.5) are certainly valid. On the other hand, if bath interactions are sufficiently strong, each of the exciton states |m> is replaced by a *group* of states with the exciton in |m> and the bath in various states, and Master equations such as (2.3) and (2.7) are considered appropriate. Let us pose two questions. First, how strong and of what nature should the bath interactions be to effect this passage to the Master equation? And second, can we find an intermediate transport instrument, valid for arbitrary strength (and nature) of bath interactions, which reduces to the Schrödinger equation in the limit of no bath and to the Master equation in an opposite limit? Clearly, these questions aim at the heart of the coherence problem. The first is linked with one of the central tasks of nonequilibrium statistical mechanics and in a different and quite general context has been tackled by many workers /2.5-11/. In examining their analyses we find an affirmative answer to the second question posed above. The intermediate transport instrument we obtain is the generalized master equation (GME):

$$\frac{dP_m(t)}{dt} = \int_0^t dt' \sum_n [\mathcal{W}_{mn}(t-t')\, P_n(t') - \mathcal{W}_{nm}(t-t')\, P_m(t')] \quad . \tag{2.11}$$

It describes motion through "memory functions" $\mathcal{W}(t)$ and is thus non-Markoffian in nature. By this phrase it is not meant that the *process* is necessarily non-Markoffian but merely that the equation is nonlocal in time /2.9,12/.

The motivation for selecting the GME (2.11) for our transport investigations stems from the fact that its non-Markoffian or memory nature is uniquely suited to the exciton coherence problem. Thus, for the two-site system described by (2.1) or (2.3) the GME is of the form

$$\frac{dP_1(t)}{dt} = \int_0^t dt'\, \mathcal{W}(t-t')\, [P_2(t') - P_1(t')] \tag{2.12}$$

and reduces immediately to the coherent and incoherent limits, (2.1) and (2.3), for extreme forms of the memory $\mathcal{W}(t)$. As in Sect. 2.1 let us consider site 1 initially occupied. The limit of a constant memory

$$\mathcal{W}(t) = 2V^2/\hbar^2 \tag{2.13a}$$

in (2.12) gives the solution (2.2), which is precisely the coherent motion consequence of (2.1). And the limit of a memory that decays infinitely rapidly

$$\mathcal{W}(t) = F\,\delta(t) \tag{2.13b}$$

yields the incoherent case (2.4). A similar demonstration for the infinite chain is also straightforward from the results of Sect. 2.6.1.

It is instructive to examine how non-Markoffian equations can unify the wave equation (2.9) and the diffusion equation (2.10). It is well known that the telegrapher's equation /2.13/

$$\frac{\partial^2 P(x,t)}{\partial t^2} + (c^2/D)\,\frac{\partial P(x,t)}{\partial t} = c^2\,\frac{\partial^2 P(x,t)}{\partial x^2} \tag{2.14}$$

has the ability to provide one way of unifying (2.9) and (2.10). It reduces to the former in the limit $c^2/D \to 0$ with c^2 finite and to the latter in the limit $D/c^2 \to 0$

with D finite. Moreover, its solutions are essentially those of the wave equation at short times and become those of the diffusion equation at long times, the transition time being D/c^2. For instance, the mean-square displacement corresponding to (2.14) goes as t^2 for $t \ll D/c^2$ but as t for $t \gg D/c^2$. These behaviours correspond, respectively, to (2.9) and (2.10). This unification ability is possessed, to an even greater degree, by the non-Markoffian equation

$$\frac{\partial P(x,t)}{\partial t} = D \int_0^t dt' \, \phi(t-t') \, \frac{\partial^2 P(x,t')}{\partial x^2} \quad . \tag{2.15}$$

The wave equation, the diffusion equation, and the telegrapher's equation, are all particular cases of (2.15) (barring some unimportant details) and are given, respectively, by the functional forms c^2/D, $\delta(t)$, and $c^2 \exp(-tc^2/D)$ of the memory $\phi(t)$.

The GME thus possesses two special characteristics, its non-Markoffian nature and the fact that it is an exact consequence of microscopic dynamics. Its non-Markoffian nature makes it ideal for coherence investigations, as has been shown above. We also stress that it is not a phenomenological construct but a direct result of dynamics and has a wide range of validity. While traditionally it has served only as an intermediate step in the derivation of the (Pauli) Master equation /2.19/, it will be put to practical use in this part of the review.

2.3 General Derivation of Exciton GME's

Beginning with the von Neumann equation for the density matrix ρ of the total exciton-bath system,

$$i \frac{d\rho(t)}{dt} = [H \, , \rho(t)] \equiv L \, \rho(t) \quad , \tag{2.16}$$

where we put $\hbar = 1$ for simplicity and where H is the Hamiltonian and L the Liouville operator, it is possible to derive, as an exact consequence under arbitrary conditions, the equation

$$\frac{dP_\xi(t)}{dt} = \int_0^t dt' \sum_\mu [\mathcal{W}_{\xi\mu}(t-t') \, P_\mu(t') - \mathcal{W}_{\mu\xi}(t-t') \, P_\xi(t')] + \mathcal{I}_\xi(t) \quad , \tag{2.17}$$

P_ξ being the diagonal element $\langle\xi|\rho|\xi\rangle$ of the density matrix.

Equation (2.17) differs from the GME (2.11) in the presence of the "initial" term $\mathcal{I}_\xi(t)$ and the fact that ξ, μ, etc. are total system states rather than pure exciton states. The basic technique for obtaining (2.17) is due to ZWANZIG /2.9/ and employs projection operators that diagonalize other operators on which they act, in the representation of eigenstates (ξ, μ, etc.) of a part H_0 of the total Hamiltonian. A generalization /2.14-16/ of this technique which employs projection operators $\mathcal{P}$ that coarsegrain as well as diagonalize, was employed by KENKRE and KNOX /2.16/ to derive exciton GME's. The incorporation of coarsegraining into (2.17) was first suggested in a different context by EMCH /2.10/. The coarsegraining consists in an *exact* elimination of bath variables and results in the following equation for P_m's which are sums $\sum_{\xi \in m} \langle\xi|\rho|\xi\rangle$ of the P_ξ's over "grains" m which represent pure exciton states:

$$\frac{dP_m(t)}{dt} = \int_0^t dt' \sum_n [\mathcal{W}_{mn}(t-t')\, P_n(t') - \mathcal{W}_{nm}(t-t')\, P_m(t')] + \mathcal{I}_m(t) \quad . \tag{2.18}$$

Equation (2.18) differs from the GME only in the existence of the "initial" term. The exact definition of $\mathcal{P}$ is that for any operator $\mathcal{O}$ in the total exciton-bath system,

$$\langle\xi|\mathcal{P}\mathcal{O}|\mu\rangle = \left[\sum_{\xi \in m} \langle\xi|\mathcal{O}|\xi\rangle\right] \left[\sum_{\xi \in m} 1\right]^{-1} Q_\xi \, \delta_{\xi\mu} \quad , \tag{2.19}$$

where Q_ξ is arbitrary except for being subject to the condition $\sum_{\xi\in m} Q_\xi = \sum_{\xi\in m} 1$ to ensure that $\mathcal{P}$ is idempotent. The choice of Q_ξ is dictated by the kind of ensemble used as well as other considerations, various possibilities having been discussed by KENKRE /2.14,15/. These and other details concerning the derivation are available in many other reviews /1.9 , 2.9 , 2.15/ and will not be repeated here. However, we display the operator form of (2.18), which may be used as a point of departure for exact or approximate evaluations of $\mathcal{W}(t)$ and $\mathcal{I}(t)$:

$$\begin{aligned} \frac{d\,\mathcal{P}\rho(t)}{dt} = &- \int_0^t dt'\ \mathcal{P}L\, e^{-i(t-t')(1-\mathcal{P})L}\, (1-\mathcal{P})L\,\mathcal{P}\rho(t') \\ &- i\mathcal{P}L\, e^{-it(1-\mathcal{P})L}\, (1-\mathcal{P})\, \rho(0) \quad . \end{aligned} \tag{2.20}$$

Equation (2.18) is thus an *exact* consequence of microscopic dynamics for arbitrary initial conditions. The derivation of the GME (2.11) now requires only that the "initial" term $\mathcal{I}_m(t)$ vanishes from (2.18). That term, the conditions under which it vanishes, and the manner in which it modifies the GME when it does not vanish, will be examined in Sect. 2.4. It will be seen that there are physical situations in which it can be dropped exactly and others in which its effects disappear rapidly so that for times which are not too short, it can be considered absent.

The derivation of the GME from microscopic dynamics, as represented by (2.16), is therefore complete. The crucial quantities are the memory functions $\mathcal{W}(t)$. They may be calculated from (2.20), in some cases exactly, in others through approximation procedures. Their form will depend on a) the Hamiltonian of the exciton-bath system, b) the partitioning of H into the part H_0, whose eigenstates are summed to obtain the "grains" m , n , etc. and the part V which describes the interaction, and c) the extent of coarsegraining. In an approximate calculation, the precise form of the Q_ξ's in (2.19) will also influence the $\mathcal{W}(t)$'s. While it is difficult to gain a further understanding of the behaviour of these $\mathcal{W}(t)$'s by an inspection of the exact expression in (2.20), considerable insight is provided by approximate expressions.

The approximation consists of the "weak-coupling" or perturbation analysis. In the exponents in (2.20) one replaces the Liouville operator L by the part L_0 corresponding to the neglect of the interaction V in the total Hamiltonian H. This procedure /2.9/ effectively retains lowest order terms in an expansion of $\mathcal{W}(t)$'s in orders of the interaction and results /2.14/ in

$$\frac{dP_m(t)}{dt} = \int_0^t dt' \sum_{\xi\in m} \sum_{all\,\mu} S_{\xi\mu}(t-t') \left[\langle\mu|P\rho(t')|\mu\rangle - \langle\xi|P\rho(t')|\xi\rangle \right] . \qquad (2.21)$$

Note that the ξ-summation is over the "grain" m, i.e., over all states corresponding to the exciton state m, but that the μ-summation is over *all* system states. In (2.21), the quantity $S_{\xi\mu}(t)$ is the weak-coupling $\mathcal{W}_{\xi\mu}(t)$ given by ZWANZIG /2.9/ for the case involving no coarsegraining:

$$S_{\xi\mu}(t) = 2\ |\langle\xi|V|\mu\rangle|^2 \cos\left[(E_\xi - E_\mu)t\right] , \qquad (2.22)$$

the E's being the eigenvalues of H_0. Splitting the μ-summation in (2.21) into one within the grain n (to be absorbed in the memory expressions) and another over the grains n (to be displayed as in (2.11)), one obtains the coarsegrained GME (2.11) with the weak-coupling, coarsegrained $\mathcal{W}(t)$'s:

$$\mathcal{W}_{mn}(t) = 2 \sum_{\xi\in m} \sum_{\mu\in n} [Q_\mu/g_n]\ |\langle\xi|V|\mu\rangle|^2 \cos\left[(E_\xi - E_\mu)t\right] , \qquad (2.23)$$

$$\mathcal{W}_{nm}(t) = 2 \sum_{\xi\in m} \sum_{\mu\in n} [Q_\xi/g_m]\ |\langle\xi|V|\mu\rangle|^2 \cos\left[(E_\xi - E_\mu)t\right] , \qquad (2.24)$$

where $g_m = \sum_{\xi\in m} 1$ and $g_n = \sum_{\mu\in n} 1$.

These coarsegraining generalizations of the ZWANZIG memory (2.22) have been obtained /2.15/ and analyzed for various levels of coarsegraining /2.14/ by KENKRE. The choice $Q_\xi = 1$ has been used by KENKRE and KNOX /2.16/ while thermal Q_ξ's involving equilibrium phonon distributions have been used by KENKRE and RAHMAN /2.17/ and KENKRE /2.18/.

Equation (2.22) shows that the memories are purely oscillatory at the microscopic level. Equations (2.23,24) show that coarsegraining introduces a summation of many such microscopic memories, corresponding to bath states that we eliminate in our description, and that the summation can result in *decaying* memories. Here we have an answer to the question concerning the passage from the Schrödinger to the Master equation that we posed in Sect. 2.2. Equations (2.23,24) can be rewritten in a way that shows that the memories $\mathcal{W}(t)$ are essentially Fourier transforms of the product of the square of the matrix element of V, the factor (Q/g), and the bath density of states. Calling this product $Y(\omega)$, we have the symbolic equation

$$\mathcal{W}(t) = \int d\omega \cos(\omega t)\ Y(\omega) \quad . \qquad (2.25)$$

When no bath exists and the pure exciton system is small, $Y(\omega)$ is a clearly recognizable sum of δ functions in ω and results in a purely oscillatory $\mathcal{W}(t)$ and, consequently, in coherent motion of the exciton. An infinite-size limit of the pure exciton system, without introduction of a bath, will lead to decaying $\mathcal{W}(t)$'s, but as is well known in statistical mechanics, this constitutes an artificial way of destroying Poincaré recurrences and of bringing in irreversibility. When we do introduce the bath, the nature of the exciton-bath interaction decides the form of $Y(\omega)$ and can make the $\mathcal{W}(t)$ decay in time, the strength and nature of the exciton-bath interaction being responsible for how rapidly, and how, this decay occurs. If the decay were infinitely fast, we would have $\mathcal{W}$'s that are delta functions in time:

$$\mathcal{W}_{mn}(t) = F_{mn}\ \delta(t) \quad . \tag{2.26}$$

The Master equation, and consequently incoherent exciton motion, would then result:

$$\frac{dP_m(t)}{dt} = \sum_n [F_{mn}P_n(t) - F_{nm}P_m(t)] \quad . \tag{2.27}$$

We have seen explicitly how the passage from the microscopic von Neumann equation (2.16) to the macroscopic Master equation for exciton motion (2.27) occurs as a result of coarsegraining. The absence of a bath, or weak exciton-bath interactions, will result in coherent motion, and strong interactions will result in incoherent motion. Every realistic situation lies in between these limits. The GME (2.11) is thus the intermediate transport instrument capable of analysing intermediate kinds of motion with arbitrary degree of coherence. The starting point for earlier standard descriptions of exciton transport /1.15-17/ used to be the Master equation (2.27). The GME approach replaces that starting point by (2.11). It consists in deriving memory functions from microscopic dynamics through formulae such as (2.23) or from exact calculations,and then employing the solutions of the GME's to explain experimental observations.

2.4 Range of Validity of Exciton GME's

We now return to the question of the validity of dropping the "initial" term $\mathcal{I}_\xi(t)$ or $\mathcal{I}_m(t)$ from (2.17) or (2.18) to obtain the GME. In the context of general nonequilibrium statistical mechanics /2.5-11/ it has been analyzed as follows. Equation (2.20) shows that the initial term vanishes if

$$(1 - \mathcal{P})\ \rho(0) = 0 \quad . \tag{2.28}$$

The GME is therefore valid if (2.28) is satisfied. In the case with no bath, i.e., when ξ, μ etc. are identical to the exciton states m, n, etc., (2.28) represents initial diagonality of the density matrix in the representation of m , n , etc. In the presence of a bath, (2.28) further means that initially the system density matrix is an outer product of the exciton ρ and the bath ρ. When, as a result of random initial phases, both these requirements are met, the passage to the GME is exact. When they are not met, it is to be considered an approximation.

The great success of the above argument of VAN HOVE, ZWANZIG, and others /2.5-11/ was in eliminating the highly objectionable repeated random-phase approximation made by PAULI /2.19/ in his derivation of the Master equation. However, exciton transport presents a special problem. Because this problem is independent of the existence of a bath, we shall consider the latter to be absent. The projection operator is, in this case, the diagonalizing operator, and (2.28) can be satisfied in one of two ways: initially a) the exciton occupies a single site, or b) it occupies many sites but has no definite phase relations over them. We now see that if the initial time is taken to be that at which the exciton is created through optical absorption, (2.28) is definitely violated. The wavelength of light is several hundreds of times as large as the intersite distance, and localized excitation in the absence of impurities does not correspond to experimental conditions. Furthermore, the creation of an exciton does involve definite phase relations over many sites and $(1-\mathcal{P})\,\rho(0)$ does not equal zero. The usual argument which serves in general nonequilibrium statistical mechanics /2.5-11/ is thus not entirely adequate here. Surely, this is not a shortcoming of the general argument but a result of the particular features of exciton transport description in *real* space. Optical absorption obeys the k-selection rule, and therefore, to an excellent approximation, the initial exciton density matrix is diagonal or "localized" in momentum space. The usual argument for dropping $\mathcal{I}(t)$ would work if the states chosen for the transport description, m , n , etc. above were *momentum* states. However, the transport description we use in molecular crystals and aggregates is that of real-space states. Initial diagonalization or random-phase condition is inapplicable in this representation, and it is necessary to inquire further into the validity range of the exciton GME.

The initial term $\mathcal{I}_m(t)$ may be rewritten as

$$\mathcal{I}_m(t) = -i \langle m \,|\, \mathcal{P}\, e^{-itL(1-\mathcal{P})}\, L(1-\mathcal{P})\,\rho(0)\,|\, m\rangle \quad , \tag{2.29}$$

where we have switched the order of the operators L and $(1-\mathcal{P})$. It follows /2.20/ that $\mathcal{I}_m(t)$ can be dropped exactly in situations which violate (2.28), provided they obey the weaker condition

$$L(1-\mathcal{P})\,\rho(0) = 0 \quad . \tag{2.30}$$

For a crystal possessing translational invariance, the matrix elements of the Hamiltonian, $\langle m|H|n\rangle$, are functions of the differences $(m-n)$. Using this property it can be shown that a sufficient condition for (2.30) is that for arbitrary m and n,

$$\sum_r e^{ikr} \langle r|(1-P)\,\rho(0)\,|n\rangle\, e^{-ikn} = \sum_r e^{ikm} \langle m|(1-P)\,\rho(0)\,|r\rangle\, e^{-ikr} . \qquad (2.31)$$

Generally m and n are vectors in the direct lattice, k is a vector in the reciprocal lattice, and km is the dot product (usually written k·m). Condition (2.31) is satisfied for initial $\rho(0)$'s whose matrix elements $\langle m|\rho(0)|r\rangle$ are functions only of the difference $(m-r)$. However, this translational invariance of the initial density matrix applies in all those cases in which no site is preferred over another in $\rho(0)$. Thus (2.30) holds for fully delocalized $\rho(0)$'s, specific examples being a single Bloch state and a thermalized state given by $\rho(0) = \text{const.}\,[\exp(-\beta H)]$. It is also possible to prove this result from simple symmetry arguments.

The GME is thus valid for initial conditions which are not only completely localized but also for those which are completely delocalized.

The above extension of the range of validity of the GME was given by KENKRE /2.20/ to answer objections to the GME raised /2.21/ on the grounds that localized excitons cannot be created experimentally in pure crystals. The GME is now known to be valid in *both* extreme limits, i.e., whenever a wave packet of very narrow *or* very wide spread is created initially. The usual case of optical absorption corresponds, to a good approximation, to the delocalized limit analyzed above. For intermediate spreads in the initial exciton wavepacket, it is possible to compute $I_m(t)$ by using techniques similar to the ones used in Sect. 2.6 for the computation of $\mathcal{W}_{mn}(t)$ and to study the effect of these driving terms on the GME. Some of this work is still under way /2.22/. As an example of the calculation of $I_m(t)$ /2.20/, we show

$$I_m(t) = (1/N)\left[C(t)\cos[m(k_1-k_2)] + S(t)\sin[m(k_1-k_2)]\right] , \qquad (2.32)$$

which holds for initial occupation of two Bloch states with wave vectors k_1 and k_2 in a crystal of N sites. For an infinite one-dimensional crystal with nearest-neighbour Hamiltonian matrix elements V and with $k_1 + k_2 = \pi$, we have, for instance,

$$C(t) = -\left\{4V\sin[(k_1-k_2)/2]\right\} J_1\left\{4Vt\sin[(k_1-k_2)/2]\right\} , \qquad (2.33)$$

$$S(t) = \left\{4V\sin[(k_1-k_2)/2]\right\} J_0\left\{4Vt\sin[(k_1-k_2)/2]\right\} , \qquad (2.34)$$

where the J's are Bessel functions.

The GME's are thus valid for initial conditions in which the total ρ is a product of the exciton ρ and the bath ρ, and the excitons are either largely localized or largely delocalized in the site representation. Situations in which it is not a good approximation to consider ρ to be a product of the exciton and bath parts can sometimes be analyzed through a transformation such as the one given by GROVER and SILBEY /2.23/ or MUNN and SILBEY /2.24/, which dresses the excitons and the bath and results in an initial product condition in the *dressed* species. Subtle issues remain to be solved in this part of the "initial" problem. They are of interest not only to the GME but to the stochastic Liouville equation and to all analyses, within exciton physics or outside, which purport to provide a microscopic analysis of interacting systems. The initial product condition has always been used in all these other analyses also /1.9,10 , 2.23-25/.

In special situations in which $I(t)$'s cannot be neglected, we may perhaps analyze exciton motion as proceeding through three time regimes: the first, when neither the GME (2.11) nor the Master equation (2.27) is valid but the *driven* GME (2.18) describes the evolution accurately; the second, when the $I(t)$'s have decayed to zero and the GME (2.11) is appropriate but the Master equation (2.27) is still not valid, and third, when the $W(t)$'s too have decayed and the Master equation is valid. Pictured thus, the GME can still address the coherence-incoherence problems in the second and third time regimes and describe details of the motion which are not accessible to the Master equation. Surely, there is no a priori guarantee that in a given system such a clear separation of time regimes exists with no overlapping.

Nevertheless, we reiterate that the vanishing of $I(t)$ is not always an approximation and that in several real situations, the GME is an exact consequence of microscopic dynamics. This point needs to be emphasized. Even now, some twenty-five years after the original derivations /2.5-11/ and in spite of clear expositions such as those of ZWANZIG /2.9/ and lucid reviews such as that of HAAKE /1.9/, it is often not realized that a closed equation for probabilities can be an exact consequence of *quantum* mechanics.

2.5 A Conceptual Application: Unification of Transfer Rates

A conceptual application of the GME, which is made possible by the very existence rather than details of the memory functions, and which illustrates the special suitability of the GME to the coherence problem, is described in this section. The question to be answered is the one posed by FÖRSTER /1.17 , 2.26/ and others /2.1,2/. Should intersite rates for excitation transfer be considered proportional to the intersite interaction V, or to its square, or to an intermediate quantity? In particular, in the singlet exciton case where V is proportional to R^{-3} with R as the intersite distance, what is the value of the exponent n in the expression for this transfer rate w? Here

$$n = \frac{d\,\ell n\,w}{d\,\ell n\,R} \quad . \tag{2.35}$$

Although it had been suggested /2.3,4,26/ that the two limits of the transfer rate could never be compared in principle, because one of them is associated with a frequency of probability oscillations and the other with a true rate of decay, a strong need for the unification of these coherent and incoherent rates had been felt /1.17/. Part of the reason for this need lies in the extensive use, and in some cases misuse, made of these rates in the interpretation of experiments. As KENKRE and KNOX have shown /2.27/, the required unification is naturally provided by the GME in the following way.

A simple unified definition of the transfer rate w may be given by identifying it with the reciprocal of the time required for the mean square displacement $\langle\langle m^2(t)\rangle\rangle$ to build from the value 0 to the value 1, for the initial condition that the exciton is localized on site 0. Clearly, the characteristics of the motion will be directly reflected in w defined in this manner. Calculations of $\langle\langle m^2(t)\rangle\rangle$ from the GME have been given by KENKRE /2.28,29/. Only the special case for an infinite linear chain with memories $\mathcal{W}_{mn}(t)$ that factor into a spatial part F_{mn}, and a temporal part $\phi(t)$ whose integral is normalized to 1, are displayed here:

$$\langle\langle m^2(t)\rangle\rangle = \langle\langle m^2(0)\rangle\rangle + \langle m^2\rangle \int_0^t dt' \int_0^{t'} dt'' \,\phi(t'') \quad . \tag{2.36}$$

The definition of $\langle\langle m^2(t)\rangle\rangle$ and of $\langle m^2\rangle$ is

$$\langle\langle m^2(t)\rangle\rangle = \sum_m m^2 P_m(t) \quad , \tag{2.37}$$

$$\langle m^2\rangle = \sum_m m^2 F_m \quad . \tag{2.38}$$

In the light of (2.36), the definition of w given above takes the quantitative form

$$\int_0^{1/w} dt \int_0^{t'} dt' \,\phi(t') = \frac{1}{\langle m^2\rangle} \quad . \tag{2.39}$$

The unification of the coherent and incoherent rates, sometimes termed fast and slow rates, is immediately apparent from (2.39). The extreme choice $\phi(t) = \delta(t)$ is representative of totally incoherent transport and results in

$$w = \langle m^2\rangle = \text{const. } V^2 = \text{const.' } R^{-6} \quad . \tag{2.40}$$

The second equality in (2.40) is based on the fact that the Golden Rule gives F_{mn} to be proportional to V^2. The third equality corresponds to singlet exciton transfer via dipole-dipole interactions. On the other hand, the extreme choice $\phi(t)$ = const. is representative of totally coherent transport and results in

$$w^2 = \text{const. } \langle m^2\rangle = \text{const. } V^2 = \text{const. } R^{-6} \quad , \tag{2.41}$$

giving a w that is proportional to V and R^{-3}. Furthermore, expanding *any* nonpathological memory as

$$\phi(t) = \phi(0) + t\,\dot{\phi}(0) + \dots \tag{2.42}$$

and retaining only the first term gives, if a *short-time* approximation can be made,

$$w^2 = [\phi(0)/2] \langle m^2 \rangle = \text{const. } V^2 = \text{const. } R^{-6} \quad . \tag{2.43}$$

On the other hand, *any* memory that eventually decays to zero gives, if a *long-time* approximation can be made,

$$w = \left[\int_0^\infty dt\ \phi(t)\right]^{-1} \langle m^2 \rangle = \text{const. } V^2 = \text{const. } R^{-6} \quad . \tag{2.44}$$

Equations (2.40-44) show how the GME, and (2.39) in particular, unify the transfer rates. Explicit expressions have been given by KENKRE and KNOX /2.27/ for the case of nearest-neighbour transfer and exponential $\phi(t)$'s, i.e.,

$$\mathcal{W}_{mn}(t) = F_{mn}\ \phi(t) = (2V^2/\alpha)(\delta_{m,n+1} + \delta_{m,n-1})\ \alpha \exp(-\alpha t) \quad . \tag{2.45}$$

The exponent α measures the rate of decay of the memory. Equation (2.39) yields the implicit equation

$$(\alpha/w) + \exp[-(\alpha/w)] - 1 = (\alpha^2/2V^2) \quad , \tag{2.46}$$

which reduces, in the coherent and incoherent limits, to $w = V$ and $w = 2V^2/\alpha$, respectively. Figure 2.1 shows the exponent of the intermolecular distance R, given by (2.35) and (2.46), plotted versus w/α to exhibit this unification graphically

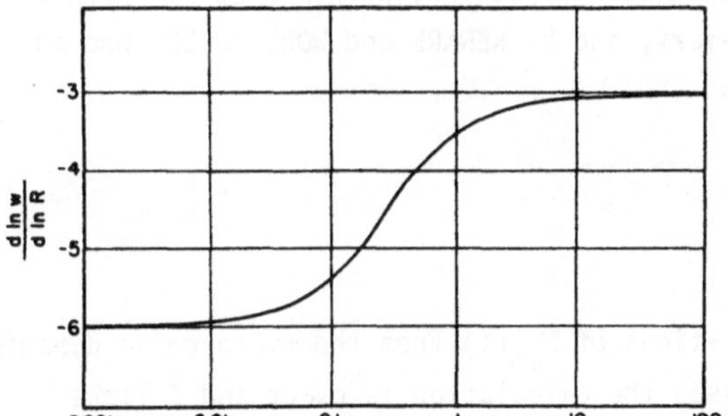

Fig. 2.1. The explicit unification of the coherent ($1/R^3$) and incoherent ($1/R^6$) limits of the transfer rate for dipolar interaction (singlets). The exponent of R, the intermolecular distance, is plotted as a function of α

/2.27/. As an example note that an application of this treatment to a bacteriochlorophyll complex studied by PHILIPSON and SAUER /2.30/ shows /2.27/ that with $V \approx 125\ \text{cm}^{-1}$ and $\alpha \approx 2.3 \times 10^{14}\ \text{s}^{-1}$, the rate $w \approx 4.8 \times 10^{12}\ \text{s}^{-1}$ with a dependence $R^{-5.9}$

on the intermolecular distance.

A word of caution is necessary. The above analysis shows that w, the reciprocal of a characteristic time for motion, goes as V in the coherent limit and V^2 in the incoherent limit. We have used the term *transfer* rate to describe w. By no means does it follow that the w can be used as a *transition* rate in a Master equation such as (2.27). The practice of taking transition rates such as F_{mn} (or inverses of hopping time for random walkers) to be proportional to V in the coherent limit is dangerous and has no foundation. Unfortunately, this point appears not to have been fully realized in the exciton literature.

The assumption (2.45) was made only /2.27/ in the interests of simplicity and is by no means crucial. The exponential memory is known to result in negative probabilities for large systems as has been shown by SILBEY /1.1/ and serves only as a pedagogical device above. The entire treatment of this section follows through unchanged in essentials for arbitrary (but physical) memories, as should be obvious by the comments made /2.15/ in conjunction with (2.40-44). Thus Fig. 2.1 will obtain for any physical memory. Changing the form of (2.45) will change only details of the shape of the curve in the transition region. These remarks should serve partly as a warning /1.1/ against careless use of simple exponential memories in extended systems and partly to emphasize the model-independent nature of the transfer rate unification treatment of /2.27/. It is also to be noted that the separation of $\mathcal{W}_{mn}(t)$ into a spatial and a temporal factor is also merely a simplification and quite unnecessary to this unification.

This treatment has been recently extended in three directions: by KENKRE /2.29/ who included the effects of *strong* intersite interaction, by KÜHNE et al. /2.31/ who added a term to the exponential memory, and by KENKRE and WONG /2.22/ who analyzed the effects of $\mathcal{I}(t)$ on the rates.

2.6 Evaluation of Memory Functions

In this section we describe the computations of $\mathcal{W}_{mn}(t)$ from the microscopic dynamics of the exciton-bath system. In some cases the calculation is exact and follows (2.19,20), while in other cases approximation procedures are used, the point of departure being (2.23,24).

2.6.1 Pure Crystal of Arbitrary Size: Nonlocal Memories

Even in the absence of bath interactions, remarkable features such as spatially nonlocal or long-range character can exist in the $\mathcal{W}(t)$'s and the associated transition rates F /2.29,32/. Consider an exciton in a pure crystal of N sites, with all site energies equal and site-to-site interaction matrix elements V_{mn}. The counterpart of the Schrödinger equation (2.1) is now

$$\frac{dc_m(t)}{dt} = -i \sum_n V_{mn}\, c_n(t) \quad . \tag{2.47}$$

Under fully localized or delocalized initial conditions, (2.47) is equivalent to (2.11). To calculate explicit $\mathcal{W}_{mn}(t)$'s, we rewrite (2.11) as

$$\frac{dP_m(t)}{dt} + \int_0^t dt' \sum_n A_{mn}(t-t')\, P_n(t') = 0 \quad , \tag{2.48}$$

where $A_{mm} = \sum_n \mathcal{W}_{nm}$ and $A_{mn} = -\mathcal{W}_{mn}$ for $m \neq n$. Translational invariance allows us to write $A_{mn} = A_{m-n}$. If we introduce discrete Fourier transforms

$$P^k(t) = \sum_m P_m(t)\, e^{ikm} \quad , \tag{2.49}$$

$$A^k(t) = \sum_m A_m(t)\, e^{ikm} \quad , \tag{2.50}$$

the solution of (2.48) gives

$$\tilde{A}^k(\varepsilon) = 1/\tilde{P}^k(\varepsilon) - \varepsilon \quad , \tag{2.51}$$

where ε is the Laplace variable and tildes denote Laplace transforms.

Equation (2.51) gives the transform of the memory in terms of the probabilities. However, these can be calculated directly via (2.47) through the Fourier transform technique and by using

$$P_m(t) = c_m^*(t)\, c_m(t) \quad . \tag{2.52}$$

The result is

$$P_m(t) = (1/N) \sum_{k,q} e^{-it(V^k - V^q)}\, e^{-im(k-q)} \quad , \tag{2.53}$$

where V^k is the transform of V_{m-n} defined as in (2.50). Using (2.53) and (2.49) in (2.51), we obtain /2.29/ the general expression for pure crystal memories as

$$\mathcal{W}_{mn}(t) = -\int d\varepsilon\, e^{\varepsilon t} \sum_k \left\{ e^{-ik(m-n)} \Big/ \sum_q \left[\varepsilon + i(V^{k+q} - V^q)\right]^{-1} \right\} \quad . \tag{2.54}$$

The ε integration is on the Bromwich contour and constitutes the Laplace inversion. The m, n, k, q are generally vectors and (2.54) is valid for an arbitrary number of spatial dimensions. It is also valid for any form and strength of the intersite interaction V_{mn}, any lattice type, and any size of the crystal, provided only that translational invariance applies. The V^k's appearing in (2.54) are essentially band energies of the Bloch states k.

Equation (2.54) may be used to investigate various particular cases. The trivial ones of a dimer (a molecule pair) and a closed trimer hardly need elaboration. The former is characterized by $\langle 1|H|2\rangle = \langle 2|H|1\rangle = V$, corresponds to (2.1) and (2.2), and has memories

$$\mathcal{W}_{12}(t) = \mathcal{W}_{21}(t) = 2V^2 \quad . \tag{2.55}$$

The trimer is characterized by $\langle 1|H|2\rangle = \langle 2|H|3\rangle = \langle 3|H|1\rangle = V$ and has

$$\mathcal{W}_{12}(t) = \mathcal{W}_{23}(t) = \mathcal{W}_{31}(t) = 2V^2 \cos(tV\sqrt{3}) \quad , \tag{2.56}$$

and obviously $\mathcal{W}_{12} = \mathcal{W}_{21}$, etc. The ring of four sites illustrates the method of working with the general result (2.54) and reveals the new phenomenon of long-range memories. We therefore exhibit its details.

The four-site ring has $\langle 1|H|2\rangle = \langle 2|H|3\rangle = \langle 3|H|4\rangle = \langle 4|H|1\rangle = V$ and $\langle 1|H|3\rangle = \langle 2|H|4\rangle = 0$. The k, q in (2.54) have the allowed values 0, $\pi/2$, π, and $3\pi/2$, the respective V^k's being $2V$, 0, $-2V$, and 0. Substituting these in (2.54) and inverting the transforms, we obtain

$$\mathcal{W}_{12}(t) = \mathcal{W}_{23}(t) = \mathcal{W}_{34}(t) = \mathcal{W}_{41}(t) = 2V^2 \cos(tV2\sqrt{2}) \quad , \tag{2.57}$$

$$\mathcal{W}_{13}(t) = \mathcal{W}_{24}(t) = 2V^2\,[1-\cos(tV2\sqrt{2})] = 4V^2 \sin^2(tV\sqrt{2}) \quad . \tag{2.58}$$

While (2.57) is an expected result, (2.58) contains a remarkable fact: memory functions $\mathcal{W}_{mn}(t)$ exist between sites for which *no intersite interaction* V_{mn} *exists*. This is the phenomenon of long-range memories /2.29,32/. It leads to long-range transition rates F_{mn} and has important consequences in the calculation of diffusion constants and the interpretation of transient grating signals. These matters will be discussed in Sects. 2.6.2 and 3.1.3.

The infinite one-dimensional crystal with nearest-neighbour interactions V provides another particular case of interest /2.29,32-35/. The allowed values of k now form a continuum of size 2π, and the V^k's are $2V\cos k$. The double transform $\tilde{\mathcal{W}}^k(\varepsilon)$ of the memories can be evaluated exactly /2.29/ as

$$\tilde{\mathcal{W}}^k(\varepsilon) = -\left[\varepsilon^2 + 16V^2 \sin^2(k/2)\right]^{1/2} \quad . \tag{2.59}$$

Using the relation

$$\tilde{A}^k(\varepsilon) = \tilde{\mathcal{W}}^0(\varepsilon) - \tilde{\mathcal{W}}^k(\varepsilon) \tag{2.60}$$

and recognizing the Laplace transform of $(1/t)\,J_1(t)$ or of $[J_0(t)+J_2(t)]$, we can write

$$A^k(t) = \left[8V^2 \sin^2(k/2)\right]\left[J_0[4Vt\sin(k/2)] + J_2[4Vt\sin(k/2)]\right] \quad . \tag{2.61}$$

The Fourier inversion can be performed with the help of Bessel function identities to yield the infinite crystal memories:

$$\mathcal{W}_{mn}(t) = 2V^2 \Big(J^2_{m-n+1} + J^2_{m-n-1} + 2J_{m-n-1}J_{m-n+1} - 2J^2_{m-n} - J_{m-n}J_{m-n+2} - J_{m-n}J_{m-n-2} \Big) . \qquad (2.62)$$

The argument of the Bessel functions in (2.62) is 2Vt, and (2.62) may also be written in the compact form

$$\mathcal{W}_{mn}(t) = \frac{1}{t}\frac{d}{dt} J^2_{m-n}(2Vt) \quad . \qquad (2.63)$$

The spatially nonlocal character is, however, particularly transparent from (2.62). The Hamiltonian has no matrix elements between nonnearest neighbours, but memories exist between every pair of sites in the crystal.

As a final example of (2.54) let us consider the case of equal interactions V among all sites. The V^k are now $NV\,\delta_{k,0}$, and we obtain

$$\tilde{\mathcal{W}}^k(\varepsilon) = -N\left\{\frac{2\varepsilon}{\varepsilon^2 + V^2N^2(1-\delta_{k,0})} + \frac{N-2}{\varepsilon}\right\}^{-1} . \qquad (2.64)$$

Simplifying (2.64) and inverting the result we obtain

$$\mathcal{W}_{mn}(t) = 2V^2 \cos\left[tV\sqrt{N(N-2)}\right] . \qquad (2.65)$$

Particular cases of (2.65) for N = 2 and N = 3 are (2.55) and (2.56), respectively. KENKRE and SILBEY /2.36/ have shown that this system possesses certain curious features, particularly in the infinite size limit. They arise from the fact that the interaction V does not decrease with distance in this system.

The special case of the infinite one-dimensional chain with short-range interactions (2.62) was treated first by SOKOLOV /2.33/. The general result (2.54), which is valid for crystals of any size, dimension, type of range of interaction, and its reduction to particular cases including the infinite-chain result were first given by KENKRE /2.29,32,34/, along with a discussion of the concept of spatial nonlocality and its effects on experiments. Subsequently, a derivation of the special case of the infinite chain was given by KÜHNE and REINEKER /2.35/ with the help of a technique different from those of /2.29,32-34/.

2.6.2 Crystal in Interaction with a Bath: Long-Range Transition Rates

Of the two factors that determine the form of the memory functions $\mathcal{W}(t)$, the crystal structure and the nature of the interaction with the bath, we have neglected the latter and focused our attention on the former in the treatment of Sect. 2.6.1. We shall now analyze the effect of bath interactions in two steps. Exact evaluations of $W(t)$'s, which continue to treat the crystal structure in detail but analyze the bath interaction in a well-known but phenomenological manner, are presented in this section. Computations which start from a microscopic system-bath Hamiltonian but employ approximations are given in Sect. 2.6.3.

A commonly used procedure, which can introduce bath interactions and the resulting damping behaviour into evolution equations for the system density matrix ρ, consists of adding certain terms to such equations. These terms describe the absolute destruction, at a rate α, of off-diagonal elements of ρ in the given representation:

$$i\,\frac{\partial \langle m|\rho(t)|n\rangle}{\partial t} = \langle m|[H,\rho(t)]|n\rangle - i(1-\delta_{m,n})\,\alpha\langle m|\rho(t)|n\rangle \quad . \tag{2.66}$$

This way of augmenting the von Neumann equation for ρ appears in many different fields of physics /2.37-39/. A lucid discussion is in the textbook by WANNIER /2.37/. The first use of this device in the exciton context is by AVAKIAN et al. /2.39/, who have also shown how diffusion equations can result from equations such as (2.66).

The methods of Sect. 2.6.1 cannot be used directly to calculate the $W(t)$'s corresponding to (2.66). Those methods are based on the solution of Schrödinger equations such as (2.47), and no clear extensions of (2.47) corresponding to (2.66) exist. However, if we rewrite (2.66) in the von Neumann form (2.16) and interpret L to have the parts L_c and L_i corresponding to the two terms in the right-hand side of (2.66), the GME in the operator form (2.20) follows immediately. Let us now use the notation

$$\mathcal{O}'' = (1-P)\,\mathcal{O} \tag{2.67}$$

for arbitrary operators $\mathcal{O}$. A key expression in (2.20) is

$$e^{-it(1-P)L} = 1 + (-it)(L_c'' + L_i'') + \frac{(-it)^2}{2!}\left[(L_c'')^2 + (L_i'')^2 + L_c''L_i'' + L_i''L_c''\right] + \ldots \quad . \tag{2.68}$$

Equation (2.66) shows that L_i'', when acting on an off-diagonal operator, merely multiplies it by $-i\alpha$. Therefore (2.68) gives

$$e^{-it(1-P)L}\,\mathcal{O}'' = \left[1 + (-it)(L_c'' - i\alpha) + \frac{(-it)^2}{2!}(L_c'' - i\alpha)^2 + \ldots\right]\mathcal{O}'' \quad . \tag{2.69}$$

This remarkable result shows that L_c'' and L_i'' commute in the expansion of the left-hand side of (2.68), with the consequence

$$e^{-it(1-P)L}\,\mathcal{O}'' = e^{-\alpha t}\,e^{-it(1-P)L_c}\,\mathcal{O}'' \quad . \tag{2.70}$$

It is clear from (2.66) that L_i contains the off-diagonalizing operator $(1-P)$. Therefore it makes no contribution to the first L and the last L in the kernel of (2.20). Equation (2.70) and (2.20) then give

$$\mathcal{W}_{mn}(t) = e^{-\alpha t}\, \mathcal{W}^{c}_{mn}(t) \quad , \tag{2.71}$$

where $\mathcal{W}^{c}_{mn}(t)$ is the "coherent" memory corresponding to no bath interactions, i.e., to the deletion of the last term in (2.66).

In (2.69) we have an exact evaluation of memory functions for exciton transport in a system whose evolution obeys the augmented von Neumann equation (2.66). The bath interactions are represented by the decaying factor $\exp(-\alpha t)$ and the details of the crystal structure are given by $\mathcal{W}^{c}_{mn}(t)$ calculated through the prescription of Sect. 2.6.1. Thus, for an arbitrary crystal, (2.71) and (2.54) give

$$\mathcal{W}_{mn}(t) = e^{-\alpha t} \int \left\{ - \sum_{k} \left\{ e^{-ik(m-n)} \Big/ \sum_{q} \left[\varepsilon + i(V^{k+q} - V^{q}) \right]^{-1} \right\} \right\} e^{\varepsilon t}\, d\varepsilon \quad . \tag{2.72}$$

The special cases (2.55-58,62,65) of the general crystal memory (2.54) are respectively augmented by the prescription (2.71) to include the effect of damping. Thus, for a dimer,

$$\mathcal{W}_{12}(t) = 2V^2 e^{-\alpha t} \quad ; \tag{2.73}$$

for a four-site ring,

$$\mathcal{W}_{12}(t) = \mathcal{W}_{23}(t) = \mathcal{W}_{34}(t) = \mathcal{W}_{41}(t) = 2V^2 e^{-\alpha t} \cos(2\sqrt{2}\,Vt) \quad , \tag{2.74}$$

$$\mathcal{W}_{13}(t) = \mathcal{W}_{24}(t) = 4V^2 e^{-\alpha t} \sin(\sqrt{2}\,Vt) \quad ; \tag{2.75}$$

and for an infinite one-dimensional crystal,

$$\begin{aligned} \mathcal{W}_{mn}(t) = 2V^2 e^{-\alpha t} \Big\{ & J^2_{m-n+1}(2Vt) + J^2_{m-n-1}(2Vt) + 2J_{m-n-1}(2Vt) \times J_{m-n+1}(2Vt) \\ & - 2J^2_{m-n}(2Vt) - J_{m-n}(2Vt)\,[J_{m-n+2}(2Vt) + J_{m-n-2}(2Vt)] \Big\} \quad . \end{aligned} \tag{2.76}$$

It is remarkable that the frequently used device (2.66) of extending the von Neumann equation to include damping results in such a simple and natural extension of the memory functions: the latter are merely multiplied by $\exp(-\alpha t)$. The demonstration displayed above, which uses simple properties of projection operators, was given by KENKRE /2.34/. A rederivation of those results was given by REINEKER and KÜHNE /2.40/ in terms of an expansion of the exponential operator and a term-by-term evaluation of the resulting series, a method which could prove to be quite powerful.

The results of this section have considerable practical use: they allow an explicit unification of the coherent and incoherent regimes through the mere assignment of the appropriate value to the bath parameter α. The coherent limit is recovered trivially when $\alpha \to 0$. The incoherent limit involves $\alpha \to \infty$ but with the well-known addition that $V \to \infty$ too, such that V^2/α = const. We then recover the (Markoffian)

Master equation. Thus, for the infinite one-dimensional crystal, the GME with the memory functions (2.76) reduces to the familiar Master equation (2.7) with

$$F = 2V^2/\alpha \tag{2.77}$$

as the prescription to calculate the transition rate F from the interaction matrix element V and the bath parameter α.

The real usefulness of the results in this section stems, however, from the *intermediate* description for arbitrary α. A detailed analysis of the memory functions, the solutions, and the inherent physical effects in the context of a ring of four sites, as well as in the infinite crystal, is available in /2.29/. Needless to say, for times small compared to $1/\alpha$, coherent effects are seen, whereas for longer times incoherent motion is recovered. It is particularly instructive, however, to understand the additonal effects of finite coherence (i.e., of the fact that $1/\alpha \neq 0$) on *long-time* motion. We review this interesting phenomenon briefly below.

For the four-site ring of (2.74,75) consider the exciton to be initially at site 1. It moves via nearest-neighbour matrix elements V and is subject to a nonzero $1/\alpha$. It is easy to solve the GME under these conditions for arbitrary time. For long times, the probability $P_3(t)$ of the occupation of the site *farthest* from the initially occupied one is /2.29/

$$P_3(t) = \frac{1}{4}\left[1 + e^{-t\zeta(2V^2/\alpha)} - 2e^{-t(V^2/\alpha)}\right] \quad , \tag{2.78}$$

where the quantity ζ is given by

$$\zeta = \frac{\alpha^2}{\alpha^2 + 8V^2} \quad . \tag{2.79}$$

If we were to employ the usual Master equation with nearest-neighbour rates $F = 2V^2/\alpha$ for the description of motion here, the quantity ζ would equal 1 in (2.78) and the exciton *would move more slowly*. Thus, if the ratio of V to α in a system is not too small, the *usual* Master equation generally overestimates the time that an exciton takes to move from one location to another.

Similar results hold for the infinite crystal /2.29/. Clearly, the motion has wave-like character at times short with respect to $1/\alpha$. Even for long times however, differences exist between the true solutions for finite α and the solutions of the Master equation (2.7). We emphasize that the form (2.8) involving modified Bessel functions is recovered *only* for sufficiently small values of V/α. Generally, the exciton moves faster than in (2.8). Thus, the general long-time solution for $P_n(0) = \delta_{n,0}$ is

$$P_m(t) = \frac{1}{2\pi}\int_0^{2\pi} dk\; e^{-ikm} \cdot \exp\left\{-t\left[(\alpha^2 + 16V^2 \sin^2(k/2))^{1/2} - \alpha\right]\right\} \quad . \tag{2.80}$$

It does not reduce to (2.8) unless $V/\alpha \to 0$ and represents motion which is faster than (2.8).

These results are the consequence of spatial nonlocality that develops in the (memory functions and) transition rates in the Master equation, even when the interaction matrix element V itself is local. More generally, the range of F_{mn} is longer than that of V_{mn}. These effects may be looked upon as arising from the necessity of considering higher perturbation terms in orders of (V/α) when the latter is not too small. In any case, the effect is very real and can make its presence felt in experiments such as transient grating observations. We shall discuss them in Chap. 3.

There are subtle questions which arise in the process of obtaining (Markoffian) Master equations such as (2.27) from the GME. Should the transition rates F_{mn} in the latter be written as integrals of the "broadened" memories $\mathcal{W}_{mn}(t)$ such as (2.72-76)? Or should we merely take the infinite size limit as in the case of (2.62) and make $\mathcal{W}_{mn}(t)$ integrable *without* the addition of any bath interactions? The latter procedure is mathematically possible, eliminates Poincaré recurrences, and results in an acceptable Master equation and physical solutions /2.29/. However, since irreversibility and the passage from the GME to the Master equation are really consequences of bath interactions, this second procedure is clearly artificial. The correct method, unlike the latter, works also for finite systems and uses the Markoffian approximation on the memories (2.72-76). For the memories of (2.72), i.e., for a crystal of arbitrary dimensions and range in V_{mn}, we then get, for the rates in (2.27),

$$F_{mn} = - \sum_k \left[e^{-ik(m-n)} \Big/ \sum_q \left(\alpha + \frac{(V^{k+q} - V^q)^2}{\alpha} \right)^{-1} \right] . \tag{2.81}$$

In the case of the one-dimensional infinite crystal, (2.81) reduces to

$$F_{mn} = \left[(-1)^{|m-n|+1} (\alpha^2 + 8V^2)^{1/2} P_{1/2}^{|m-n|}(\beta') \, \Gamma(3/2) \right] \times \left[\beta'^{1/2} \, \Gamma(|m-n| + \tfrac{3}{2}) \right]^{-1} \tag{2.82}$$

where $\beta' = [\alpha(\alpha^2 + 16V^2)^{1/2}]^{-1} (\alpha^2 + 8V^2)^{1/2}$ and Γ and $P_{1/2}^m$ are, respectively, the gamma function and the Legendre function of fractional order /2.29/.

A more extensive discussion of this problem of deriving expressions for rates F_{mn} as well as of the phenomenon of spatial nonlocality of memories and rates can be found in the analysis of KENKRE /2.29/. The first of these has also been commented on by SUBRTA and CAPEK /2.41/.

2.6.3 Linear Exciton-Phonon Coupling

Exciton interaction with the bath was taken into account phenomenologically in the last section. We shall now demonstrate the method of microscopic evaluation of memory functions from a given Hamiltonian which includes interactions with the bath explicitly. The model considered has been treated by a number of authors /2.23-25,

42,43/ for a variety of purposes. It has the Hamiltonian

$$H = E_0 \sum_m a_m^\dagger a_m + \sum_{m \neq n} V_{mn} a_m^\dagger a_n + \sum_q \omega_q b_q^\dagger b_q + \sum_{m,q} X_q^m \omega_q (b_q + b_{-q}^\dagger) a_m^\dagger a_m \quad , \tag{2.83}$$

and it describes excitons with site energy E_0, moving with matrix elements V_{mn} among sites m , n , etc. and interacting with phonons of energy (or frequency) ω_q through a term which is site-diagonal, linear in the phonon coordinate, and has the coupling constant X_q^m.

The evaluation of the memory functions is a trivial exercise if the last term in (2.83) is absent. A Bloch transformation, as for electrons in a metal, solves the problem immediately. That transformation would therefore be called for if the exciton-phonon interaction were small. In systems of interest to this review, it is, however, not negligible. The evaluation of $\mathcal{W}(t)$'s is a straightforward exercise also when the second term in (2.83), describing exciton motion, is absent, the useful device being now the displaced-oscillator tranformation /2.23-25,42-44/. If neither the second term nor the last term is small, but if the coupling constant X_q^m is independent of m, i.e., of the site, diagonalization and therefore the evaluation of $\mathcal{W}(t)$'s is again straightforward. The method in this case employs /2.36/ the Bloch and the displaced-oscillator transformations successively. However, when none of the above three simplifications is applicable, we must evaluate $\mathcal{W}(t)$'s through the approximation method discussed in Sect. 2.3.

The procedure consists in selecting a convenient partitioning of H into an unperturbed H_0 and a perturbed V, choosing the "grains" of coarsegraining operators to sum over all phonon states, selecting an appropriate factor Q, and then applying directly (2.23,24). We exhibit the results of this procedure carried out for thermal Q's. However, we first apply the displaced-oscillator transformation to (2.83) and reexpress it as

$$H = \sum_m (E_0 - \sum_q |X_q^m|^2 \omega_q) A_m^\dagger A_m + \sum_q \omega_q B_q^\dagger B_q + \sum_{m \neq n} V_{mn} \, e^{\alpha_m^\dagger} \, e^{\alpha_n} \, A_m^\dagger A_n \quad . \tag{2.84}$$

Following the usual method /2.23-25/, we have dropped a term involving exciton-exciton interactions and have introduced dressed exciton operators A , $A^\dagger$ and dressed phonon operators B , $B^\dagger$. They are given, for instance, by

$$A_m = a_m \, e^{\alpha_m^\dagger} \quad , \tag{2.85}$$

$$\alpha_m = \sum_q X_q^m \, (b_q - b_{-q}^\dagger) \quad . \tag{2.86}$$

The operator $B^\dagger$ creates an excitation of the "displaced oscillator", and $A^\dagger$ creates an excitonic polaron. The reason for applying the transformation to (2.83) is that we may now partition the Hamiltonian (2.84) into an H_0, which includes its first and second terms, and a "small" site-to-site interaction V described by the third term. The site energies are lowered by polaronic binding energy terms and the motion interaction is phonon-influenced unlike in (2.83).

We consider a dimer for simplicity: the m , n take only the values 1 and 2. The application of (2.23,24) to (2.84) gives /2.17,18/

$$\mathcal{W}_{mn}(t) = 2\,|V_{mn}|^2 \exp\left\{-\sum_{r,s}[h_{rs}(t) - h_{rs}(0)]\right\} , \tag{2.87}$$

where each r and s takes values m and n, and

$$h_{rs}(t) = -\left(|X_q^r|^2 + |X_q^s|^2 - X_q^r X_{-q}^s - X_q^s X_{-q}^r\right) \times \left(N_q\, e^{i\omega_q t} + (N_q + 1)\, e^{-i\omega_q t}\right) , \tag{2.88}$$

where $\beta = 1/k_BT$, k_B is the Boltzmann constant, and N_q is the Bose distribution $N_q = [\exp(\beta\omega_q) - 1]^{-1}$. Equation (2.87) thus gives the explicit memories in terms of microscopic quantities: phonon frequencies ω_q, coupling constants X_q^m, and interaction matrix elements V_{mn}. It also gives the explicit dependence of $\mathcal{W}(t)$'s on the temperature T.

The memory in (2.87) should be compared to the phenomenological dimer memory (2.73). The simple single exponential term in the latter is replaced by a much richer one in the microscopic evaluation. This result was obtained by KENKRE and RAHMAN /2.17/. A similar result has appeared, more recently, in the work of CAPEK and RIPS /2.45/.

In analyzing the physical meaning of the memory (2.87) one finds /2.18/ an unexpected feature. The functions h(t) of (2.88), which are effectively Fourier transforms of products of the phonon density of states, coupling constants, and related factors, have appeared in other contexts such as the analysis of optical spectra by LAX /2.46/ and MARADUDIN /2.47/. They are known, or at least expected, to vanish for long times for all nonpathological systems. But if the h(t)'s vanish for long times, the memories $\mathcal{W}(t)$, which contain the h(t)'s in the exponent, do *not* vanish as $t \to \infty$. Indeed, they decay from the value $2|V_{mn}|^2$ to the value $2|\tilde{V}_{mn}|^2$, where

$$|\tilde{V}_{mn}|^2 = |V_{mn}|^2 \exp\left\{-\sum_q \left(|X_q^m|^2 + |X_q^n|^2 - X_q^m X_{-q}^n - X_q^n X_{-q}^m\right) \coth\,(\beta\omega_q/2)\right\} . \tag{2.89}$$

We have retained SILBEY's usage /2.23-25/ of the tilde to avoid too much variety in notation but point out that here, unlike everywhere else in this review, the tilde does *not* denote Laplace transforms.

SILBEY /1.1/ has expressed the opinion that this feature of the memory is an artifact of the perturbation technique used by KENKRE and RAHMAN /2.17/ and has suggested that the thermalized portion of the third term in (2.84) should be removed from V and included in H_0. While this procedure does make the $\mathcal{W}(t)$ decay to zero, there is little guarantee that the resulting approximation is any better than that in /2.17/. In fact, KENKRE /2.18/ has argued on the basis of model calculations that the approximation procedure of SILBEY could well be considerably worse than that in /2.17/, as far as the solutions for the probabilities are concerned, and that the procedure of extracting the thermalized portion of V is itself an artificial and rather violent manner of replacing $W(t)$ by $[W(t) - W(\infty)]$.

One is faced with the following question: should memory functions always decay to zero as $t \to \infty$? It would appear that they should for real systems. However, we know that for a two-site system with no bath, $W(t)$ is a constant. In fact, if we accept $W(t)$ as being of the form $\sum_i f(z_i) \cos z_i t$ [see the general fomulae (2.23, 24)], we see that $W(t)$ will decay from the value $\sum_i f(z_i)$ at $t = 0$ to the value $f(0)$ at $t \to \infty$. A nonzero $W(\infty)$ is thus a natural consequence of (2.23) unless $f(0)$ is itself zero. For real systems, interaction with the bath makes $f(0) = 0$. The model of (2.84), however, does not, even in the thermodynamic limit. We believe this is an artifact of the model and not of the perturbation approximation in /2.17/ and that it corresponds to the well-known fact /2.46,47/ that zero-phonon lines do not get broadened by linear interactions. The persistence of the zero-phonon lines is intimately connected to the persistence of the memory functions as will become clearer in Sect. 2.6.4 below.

The situation is, thus, that it is not known which of the approximations is preferable for the model of (2.84). If an exact diagonalization of the Hamiltonian in (2.84) were possible, the issue would be immediately settled. Models of the model, such as those mentioned in /2.18/ serve as no more than indicators. It is hoped that future work will clarify this problem.

Fortunately, the disagreement exists only for the singular model of (2.84) and disappears for all real systems. As soon as the model is considered to be open, i.e., in interaction with a further bath, the latter can cause a decay of the undecaying component of $W(t)$. This further bath may represent, for instance, nonlinear interactions in the phonon coordinate, which have been neglected in (2.83,84). It is well-known that nonlinear interactions can broaden zero-phonon lines /2.46,47/. That they impose additional decays on $W(t)$ can be shown easily and is particularly clear from the work of ABRAM and SILBEY /2.48/.

The $\tilde{V}_{mn}$ in (2.89) is the bandwidth of the excitonic polaron, while V_{mn} is the bandwidth of the bare exciton. The decay of the memory function from $2|V_{mn}|^2$ to $2|\tilde{V}_{mn}|^2$, given by the expression of KENKRE and RAHMAN /2.17/, has thus a natural physical meaning: the evolution of the process of dressing the exciton with virtual phonons. KENKRE has shown /2.18/ that the memory in (2.87) results in the coexistence of two types of motion: "tunneling" or "band motion" with the polaronic bandwidth $\tilde{V}_{mn}$ and "hopping" motion with a hopping rate which equals the integral from $t = 0$ to $t = \infty$ of $[W(t) - W(\infty)]$. The second type has a hopping character only for times much larger with respect to the decay time of $[W(t) - W(\infty)]$, and can certainly exhibit band character, i.e., coherence, at shorter times. Similarly, coherent motion with the polaronic transfer $\tilde{V}_{mn}$ will become incoherent for times long with respect to the time for any additional scattering imposed on the model. It is interesting that even in the absence of such scattering, no unphysical behaviour is found in the solution of the GME, although $\mathcal{W}(t)$ is not integrable from $t = 0$ to $t = \infty$ /2.18/.

The model of (2.83,84) is simple enough, familiar enough, and even rich enough to deserve serious treatment in transport theory. It appears, however, that it has the artificial feature that its $\mathcal{W}(t)$ does not decay to zero. Even if one takes issue with this assertion, one has to accept the fact that for this model, the corresponding monomer zero-phonon line is an unbroadened, i.e., unphysical, δ-function even in the thermodynamic limit. One must therefore find a way of using the insights provided by this model without being hindered or distracted by its unphysical features. We suggest that the proper way consists of combining the results of this section and those of the phenomenological treatment of the last section and writing

$$\mathcal{W}_{mn}(t) = 2|V_{mn}|^2 \, e^{-\alpha t} \exp\left\{-\sum_{r,s} [h_{rs}(t) - h_{rs}(0)]\right\} , \tag{2.90}$$

where α represents all the important bath interactions left out of the Hamiltonian in (2.83,84).

2.6.4 Generalization of Förster's Spectral Prescription

A major contribution made by FÖRSTER /1.15,17/ to the theory of excitation transfer in molecular aggregates consists in his prescription for calculating the transition rates F_{mn} directly from optical spectra of monomers. In this section we describe a generalization of the FÖRSTER prescription /1.15,17/ to obtain the memory functions $\mathcal{W}_{mn}(t)$ directly from monomer spectra /2.16,49/.

We begin with (2.23,24) for the memory functions $\mathcal{W}(t)$ and merely rewrite the quantities therein in the notation of FÖRSTER /1.15,17/. The result is

$$\mathcal{W}_{mn}(t) = 2 \int_{W=0}^{\infty} \int_{\omega_n''=0}^{\infty} \int_{\omega_m'=0}^{\infty} \int_{W=-\infty}^{+\infty} d\omega_m' \, d\omega_n'' \, dW \, d(\Delta W) \, \cos(\Delta W t) \, g'(\omega_m') g''(\omega_n'')$$
$$\times \left| u_{mn}\left(\omega_m', \omega_n'', W_0 - W + \omega_m' + \frac{\Delta W}{2}, W - W_0 + \omega_n'' + \frac{\Delta W}{2}\right)\right|^2 \quad . \tag{2.91}$$

Most of the notation of /1.15,17/ is followed here. The labels m and n represent molecular sites, vibrational contributions to the energy are denoted by ω' for the (electronically) excited state and ω'' for the ground state. The energy of the purely electronic excitation on either site m or n is W_0, and W and ΔW are given by

$$W = W_0 + \frac{1}{2}(\omega_m' - \omega_m'' + \omega_n' - \omega_n'') \quad , \tag{2.92}$$

$$\Delta W = \omega''_m + \omega'_n - \omega'_m - \omega''_n \quad . \tag{2.93}$$

The g' and g" are products of thermal factors and density of states, and the Coulomb interaction matrix elements u_{mn} are evaluated partially in terms of energy-normalized states as in /1.15/.

As in the treatment of /1.15-17/, we introduce the solvent refractive index n_s and rewrite (2.91) in terms of the intermolecular separation R, the speed of light $\bar{c}$, the number of molecules per millimole $N' = 6.02 \times 10^{20}$, and the matrix element M(W), whose square is proportional to the transition probability density on an energy scale. Finally, using the connection between the latter and the fluorescence intensity per molecule $\bar{A}(W)dW$ in the range dW on the one hand, and the extinction coefficient $\bar{\varepsilon}(W)$ on the other, we obtain

$$\mathcal{W}_{mn}(t) = \frac{3[\ell n(10)]\hbar^4\bar{c}^4}{4\pi^2 n_s^4 N'} \frac{1}{R_{mn}^6} \int_{W=-\infty}^{W=+\infty} d(\Delta W/\hbar) \cos[(\Delta W/\hbar)t]$$
$$\times \int_{W=0}^{\infty} dW \left[\frac{\bar{A}(W - \Delta W/2)\,\bar{\varepsilon}\,(W + \Delta W/2)}{(W - \Delta W/2)^3 \ (W + \Delta W/2)}\right] \quad . \tag{2.94}$$

In (2.94) we have reintroduced 2π times the Planck's constant for convenience in comparison to expressions appearing elsewhere in the literature.

Equation (2.94) is the explicit prescription for calculating memory functions from modified optical spectra $\bar{A}(W)$ and $\bar{\varepsilon}(W)$. By the word "modified" we describe the fact that $\bar{A}(W)$ and $\bar{\varepsilon}(W)$ are to be divided by W^3 and W, respectively, before their overlap integrals are used to calculate $\mathcal{W}(t)$. The prescription is as follows. One calculates F_{mn} as in FÖRSTER's theory /1.15/ and renames it $F_{mn}(0)$. One then shifts the two modified spectral curves $W^{-3}\bar{A}(W)$ and $W^{-1}\bar{\varepsilon}(W)$ on the frequency axis, the former by $+(1/2)(\Delta W)$ and the latter by $-(1/2)(\Delta W)$ and redoes the overlap integral, calling it $f(\Delta W)$. This procedure is repeated for all values of ΔW. The resulting curve $F_{mn}(\Delta W)$ is then Fourier-transformed to yield the memory function:

$$\mathcal{W}_{mn}(t) = \frac{1}{\pi} \int_{-\infty}^{+\infty} d(\Delta W)\ F_{mn}(\Delta W) \cos(\Delta Wt) \quad . \tag{2.95}$$

Well-known theorems of Fourier analysis result in several immediate insights. How accurate the calculated $\mathcal{W}(t)$ is at short times will depend on how accurately the large ΔW separations have been taken into account. The $t = 0$ limit of $\mathcal{W}_{mn}(t)$ will be given by the integral of $F_{mn}(\Delta W)$ over the entire frequency range. Unless $F_{mn}(\Delta W)$ has a δ-function behaviour at $\Delta W = 0$, $d(\Delta W)F_{mn}(\Delta W)$ will be a true infinitesimal and the memory function will decay to zero at long times. It is here that we see that the model of (2.83,84) violates this condition on account of its inherent spectral zero-phonon lines. Fourier analysis also shows that the Markoffian approximation (2.26), whereby $\mathcal{W}(t)$ is replaced by the product of its integral from $t = 0$

to ∞ and of $\delta(t)$, allows us to recover FÖRSTER's prescription /1.15/ for transition rates F_{mn} as a particular approximation.

This treatment has been applied by KENKRE and KNOX /2.16,49/ to calculate memory functions for a number of systems directly from their monomer spectra: anthracene in cyclohexane, bacteriochlorophyll, adenosine monophosphate which all show incoherent behaviour, and F_3^+ centers in NaF and MgO:V^{2+}, which show some coherence in the memory functions corresponding to sharp peaks in the optical spectra.

While the notation used in (2.91-95) has been chosen to facilitate comparison to the FÖRSTER analysis /1.15/ and the relevant literature, it is possible that the details shown there may obscure the simplicity of the generalization that (2.94) provides. To appreciate the essentials, the memory function expression (2.23) should be compared to a standard Fermi Golden Rule expression such as

$$F_{mn} = 2\pi \sum_{\xi \in m} \sum_{\mu \in n} [Q_\mu / g_n] \, |\langle \xi | V | \mu \rangle|^2 \, \delta(E_\xi - E_\mu) \quad . \tag{2.96}$$

The basic difference is the $\cos[(E_\xi - E_\mu)t]$ term in (2.23) in contrast to $\delta(E_\xi - E_\mu)$ in (2.96). The separability of the dipole-dipole interaction inherent in the FÖRSTER theory /1.15/ results in F_{mn} being given by the overlap of the (modified) absorption and emission spectra. Whereas the factor $\delta(E_\xi - E_\mu)$ corresponds to a fixed overlap, the transform of $\cos[(E_\xi - E_\mu)t]$ corresponds to the variable shift by ΔW discussed above in the construction of the function $F_{mn}(\Delta W)$, whose transform yields $\mathcal{W}_{mn}(t)$. An appropriate physical description is that while energy conservation demanded by $\delta(E_\xi - E_\mu)$ is essential in the FÖRSTER prescription, the variable shift in the KENKRE-KNOX prescription corresponds to virtual processes at short times that do not require energy conservation. The practical question of how far the variable shift (ΔW) of the spectra should be carried out numerically, is therefore immediately answered: it is decided, by a simple reciprocity relation, by how small a time resolution is required in the $\mathcal{W}(t)$'s.

The connection between transport as described by $\mathcal{W}(t)$'s and optical spectra as described by $\bar{A}(W)$ and $\bar{\varepsilon}(W)$ becomes particularly transparent /2.18/ in the context of the model of Sect. 2.6.3. Let us first rewrite (2.94) or (2.95) in the form

$$\mathcal{W}_{mn}(t) = \text{const.} \int_{z=-\infty}^{+\infty} dz \cos zt \int_{\omega=0}^{\infty} d\omega \, I_m^a\left(\omega - \frac{z}{2}\right) I_n^e\left(\omega + \frac{z}{2}\right) \quad , \tag{2.97}$$

where I_m^a and I_n^e are the modified absorption and emission spectra, and then apply trivial Fourier arguments to reexpress (2.97) as

$$\mathcal{W}_{mn}(t) = \text{Re}\left\{ [\mathcal{I}_m^a(t)] \, [\mathcal{I}_n^e(t)]^* \right\} \quad . \tag{2.98}$$

Equation (2.98) provides another useful prescription to obtain memories from spectra. It involves $\mathcal{I}_m^a(t)$ and $\mathcal{I}_n^e(t)$, the Fourier transforms of the modified spectra $I_m^a(\omega)$

and $I_n^e(\omega)$. The superscripts a and e refer to absorption and emission, respectively, and m and n to molecular sites: the above prescription can be used also for obtaining $W(t)$'s for unlike molecules /2.16/.

The functions $I_m^a(t)$ and $I_n^e(t)$ are precisely the "characteristic functions" of LAX /2.46/. They describe spectra in the model with the Hamiltonian

$$H = E_0 a_m^\dagger a_m + \sum_q \omega_q b_q^\dagger b_q + \sum_q X_q^m \omega_q (b_q + b_{-q}^\dagger) a_m^\dagger a_m \tag{2.99}$$

and are given, for instance, by

$$I_m^a(t) = \text{const.} \exp\{-[h_{mm}(t) - h_{mm}(0)]\} \quad . \tag{2.100}$$

The intimate relationship between spectra and transport should be quite clear in terms of this analysis. If we compare the Hamiltonians in (2.99) and (2.83), we see that they are identical except for the fact that the former describes monomer and the latter a dimer, crystal, or aggregate. If we compare (2.100) and (2.87), we discover the precise form of the close connection between spectra and transport. Exciton-phonon interactions, temperature, and other microscopic parameters determine the functions $h_{mm}(t)$, which in turn decide the form of the monomer spectra in (2.100) and the memory functions in (2.87). A broad spectral lineshape corresponds to fast decay of $W(t)$'s and thus to incoherent motion. Sharp peaks in the lineshape should warn us of coherent transport. These conclusions should not be surprising since it is generally the same agency that broadens lines and randomizes the phase of the moving exciton. It is important to stress here that inhomogeneous broadening is entirely neglected in this discussion. The various spectral prescriptions analyzed above assume that the line is homogeneously broadened. A usable prescription for connecting $W(t)$'s to inhomogeneous lineshapes is being worked on but is not yet available in a satisfactory form.

Sharp spectral lines and coherent motion go together, as we have seen above. The zero-phonon line of (2.99), which refuses to be broadened unless additional interactions are included, has been well-known /2.46/ to produce the singularity in $I_m(\omega)$, the transform of $I_m^a(t)$. We see from (2.100) that this behaviour corresponds to the fact that as $t \to \infty$, $h_{mm}(t) \to 0$. Therefore, that $I_m^a(t)$ in (2.100), and consequently the memory functions $W(t)$ in (2.98), do *not* tend to 0 as $t \to \infty$, is a result of the zero-phonon line singularity of the model. It is quite reasonable to maintain, therefore, that the nondecaying nature of $W(t)$'s for the model of (2.83) is a property of the model rather than of the perturbation approximation used to obtain (2.87).

We emphasize, as HOCHSTRASSER and PRASAD have also done /2.50/, that very useful insights into motion can be gained from a study of spectral lineshapes. The GME-spectra connections described here show the quantitative relationship between coherence and broadening. Since these various connections are especially attractive to the experimentalist, (they appear to connect observations in one domain to those in another without the need of specific models), it is important to point out some of the pitfalls. One of these has been mentioned above: inhomogeneous broadening. Another is the effect of phonon correlations. The breakdown of the FÖRSTER-DEXTER spectral prescriptions /1.15-17/ has been carefully analyzed by SOULES and DUKE /2.43/ in this context. Since the prescription discussed in this section is a generalization of that in /1.15-17/, it shares the essential drawbacks as well as the advantages of the latter. However, it is possible to write down an explicit correction factor for the spectrally obtained GME memory, to take into account the issues raised by SOULES and DUKE /2.43/. The expression for $\mathcal{W}(t)$ in (2.87) may be rewritten in the form

$$\mathcal{W}_{mn}(t) = [\mathcal{W}^{sp}_{mn}(t)]\,[\mathcal{W}^{co}_{mn}(t)] \quad , \tag{2.101}$$

where $\mathcal{W}^{sp}_{mn}(t)$ is given by the spectral prescription (2.94,95,97, or 98) and the correction factor is multiplicative:

$$\mathcal{W}^{co}_{mn}(t) = \exp\left\{-[h_{mn}(t) + h_{nm}(t) - h_{mn}(0) - h_{nm}(0)]\right\} \quad . \tag{2.102}$$

We see from (2.88) that the correction factor is 1 if the two molecules m and n can be assumed to be connected to separate phonon baths. The usual spectral prescription is recovered in this case. However, phonon correlations will generally make $h_{mn}(t)$ nonzero for $m \neq n$. Equations (2.101,102) find nontrivial use in such a case.

Much can be learnt about exciton motion through spectral gateways along the lines discussed in this section. KENKRE /2.18/ has shown what kinds of motion are related to what kinds of spectra by solving the GME for several representative spectra. The coexistence of coherent and incoherent motion has been shown to correspond to the coexistence of sharp (zero-phonon) lines and broad bands and the effect of polaronic dressing has been discussed. Figure 2.2, which emerges from that analysis, shows the multiple-time-constant nature of $\mathcal{W}(t)$ arising from structured spectra. Oscillations arising from the Stokes shift have been suppressed for the sake of clarity, and mirror symmetry and Lorentzian lineshapes have been assumed in the spectral lines for simplicity. The rapidly decaying part of $\mathcal{W}(t)$ arises from the broad band and the slow component from the sharp peak.

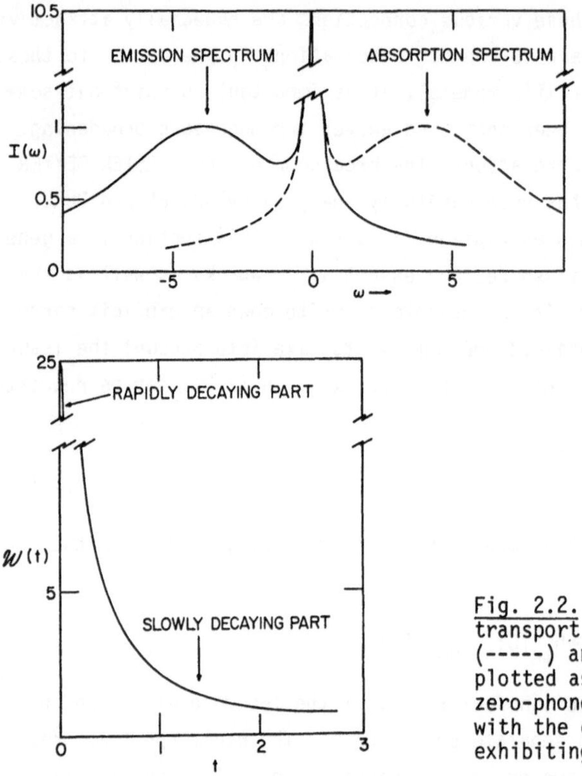

Fig. 2.2. Connection between spectra and transport coherence. "Modified" absorption (-----) and emission (———) spectra are plotted as functions of frequency, showing zero-phonon peaks and sidebands, along with the corresponding memory function, exhibiting multiple time constants

2.6.5 Present Status of Available Memories

In Sects. 2.6.1-4 we have evaluated memory functions for a variety of systems. Exact calculations for pure crystals as well as crystals interacting with a phenomenologically treated bath are thus available. The crystal can have arbitrary structure and dimension. Translational invariance has allowed us to carry the analysis through without approximations. The inversion of transforms in expressions such as (2.54) is analytically possible for several cases such as the one-dimensional infinite crystal and the equal interaction n-mer. Numerical inversion is also straightforward for other cases since numerical Fourier techniques are well established. We have also shown how to calculate $\mathcal{W}(t)$'s microscopically from exciton-phonon interaction Hamiltonians through the use of perturbation procedures. Finally, we have given explicit prescriptions to obtain them from optical spectra,bypassing model calculations.

A wealth of information on memory functions is thus available and will be put to practical use in Chap. 3. However, efforts to obtain usable $\mathcal{W}(t)$'s must be continued. We know how to treat pure systems, phenomenological baths, dimers, and how to use spectral information directly. We also know how to combine all these sources through physical arguments. We thus can construct memory functions that possess the spatially long-range character arising from the analysis of strong inter-site coupling in Sect. 2.6.1 and 2.6.2, the microscopically accessible decay information arising from the theory in Sect. 2.6.3, and the spectral details from the arguments of Sect. 2.6.4. However, calculations are still being done to unify all these aspects into a single picture and procedure and to include further features such as the effects of phonon correlations.

2.7 Relation of GME's to Other Transport Entities

Thus far in this chapter, generalized master equations have been motivated, formulated, analyzed, and utilized, and their memory functions have been computed, evaluated, extracted, and investigated. Since the GME is a relatively new transport instrument in the exciton field, it is of interest to understand the relations that it bears to other transport entities already existing in this and similar fields. We shall discuss these relations to continuous-time random walks, scattering functions, velocity autocorrelation functions, stochastic Liouville equations, and the (Pauli) Master equation.

2.7.1 Relation of GME to the Pauli Master Equation

The Master equation is formally the Markoffian limit of the GME. This connection has been amply discussed earlier in this chapter. The precise form of the Markoffian assumption is (2.26). The physical meaning of replacing the memory functions $\mathcal{W}(t)$ by a δ-function in time [multiplied by $\int_0^\infty dt'\ \mathcal{W}(t')$] is that the relevant $\mathcal{W}(t)$'s decay very rapidly on the time scale of interest. In other words, the solutions of the GME tend to those of the Master equation at sufficiently long times. Expressions such as (2.96) for the rates F_{mn} of the Master equations should be compared to expressions such as (2.23,24) for the memory functions $\mathcal{W}_{mn}(t)$ of the GME. The latter clearly contain much more information than the former.

If the memory functions in a given GME do decay very rapidly on the time scale of an experimental probe time, the GME has no advantage over the Master equation. By the same token, the GME is able to analyse coherence and other details of motion at short times, which are not accessible to the Master equation. There is a general theorem, sometimes called /2.51/ the BALESCU-SWENSON theorem /2.52/, which states that a non-Markoffian equation gives results for steady-state quantities which are identical to those given by the Markoffian Master equation. The essential argument is as follows. Quantities calculated from the GME may be expressed in the Laplace domain in terms of the Laplace transform of the memories. Steady-state observables involve integrals from $t = 0$ to $t = \infty$ or, equivalently, the limit $\varepsilon \to 0$ of the

Laplace transform. Therefore, the use of $\tilde{W}(\varepsilon)$ followed by the eventual limit $\varepsilon \to 0$ gives no more information than the use of $\tilde{W}(0)$ from the very beginning. And to replace $\tilde{W}(\varepsilon)$ by $\tilde{W}(0)$ for all ε's is precisely to replace the GME by the Master equation.

To make this more concrete, let us write, from (2.48,49)

$$\tilde{P}^k(\varepsilon) = \frac{P^k(0)}{\varepsilon + \tilde{A}^k(\varepsilon)} \quad . \tag{2.103}$$

If a steady-state quantity involves $\int_0^\infty dt\, P^k(t)$, i.e., $\tilde{P}^k(0)$, we see from (2.103) that we might as well have used the Master equation which replaces $A(t)$ by $\delta(t)\cdot[\int_0^\infty dt'\, A(t')]$ in (2.48).

This argument, while valid in the analysis of BALESCU and SWENSON /2.52/, fails in the exciton context. Excitons generally obey transport equations such as the GME but in a form augmented to include radiative (or nonradiative) decay processes. Thus, if we add a term $P_m(t)/\tau$ to (2.48), we must replace (2.103) by

$$\tilde{P}^k(\varepsilon) = \frac{P^k(0)}{\varepsilon + \frac{1}{\tau} + \tilde{A}^k(\varepsilon + \frac{1}{\tau})} \quad . \tag{2.104}$$

We now find that a steady-state quantity involving $\tilde{P}^k(0)$ as before, is sensitive to $\tilde{A}^k(1/\tau)$ rather than to $\tilde{A}^k(0)$. Unless $\tau \to \infty$, the results of the GME are thus different from those of the Master equation *even for steady-state* quantities.

We shall explore the above issue in Chap. 3 in the discussion of the effects of coherence on quantum yield. The message here is that although the GME reduces to the Master equation at sufficiently long times, probe times (τ in the above case), if short enough, can make GME predictions different from those of the Master equation. Nevertheless, many situations exist, particularly in complex, dirty, or high-temperature systems wherein the memory functions are so short-lived, often less than a tenth of a picosecond, that they are $\delta(t)$'s for all practical purposes. In such cases the GME is identical to the Master equation.

2.7.2 Relation of GME to the Stochastic Liouville Equation

The first unified treatment of coherent and incoherent motion of excitons was given in terms of the stochastic Liouville equation (SLE). The study of the relations the GME bears to the SLE is therefore particularly interesting. The SLE was first introduced by HAKEN and STROBL /2.53,54/ on the basis of semistochastic considerations: they treated exciton motion in two parts, the coherent one through a Hamiltonian and the incoherent one as a random process. The treatment was greatly extended by HAKEN and REINEKER /2.55/ and others. GROVER and SILBEY /2.23/ began the analysis of exciton motion from an entirely different starting point, the linearly interacting exciton-phonon model of Sect. 2.6.3, and used nonstochastic arguments. But they too arrived at a transport equation which is formally identical to that of HAKEN and STROBL. The transport equation has the essential features of what KUBO /2.56/ and others have called the stochastic Liouville equation. We therefore use the term SLE to refer to the result, irrespective of whether it arises from a HAKEN-STROBL-REINEKER treatment or a GROVER-SILBEY one. Exciton transport analysis from the SLE viewpoint has been developed in the second article in this book; here we shall be interested primarily in the formal structure of the SLE. It describes the evolution of the elements ρ_{mn} of the exciton density matrix:

$$\frac{\partial \rho_{mn}}{\partial t} = -i \sum_r (\tilde{V}_{mr}\rho_{rn} - \tilde{V}_{rn}\rho_{mr}) + 2\delta_{m,n} \sum_r (\gamma^s_{mr}\rho_{rr} - \gamma^s_{rm}\rho_{mm})$$
$$- (1-\delta_{m,n})\, 2\Gamma^s \rho_{mn} + (1-\delta_{m,n})\, 2\bar{\gamma}_{m-n}\rho_{nm} \quad . \tag{2.105}$$

The superscripts s on γ and Γ refer to the SLE and distinguish γ^s and Γ^s from other γ's and Γ's used in this article. The first term on the right side of (2.105) describes coherent motion. We have used the notation $\tilde{V}_{mr}$ to denote the matrix elements because, in the SILBEY treatment, they are identical to the polaronic bandwidth quantities discussed in (2.89) and Sect. 2.6.3. Needless to say, here too the tildes do not refer to Laplace transforms. The second term in (2.105) describes incoherent or "hopping" motion with transition rates $2\gamma^s_{mr}$. Note that the term involves only *probabilities* $\rho_{mm} \equiv P_m$ and $\rho_{rr} \equiv P_r$. The last two terms describe the destruction of off-diagonal elements of ρ at rates $2\Gamma^s$ and $2\bar{\gamma}_{m-n}$, which is indicative of bath interactions.

The first term on the right side of (2.105), if present alone, makes (2.105) the same as the von Neumann equation describing pure coherent motion. If the third term is added, we have the augmented von Neumann equation (2.66) which was given for excitons by AVAKIAN et al. /2.39/ and which we have analyzed in Sect. 2.6.2. If the right side of (2.105) has only the second term, we have the Master equation (2.27). It is thus clear that the SLE achieves its unification of coherent and incoherent motion by straightforward addition of terms corresponding to the two kinds of motion. The GME, however, has the combined behaviour built into its memory functions. To understand the relations between the GME and the SLE, we must therefore recast one in the form of the other.

For simplicity, we begin with a dimer. The SLE now has the form

$$\frac{d\rho_{11}}{dt} = -i\tilde{V}(\rho_{21} - \rho_{12}) + A'(\rho_{22} - \rho_{11}) \quad , \tag{2.106}$$

$$\frac{d\rho_{12}}{dt} = -i\tilde{V}(\rho_{22} - \rho_{11}) + B'(\rho_{21} - \rho_{12}) \tag{2.107}$$

with two other similar equations. Here V , A', and B' are derived in a straightforward fashion from the quantities in (2.105). Laplace transforms allow us to rewrite /2.57/ the GME (2.106,107):

$$\frac{d\rho_{11}(t)}{dt} = \int_0^t dt' \, \mathcal{W}(t-t') \, [\rho_{22}(t') - \rho_{11}(t')] \tag{2.108}$$

and a similar equation for $\rho_{22}(t)$, the memory $\mathcal{W}(t)$ being given by

$$\mathcal{W}(t) = 2\tilde{V}^2 \, e^{-2B't} + A' \, \delta(t) \quad . \tag{2.109}$$

Equation (2.109) shows the relation between the GME (2.108) and the SLE (2.106, 107) for the dimer: the SLE is equivalent to a GME with a memory function which is a sum of an exponential and a δ-function term. The strength of the $\delta(t)$ term is the "hopping" rate A', the decay constant of the exponential term is essentially the rate of destruction of off-diagonal terms of ρ in (2.107), and the initial value of the exponential term is $2\tilde{V}^2$.

To study the GME - SLE relationship in the dimer, the above simple substitution of the Laplace transform of (2.107) in (2.106) suffices /2.57/. For the extended crystal, it is necessary to use a systematic method /2.34/. One applies ZWANZIG projection operators to the SLE and uses simple properties of those operators. The development is similar to that in Sect. 2.6.2 and includes the latter as a special case. We shall therefore explain it only briefly, with attention on features not shown in Sect. 2.6.2. If we cast the SLE (2.105) in the form (2.16), neglect the last term in (2.105), and apply diagonalizing projection operators, we obtain (2.20) with an additional term in its right-hand side. This term is $\mathcal{P}L\mathcal{P}\rho(t)$. In ZWANZIG's treatment /2.9/, it is identically zero. Here, however, L is not merely a commutator operator but has also the "stochastic" or "hopping" part corresponding to the second term in (2.105). It is amusing that the term $\mathcal{P}L\mathcal{P}\rho(t)$, which in most applications of projection techniques vanishes identically, makes an important contribution here. That term and others from the analysis given in Sect. 2.6.2 result in the following memory expression:

$$\mathcal{W}_{mn}(t) = e^{-2\Gamma^s t} \, \mathcal{W}^c_{mn}(t) + 2\gamma^s_{mn} \, \delta(t) \quad . \tag{2.110}$$

Comparison to (2.71) shows that there is now a new part in the memory, which is proportional to $\delta(t)$, and that $2\Gamma^s$ takes the place of α.

The SLE (2.105), without its last term, is equivalent to the GME with memory functions given by (2.110). We see here a generalization of the dimer result (2.109). The first terms in the right-hand side of both (2.109) and (2.110) represent "band motion" with scattering controlled by the exponent in the exponential. The second term represents "hopping motion". The SLE, therefore, corresponds to a GME with memories that have two additive parts, one of which has an infinitely small time constant.

It is possible to derive various special cases of (2.110). Thus, for the infinite one-dimensional crystal with nearest-neighbour interaction matrix elements $\tilde{V}$ and hopping rates γ^s, we have /2.34/

$$\begin{aligned} \mathcal{W}_{mn}(t) = 2\tilde{V}^2\, e^{-2\Gamma^s t} \Big\{ & J^2_{m-n+1}(2\tilde{V}t) + J^2_{m-n-1}(2\tilde{V}t) + 2J_{m-n-1}(2\tilde{V}t)\, J_{m-n+1}(2\tilde{V}t) \\ & - 2J^2_{m-n}(2\tilde{V}t) - J_{m-n}(2\tilde{V}t)\, [J_{m-n+2}(2\tilde{V}t) + J_{m-n-2}(2\tilde{V}t)] \Big\} \\ & + 2\gamma^s\, \delta(t)\, [\delta_{m,n+1} + \delta_{m,n-1}] \quad . \end{aligned} \tag{2.111}$$

The essential formal relationship of the SLE and the GME is given by (2.109-111). The SLE contains the inherent assumption that exciton motion occurs simply through two additive channels: a "band" channel and a "hopping" channel. It is not possible to use the SLE in the above form to analyze exciton motion with arbitrary kind of coherence. We have seen that the actual memory functions can have involved behaviour which may or may not be appropriately represented by the sum of an exponential and a δ-function. This study of the relationship of the GME and the SLE focuses attention on both the limitations and the power of the SLE. While unable to tackle transport corresponding to arbitrary memories, the SLE is especially useful when there do exist two sharply differentiated channels of motion. The spectral relations discussed in Sect. 2.6.4 suggest that the SLE is particularly appropriate for systems with monomer spectra containing sharp zero-phonon lines in addition to broad bands. An example is provided by Fig. 2.2. The sharp line would correspond to the first and the broad band to the second term in (2.110).

Many more interesting features of the SLE - GME relationship exist, particularly in the spectral context. They concern polaron or dressing effects and spectral connections. For space reasons we shall not discuss them but refer the interested reader to /2.18,57/.

The partial polaronic dressing treatment of YARKONY and SILBEY /2.25/ and recent work of SILBEY and MUNN /2.24/ have made the SLE a much richer transport entity. The work of ABRAM and SILBEY /2.48/ has shown further subtleties existing in transport-spectra connections. Although in the form discussed above the SLE does correspond to a fixed form of memory with one part a δ-function, it can be generalized /2.57/ very simply to remove these limitations. The δ-function part in (2.110) arises from what is in effect a partial Markoffian approximation on $\mathcal{W}_{mn}(t)$. It appears in slightly different forms in the treatment of HAKEN-STROBL-REINEKER /2.53-55/ and of SILBEY and collaborators /2.23-25/. It is possible to undo the partial Markoffian approximation and use a generalized SLE which has a non-Markoffian part. Indeed, such non-Markoffian terms already exist in the analysis of SILBEY and collaborators

/2.23-25/ and in the treatment of HAKEN and STROBL /2.53,54/ before the Markoff process assumptions are made. KENKRE has shown /2.57/ that unphysical effects can arise in certain applications if this generalization is not used. SUMI /2.58/ and KOHNE et al. /2.59/ have used such generalized SLE's.

We have discussed some of the limitations of the SLE above. It is important to point out one significant advantage that it possesses over the GME. It describes the evolution of off-diagonal as well as diagonal elements of the exciton density matrix. For the analysis of observables requiring ρ_{mn} with $m \neq n$, the GME is of little use since it is constructed especially to eliminate the off-diagonal elements of ρ. The SLE must be employed in such cases. An example of this situation is provided by the theory of depolarization of fluorescence given by RAHMAN et al. /2.60/. Other examples can be found in the work of REINEKER and others, described in the second article in this book.

Finally, we comment on the neglect of the last term in (2.105). That term is perhaps the least physical part of the SLE but provides the most severe obstacle in mathematical manipulation. Several authors have dropped it /2.34 , 2.61/ on physical grounds. KOHNE and REINEKER /2.40/ have calculated the $\mathcal{W}_{mn}(t)$ for the full SLE including the troublesome term. But the computational exercise is quite complicated and novel physical behaviour does not yet seem to have emerged from the additional memory. While that term undoubtedly possesses additional information, it is likely that the essential physics is contained in the simplified SLE.

2.7.3 Relation of GME to Continuous-Time Random Walks

The continuous-time random walk (CTRW) consists of a generalization of the ordinary discrete-time random walk. The generalization is due to MONTROLL and WEISS /2.62/. The CTRW has found wide use in recent times, particularly as a result of the work of SCHER et al. /2.63-65/ in charge carrier transport. Whereas the ordinary random walk is typified by

$$P_m(r\tau') = \sum_n Q_{mn}\, P_n(r\tau' - \tau') \quad , \tag{2.112}$$

where the probability P_m of the walker after the rth time step of fixed duration τ' is determined by the probabilities P_n at the immediately earlier time step,

$$P_m(t) = P_m(0)\left[1 - \int_0^t dt'\, \psi^P(t')\right] + \int_0^t dt' \sum_n Q_{mn}\psi^P(t-t')\, P_n(t') \tag{2.113}$$

is a generalization which is equivalent to the original formulation of the CTRW. The "pausing-time distribution function" $\psi^P(t)$ describes how long the walker may pause at a site before taking the next step. The first term in (2.113) represents the probability that the walker has continued to linger at site m.

The CTRW had been proposed /2.62/ as only a richer version of the ordinary random walk. However, as KENKRE et al. /2.66/ showed, it is exactly equivalent to a simplified form of the GME. What is perhaps more interesting is that the *general* GME can be shown to be equivalent to a natural extension of the CTRW of (2.113). This demonstration, given by KENKRE and KNOX /2.16/, consists merely of rewriting the GME (2.11) in the Laplace domain

$$\varepsilon \tilde{P}_m(\varepsilon) - P_m(0) = \sum_n \tilde{\mathcal{W}}_{mn}(\varepsilon)\, \tilde{P}_n(\varepsilon) - \left[\sum_n \tilde{\mathcal{W}}_{nm}(\varepsilon)\right] \tilde{P}_m(\varepsilon) \quad , \tag{2.114}$$

rearranging it as

$$\tilde{P}_m(\varepsilon) = P_m(0)\left[\varepsilon + \sum_r \tilde{\mathcal{W}}_{rm}(\varepsilon)\right]^{-1} + \left[\varepsilon + \sum_r \tilde{\mathcal{W}}_{rm}(\varepsilon)\right]^{-1} \sum_n \tilde{\mathcal{W}}_{mn}(\varepsilon)\, \tilde{P}_n(\varepsilon) \quad , \tag{2.115}$$

and inverting it immediately into

$$P_m(t) = P_m(0)\left[1 - \int_0^t dt'\, \psi_m^p(t')\right] + \int_0^t dt' \sum_n \mathcal{Q}_{mn}(t-t')\, P_n(t') \quad . \tag{2.116}$$

Equation (2.116) is completely equivalent to the GME and is a natural generalization of (2.113) to include $\psi^p(t)$'s which depend on m. The relation of the generalized CTRW of (2.116) and the GME (2.11) is

$$\tilde{\mathcal{Q}}_{mn}(\varepsilon) = \left[\tilde{\mathcal{W}}_{mn}(\varepsilon)\right]\left[\varepsilon + \sum_r \tilde{\mathcal{W}}_{rm}(\varepsilon)\right]^{-1} \quad , \tag{2.117}$$

$$\tilde{\psi}_m^p(\varepsilon) = \left[\sum_r \tilde{\mathcal{W}}_{rm}(\varepsilon)\right]\left[\varepsilon + \sum_r \tilde{\mathcal{W}}_{rm}(\varepsilon)\right]^{-1} \quad . \tag{2.118}$$

This relation does *not* assume $\mathcal{W}_{mn}(t)$'s that are separable as $F_{mn}\phi(t)$ nor does it make any assumption about the set of states m , n. In particular, no translational invarience or crystal symmetries are assumed. In the light of this equivalence, the CTRW becomes a microscopically derivéd transport instrument. Evaluation of the ψ^p's and $\mathcal{Q}$'s may be carried out through the methods discussed in Sect. 2.6 for $\mathcal{W}(t)$'s and the relations (2.117,118). Further elaboration may be found in /2.15/ and various cases and applications in the work of SHUGARD and REISS /2.67/, LANDMANN et al. /2.68/, and KLAFTER and SILBEY /2.69/.

2.7.4 Relation of GME to Velocity Autocorrelation Functions

Correlation functions are central in KUBO's development /2.70/ of linear response theory. Of special importance to mobility and other transport phenomena are velocity autocorrelation functions <v(t) v>. It is well known that mobility and conductivity are directly related to <v(t) v>. If a particle moving in a band is scattered only infrequently, we know that its $\langle v(t)\, v\rangle$ is long-lived. It appears reasonable that such a situation represents coherent motion in real space. It is interesting to ask, therefore, whether or not long-lived $\langle v(t)\, v\rangle$'s correspond to long-lived $\mathcal{W}(t)$'s and, more generally, to seek whatever connections that may exist between velocity autocorrelation functions and memory functions. The question has been recently answered by KENKRE et al. /2.71/ and then used /2.59/ to calculate the frequency-dependent mobility of charge carriers in molecular crystals.

The answer is based on the theory of SCHER and LAX /2.63/ who have shown that the symmetrized velocity autocorrelation function, written below in the standard notation of KUBO /2.70/, is related simply to a *generalized* mean square displacement,

$$\frac{1}{2}\langle v(t)\, v + v\, v(t)\rangle = \frac{1}{2}\frac{d^2}{dt^2}\langle x(t)\, x + x\, x(t)\rangle \quad , \tag{2.119}$$

where $\langle\ \rangle$ signifies trace with the equilibrium density matrix ρ^e, and for any $\mathcal{O}$, $\mathcal{O}(t)$ is given by $e^{itL}\mathcal{O}$. If we follow SCHER and LAX /2.63/ in approximating (2.119) by

$$\frac{1}{2}\langle v(t)\, v + v\, v(t)\rangle = \frac{1}{2}\frac{d^2}{dt^2}\sum_{m,n}(m-n)^2\, P(m,t|n,0)\, \langle n|\rho^e|n\rangle \quad , \tag{2.120}$$

where $P(m,t|n,0)$ is the probability that the moving particle would be at m at time t if it occupied site n at time t, the connection between velocity autocorrelation functions and memories is obtained immediately through an intermediate connection of the latter and the mean-square displacement. This second connection has been given by KENKRE /2.28,29/ for a periodic crystal:

$$\frac{d}{dt}\langle\!\langle m^2(t)\rangle\!\rangle = -\int_0^t dt' \left[\frac{\partial^2 \mathcal{W}^k(t')}{\partial k^2}\right]_{k=0} \quad . \tag{2.121}$$

For a translationally invariant (periodic) system such as a molecular crystal, $\langle n|\rho^e|n\rangle$ equals $1/N$ where N is the number of crystal sites, the right-hand side of (2.120) is half the second derivative of the mean-square displacement $\langle\!\langle m^2(t)\rangle\!\rangle$, and we have

$$\frac{1}{2}\langle v(t)\, v + v\, v(t)\rangle = -\frac{1}{2}\left[\frac{\partial^2 \mathcal{W}^k(t)}{\partial k^2}\right]_{k=0} \quad . \tag{2.122}$$

We have here a simple relation between $\langle v(t)\, v\rangle$ and $\mathcal{W}(t)$. Equation (2.122) states that their time dependence is essentially the same. Indeed, if the $\mathcal{W}_{mn}(t)$'s were separable into a space part and a time part, i.e., if $\mathcal{W}_{mn}(t)$ were to equal $F_{mn}\phi(t)$, then the memory function and velocity autocorrelation function would be identical to each other except for a constant factor:

$$\frac{1}{2}\langle v(t)\, v + v\, v(t)\rangle = \frac{1}{2}\langle m^2\rangle\, \phi(t) \quad . \tag{2.123}$$

The constant $\langle m^2 \rangle$, defined in (2.38), is m^2 weighted with the transition rates F. We see from (2.123) that a long-lived velocity autocorrelation function does indeed correspond to a long-lived memory function and vice versa. The concept of transport coherence may thus be related to the decay time of $\langle v(t)\, v \rangle$ in a similar way to that of $\phi(t)$. We know from calculations such as that leading to (2.62) that the separation of $\mathcal{W}_{mn}(t)$ into factors F_{mn} and $\phi(t)$ can be generally only an approximation. In the general case we must therefore use (2.122).

As an example of the full relation (2.122), we treat the system described by the simplified SLE of Sect. 2.7.2. The application of (2.122) to (2.111) and the use of (2.59) for $\mathcal{W}^k$ give

$$\frac{1}{2} \langle v(t)\, v + v\, v(t) \rangle = 2V^2\, e^{-2\Gamma^s t} + 4\gamma^s\, \delta(t) \quad . \tag{2.124}$$

Thus the SLE corresponds to an autocorrelation function that consists of a sum of an exponential and a δ-function. Equations (2.124,111,109) should be compared to one another. Essential coherence information may be accessed through the $\langle v(t)\, v \rangle$ expression (2.124) or the dimer memory expression (2.109). The full crystal memory expression contains the additional complications seen in (2.111) (which corresponds to Bessel functions) because of spatial details. These spatial details are washed out in computing mean-square displacements or velocity autocorrelation functions. While such elimination of information is undesirable for calculations of quantities such as probabilities $P_m(t)$, it is very desirable for obtaining essential coherence information. This analysis thus answers the question one might raise /2.72/ about the importance of studying the mean-square displacement in coherence investigations.

The relation (2.122) is an approximation because (2.120) is an approximation to (2.119). It is possible to show /2.71/ that the exact relation states the proportionality of the velocity autocorrelation function to the second time derivative of the mean-square displacement calculated for a rather unusual initial condition. One obtains

$$\frac{1}{2} \langle v(t)\, v + v\, v(t) \rangle = \frac{1}{2} \frac{d^2}{dt^2} \sum_{m,n} (m-n)^2\, p_m^{(n)}(t) \quad , \tag{2.125}$$

where $p_m^{(n)}(t)$ is the probability $P_m(t)$ for an initial density matrix $\rho(0)$ given by

$$\langle a \,|\, \rho(0) \,|\, b \rangle = \frac{1}{2} \langle a \,|\, \rho(0) \,|\, b \rangle \, (\delta_{a,n} + \delta_{b,n}) \quad . \tag{2.126}$$

KENKRE et al. /2.71/ have recovered the result of SCHER and LAX /2.63/, i.e., (2.120) above, in the high-temperature approximation. With the help of SLE calculations for $\langle\langle m^2(t) \rangle\rangle$ carried out by REINEKER /2.73/ for arbitrary initial conditions, the analysis from /2.71/ immediately gives the $\langle v(t)\, v \rangle$ as well as the frequency-dependent mobility or conductivity arising from the SLE /2.59/.

That the simple connection (2.122) between $\langle v(t)\, v \rangle$ and $\mathcal{W}(t)$ is not completely general should not come as a surprise. The correlation function should indeed contain more information than merely the evolution of the *site-diagonal* elements of the density matrix. The latter is the sole concern of the $W(t)$'s. This statement should not, however, preclude the possibility that future investigations might uncover further connections between $\langle v(t)\, v\rangle$ and $W(t)$.

2.7.5 Relation of GME to the Scattering Function

In the treatment /2.74/ of the scattering of probe particles (such as neutrons) by moving particles (such as light atoms), attention is focused on the scattering function $S(q,\omega)$ for momentum transfer q and energy transfer ω. This quantity is connected to the self-correlation function through the well-known van Hove relation and is essentially its space-Fourier *and* time-Fourier transform. The self-correlation function is, however, closely related to the propagator, i.e., to the solution $P_m(0)$ of an evolution equation such as the Master equation or the GME corresponding to an initial localized condition $\delta_{n,0} = 0$. Therefore, there exists a simple relation between the scattering function $S(q,\omega)$ and memory functions $W(t)$ of the GME.

The relation of $S(q,\omega)$ to the self-correlation function $G_s(m,t)$ is

$$S(q,\omega) = \frac{1}{2\pi}\int_{-\infty}^{+\infty} dt\; e^{-i\omega t} \sum_m e^{iqm}\, G_s(m,t) \quad , \tag{2.127}$$

whereas the self-correlation function is given, for high temperatures, by

$$\sum_m \tilde{G}_s(m,\varepsilon)\, e^{iqm} = [\varepsilon + \tilde{A}^q(\varepsilon)]^{-1} \quad . \tag{2.128}$$

For arbitrary temperatures a mere multiplicative factor appears in the right side of (2.128) /2.75/. Here $\tilde{A}^q(\varepsilon)$ is the Laplace and Fourier transform of memory functions $A_m(t)$. We give an example of the relation given in the conjunction of (2.127,128). If we consider particles moving as in the GME with memories $W_{mn}(t)$ given by (2.72), corresponding to motion of arbitrary degree of coherence in a one-dimensional crystal, we obtain, in a straightforward fashion,

$$\sum_m \tilde{G}_s(m,\varepsilon)\, e^{iqm} = \left\{\left[(\varepsilon+\alpha)^2 + 16V^2 \sin^2(q/2)\right]^{1/2} - \alpha\right\}^{-1} \quad , \tag{2.129}$$

which leads to, with $V_q = 4V\sin(q/2)$,

$$\begin{aligned} S(q,\omega) = {} & \left\{\left[(1/2)\left(\sqrt{(\alpha^2+V_q^2-\omega^2)^2 + 4\alpha^2\omega^2} + \alpha^2 + V_q^2 - \omega^2\right)\right]^{1/2} - \alpha\right\} \\ & \times\left\{\sqrt{(\alpha^2+V_q^2-\omega^2)^2 + 4\alpha^2\omega^2} + \alpha^2 \right. \\ & \left. - 2\alpha\left[(1/2)\left(\sqrt{(\alpha^2+V_q^2-\omega^2)^2 + 4\alpha^2\omega^2} + \alpha^2 + V_q^2 - \omega^2\right)\right]^{1/2}\right\}^{-1} \end{aligned} \tag{2.130}$$

BROWN and KENKRE /2.75/ have used (2.130) to study an anomalous variation of the halfwidth of the scattering function for hydrogen diffusion in metals discussed by SKÖLD /2.74/. The relation given above is, however, not restricted to hydrogen diffusion. In the exciton context it is closely connected to our theory of transient grating experiments which we shall develop in Chap. 3. It allows us to study the effect of coherence, as described by the GME, on scattering observables. The lineshape given by (2.130) shows all the expected behaviour characteristic of band motion in the coherent limit $\alpha \to 0$, and hopping motion in the incoherent limit, including the phenomenon of motional narrowing /2.75/. It is straightforward to generalize (2.130) to higher dimensions and specific lattice structures, to analyze phonon-*assisted* motion, and to incorporate additional effects that appear at low temperatures.

3. Coherence and the Generalized Master Equation Approach: Application to Experiments

3.1 Introduction

The formalism of generalized master equations, developed in Chap. 2, will be applied to experiments in the present chapter. HARRIS and ZWEMER /1.4/ have classified experiments in this subject into the following categories: optical lineshape, magnetic resonance, and direct migration experiments. The third category is best suited to being analyzed in terms of GME's. The applications discussed below refer to three kinds of direct migration experiments. Transient grating observations /3.1-4/ constitute a relatively novel and powerful method of probing exciton motion. Sensitized luminescence /3.5-10/ and annihilation phenomena /3.11-14/ are veteran fields but several old, but unresolved, issues have been brought to the surface recently. This is a result both of new experimental techniques such as picosecond spectroscopy and new theoretical developments such as the GME approach. It will be seen that a common theme, the effect of coherence on observable phenomena, underlies the development below. The GME approach has been sometimes criticized as being rather formal and removed from experiment. Nothing could be further from the truth, as the next sections will show.

3.2 Transient Grating Observations

Pioneered by FAYER and collaborators /3.1/ in the present context, transient grating observations /3.1-4/ consist of an extremely direct method of studying exciton motion. Two time-coincident picosecond excitation pulses are crossed in the bulk of the crystal at a definite, but variable, angle. Optical absorption results in a sinusoidal inhomogeneity in the exciton density. When the exciting pulses are switched off, the inhomogeneity or "grating" decays in time because of the combined effect of radiative emission and exciton motion. The grating is thus a transient one. A new picosecond pulse is diffracted off the transient grating to measure its amplitude. The signal gives the time evolution of what is essentially the square of the grating amplitude. Characteristics of exciton motion can therefore be deduced directly from these observations without the need for (and without the theoretical complications introduced by) such material detectors as guest molecules. The analysis of these experiments in terms of the GME has shown that a very general connection exists between the observables and memory functions. This connection /3.15/ is akin to that between the VAN HOVE /3.16/ self-correlation function and scattering observables, and it is discussed in Sect. 3.2.1 below. The effect of coherence on the signal is studied with a model memory in Sects. 3.2.2 and 3, and further developments are described in Sect. 3.2.4.

3.2.1 A Universal Connection

Elementary diffraction considerations show that the dimensionless wave vector η of the transient grating is given by

$$\eta = (4\pi a/\lambda)\,\sin(\theta/2) \quad , \tag{3.1}$$

where a is the lattice spacing, λ is the wavelength of excitation, and θ is the angle of crossing of the exciting pulses. To analyze the experiment with the GME, we first notice that the experimental signal $S^{\eta}(t)$ is essentially the Fourier component $P^{\eta}(t)$ of the probabilities [see (2.49)]

$$S^{\eta}(t) = \text{const.} \sum_m P_m(t)\, e^{i\eta m} \quad . \tag{3.2}$$

We have used the word "essentially" here because the actual signal is proportional to the convolution with pulse excitation functions of the square of the amplitude, whereas for the sake of simplicity (in this review) we call the amplitude the signal S(t). The theoretical problem is therefore the simple one of the evaluation of the solution of the GME (2.11) in the *Fourier domain*. If we cast (2.11) in the form (2.48) and notice that the initial condition is

$$P_m(0) = \text{const.}\ [1 + \cos(\eta m)] \quad , \tag{3.3}$$

which corresponds to only three Fourier components P^0, P^{η}, and $P^{-\eta}$ being excited, a universal model-independent connection between the signal S(t) and the memory function $A^{\eta}(t)$ [see (2.50)] is obtained immediately:

$$\frac{dS^{\eta}(t)}{dt} + \frac{S^{\eta}(t)}{\tau} + \int_0^t dt'\, A^{\eta}(t-t')\, S^{\eta}(t') = 0 \quad . \tag{3.4}$$

Here τ is the exciton lifetime. Relation (3.4) has an even more direct appearance in the Laplace domain:

$$\tilde{S}^{\eta}(\varepsilon)/S^{\eta}(0) = [\varepsilon + \frac{1}{\tau} + \tilde{A}^{\eta}(\varepsilon)]^{-1} \quad , \tag{3.5}$$

and allows one to build a chart (Fig. 3.1) showing the connection between grating signals and memory functions. As examples we see that incoherent motion corresponding

MEMORY FUNCTION	GRATING SIGNAL
$\mathcal{A}^{\eta}(t)$	$\int d\epsilon\, e^{\epsilon t}\left[\epsilon + \int_0^{\infty} d\tau\, e^{-\epsilon\tau}\mathcal{A}^{\eta}(\tau)\right]^{-1}$
$R\delta(t)$	e^{-tR}
J^2	$\cos Jt$
$J^2 e^{-\alpha t}$	$e^{-\alpha t/2}\left[\cos \Omega t + (\alpha/2\Omega)\sin \Omega t\right]$ with $\Omega = \sqrt{J^2-(\alpha^2/4)}$
$(2J^2 e^{-\alpha t})(4\sin^2\frac{\eta}{2})\left[J_0(4Jt\sin\frac{\eta}{2}) + J_2(4Jt\sin\frac{\eta}{2})\right]$	$1-e^{-\alpha t} b\int_0^t du\, e^{\alpha(t^2-u^2)^{1/2}} J_1(bu)$ with $b = 4J\sin\frac{\eta}{2}$

Fig. 3.1. Correspondence chart showing the direct connection between transient grating signals and memory functions. By "signal" is meant $e^{t/\tau}\, S^{\eta}(t)$ and by "memory" is meant $A^{\eta}(t)$

to a δ-function memory $A^{\eta}(t) = R\delta(t)$ results in an exponential signal $[S^{\eta}(t)/S^{\eta}(0)] = \exp(-tR)$ and that if there is a constant memory (in an extended crystal such a memory is unphysical /1.1/), i.e. $A^{\eta}(t) = u^2$, then the signal would be oscillatory $[S^{\eta}(t)/S^{\eta}(0)] = \cos(ut)$. Other examples will be discussed in Sect. 3.2.2.

The generality and practical usefulness of the above connection is worth emphasizing. It is independent of all assumptions about detail and requires only the validity of the GME and the applicability of the initial condition used. The former has been discussed in Chap. 2 and requires no further comment. The initial condition used is that suggested by the experimentalists /3.1/. Relations (3.4,5) allow one to determine, at least in principle, the entire dynamics of excitons through transient grating experiments. Knowing all the memory functions $\mathcal{W}_{mn}(t)$ or $A_{mn}(t)$ is equivalent to knowing the entire dynamics. But all these $\mathcal{W}$'s or A's can be found directly from experiment through the inverse transform of the above relations. Experiments that span the range of η by varying the angle of crossing θ and/or the excitation wavelength λ are thus required. They will provide $S^{\eta}(t)$'s for various (in principle all) η's. The explicit prescription to extract all information from dynamics is

$$A_{mn}(t) = -\frac{1}{N}\sum_{\eta} e^{-i\eta(m-n)} \int_{c} d\varepsilon\, e^{\varepsilon t}\left[\varepsilon + \frac{1}{\tau} - S^{\eta}(0)\left(\int_{0}^{\infty} dt'\, e^{-\varepsilon t'}\, S^{\eta}(t')\right)^{-1}\right] . \quad (3.6)$$

Experimental signal data are to be fed into the right-hand side of (3.6) and the left-hand side gives the exciton dynamics.

The above prescription is capable of giving detailed information about not only the degree of coherence but also many other transport characteristics such as transfer rates, their anisotropy, and their temperature dependence /3.15/. The universal character of the relation can be appreciated from the fact that it is of the same form as the connection between the VAN HOVE self-correlation function /3.16/ and the scattering function. A wealth of information about excitons can thus be deduced experimentally through (3.6) in the same manner that experiments on scattering of neutrons and other probes have done in other fields through the VAN HOVE relation.

3.2.2 Nonexponential Signals and Coherence

We now exhibit explicit predicted signals showing the effect of coherence in terms of a specific GME. The analysis is based on the memory function (2.76). Exciton motion is characterized by nearest-neighbour matrix elements V for transfer and by the randomizing bath parameter α. As KENKRE /3.15/ has shown, when (3.5) is Laplace inverted,(2.54,59-62) and (3.5) give

$$S^{\eta}(t) = S^{\eta}(0)\, e^{-(\alpha+1/\tau)t} \int_{c} d\varepsilon\, e^{\varepsilon t}\left\{[\varepsilon^2 + 16V^2 \sin^2(\eta/2)]^{1/2} - \alpha\right\}^{-1} , \quad (3.7)$$

the contour c being the Bromwich contour. The extreme limits of (3.7) are obtained immediately /3.17/. For purely coherent motion, the bath is absent, $\alpha = 0$, and

$$S^{\eta}(t) = S^{\eta}(0)\, e^{-t/\tau}\, J_0[4Vt \sin(\eta/2)] . \quad (3.8)$$

For completely incoherent motion with small intersite interactions $\alpha \to \infty$, $V \to \infty$, $2V^2/\alpha = F$ and we get, from (3.7), a purely exponential signal

$$S^{\eta}(t) = S^{\eta}(0)\, e^{-t/\tau} \exp\{-t[4F \sin^2(\eta/2)]\} . \quad (3.9)$$

If we take the limit $\eta \to 0$ in (3.9), we obtain a simplified expression $(1/\tau + \eta^2 F)$ for the exponent in (3.9). The continuum limit, i.e.,

$$a \to 0 \ , \quad (1/a)P_m(t) \to p(x,t) \ , \quad F \to \infty \ , \quad Fa^2 \to D \ , \tag{3.10}$$

finally gives the diffusion-equation results of SALCEDO et al. /3.1/:

$$S^\eta(t) = S^\eta(0)\ e^{-t/\tau} \exp(-t\Delta^2 D) \ , \tag{3.11}$$

where $\Delta = \eta/a$ and D is the diffusion constant.

The exact signal (3.7) thus reduces to a Bessel function (3.8) in the purely coherent limit and to an exponential, (3.9) or (3.11), in the opposite limit of extreme incoherence. The evaluation of (3.7) in the intermediate domain /3.17/ permits one to see explicitly the effect of coherence over the entire range:

$$S^\eta(t) = S^\eta(0)\ e^{-t/\tau} \left[1 - e^{-\alpha t}\ y \int_0^t du\ e^{\alpha(t^2-u^2)^{1/2}}\ J_1(yu)\right] \ , \tag{3.12}$$

where y equals $4V \sin(\eta/2)$ und J_1 is the Bessel function. Figure 3.2 compares the signal [actually the square of $S(t)/S(0)$] for the arbitrary values $\alpha = 1 = 4V \sin(\eta/2)$ to its coherent and incoherent limits. It is not significant that the exact signal in the figure exhibits no oscillations. For other values of α it can obviously show much greater similarity to the coherent limit. These and other details have been explored by WONG and KENKRE /3.17/.

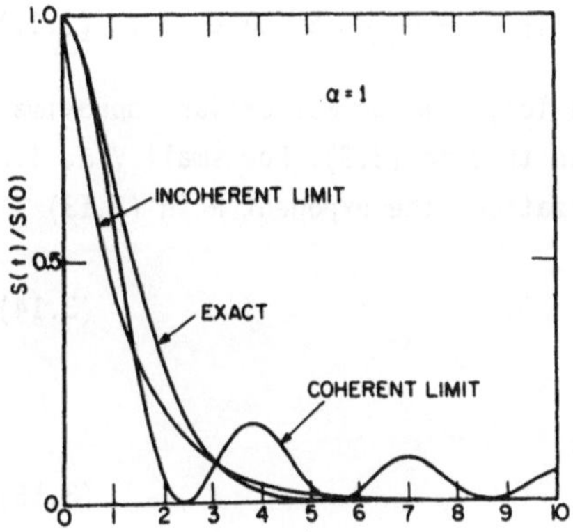

Fig. 3.2. Effect of coherence on transient grating signals. The signal, plotted as a function of time, is the square of $[S^\eta(t)/S^\eta(0)]$. The coherent and incoherent limits and a particular intermediate case are shown

The GME approach is thus seen to lend itself naturally to the analysis of transient grating experiments. The physics underlying the dynamics in this section is that of the memory function (2.72) or, perhaps more transparently, that of the transport equation (2.66) given, e.g., by AVAKIAN et al. /2.39/. The degree of coherence can be extracted from experiments by comparison to (3.12) and plots such as that in Fig. 3.2. It is straightforward to generalize this analysis to include phonon-assisted terms and microscopic memory functions by replacing the memory function used above, (2.76), by more detailed ones, such as (2.111) and (2.90).

3.2.3 Exponential Signals and Coherence

We have seen above that coherence manifests itself in the grating signal by making its time-evolution nonexponential, the purely coherent limit being represented by the Bessel function J_0. However, the intermediate coherence signal, given by (3.12), does look exponential for times large with respect to $1/\alpha$. This is, of course, a consequence of the general feature of the GME: its solutions essentially coincide with those of the Pauli Master equation for large times. If the time resolution in the experiment is much coarser than $1/\alpha$, the signal will appear to be exponential and the rich structure of (3.12) will be entirely hidden. We analyze such a situation in this section.

Are there any vestiges of coherence left in the grating signal when it appears exponential? This is an interesting question whose answer is, perhaps surprisingly, in the affirmative. In a distant way it is related to a completely different questio answered in Sect. 2.5: is there any effect of coherence on transfer *rates*? Although the physics and the method of answering the question are both quite different in these two contexts, a striking similarity will be noticed in the final result. To answer the present question we take the Markoffian limit of the GME and use (2.27) with (2.82) rather than (2.76) for the memory function. It should be noted that although (2.27) with (2.82) approximates (2.76) by a δ-function, it does *not* involve the limit which gives (3.9) for the grating signal. Instead we obtain

$$S^{\eta}(t) = S^{\eta}(0)\ e^{-t/\tau}\ \exp\left(-t\left\{\left[\alpha^2 + 16V^2\ \sin^2(\eta/2)\right]^{1/2} - \alpha\right\}\right) \equiv S^{\eta}(0)\ e^{-tR}\ . \quad (3.13)$$

The signal is exponential since we have made the long-time or Markoffian approximation. However, the exponent has more content than that in (3.9). For small V/α, i.e. for motion which is slow with respect to randomization, the exponent R in (3.13)

$$R = \frac{1}{\tau} + \left[\alpha^2 + 16V^2\ \sin^2(\eta/2)\right]^{1/2} - \alpha \quad (3.14)$$

reduces, through a binomial expansion, to

$$R = \frac{1}{\tau} + \frac{8V^2}{\alpha}\ \sin^2(\eta/2)\ , \quad (3.15)$$

which is the result of (3.9) with the well-known correspondence $F = 2V^2/\alpha$.

We thus see that even when the Markoffian approximation is made in the GME becaus the time resolution is coarse, the exponent of the exponential signal thus obtained can be used to extract information about the degree of coherence. In the present context the degree of coherence means the relative value of V/α. When it is small, the familiar result (3.15) is obtained. However, when it is large, completely different behaviour results from (3.14): in the opposite extreme limit of coherent motion, when α may be neglected with respect to $4V\ \sin(\eta/2)$, we get

$$R = \frac{1}{\tau} + 4V \sin(\eta/2) \quad , \qquad (3.16)$$

which never reduces to the diffusion equation result (3.11) even when the continuum limit $a \to 0$ is taken.

The diffusion equation result /3.1/ is thus not an automatic consequence of the Markoffian approximation (exponential signal) and the continuum limit. It requires the further limit $V/\alpha \to 0$ (small intersite interaction). These are the experimentally important manifestations of spatially long-range memories and rates appearing in transport in the strong-coupling limit. This theory was given by KENKRE /2.29/, who also showed that the exponent R when plotted against V shows a proportionality to V^2 in the incoherent limit but a linear dependence V in the coherent limit. Here we see the similarity to the discussion in Sect. 2.5 and to the behaviour there of the rate w as a function of V in the two limits of (2.40,41).

3.2.4 Further Calculations

Observations reported thus far in the literature /3.1-4/ appear to contain only exponential signals. Does this mean that in the systems studied, probe times are definitely larger than coherence times? This is a difficult question to answer at the moment because the systems studied are mixed crystals, whereas the theory reviewed deals with pure crystals. In the experimental systems the excitons thus move along a disordered array. As is well known, dynamics in random media /3.18,19/ is extremely difficult to analyze properly. While standard treatments of motion in disordered systems may be appended to the transient grating theory reviewed above, the degree of confidence in the results is not the same as in those for pure crystals. While we wait for more and better experiments on pure crystals, to which this theory is immediately applicable, we hope that theoretical advances will be made which will allow an unambiguous interpretation of observations in mixed crystals. One particular direction for such advances that we foresee is that based on the replacement of a disordered array by an ordered one but with a pausing-time distribution function as in the treatment of SCHER and LAX /2.63/ and SCHER and MONTROLL /2.64/. It is clear from the equivalences between continuous-time random walks and GME's established in Sect. 2.7.3 /2.15,16,66/, that such a replacement would result in a GME whose memory functions are a composite of the natural dynamic ones and those introduced specifically by the replacement of the disordered by the ordered array. That GME's are obtained by such a replacement is particularly clear from the work of KLAFTER and SILBEY /2.69/. If the ideas of /2.63,64,69/ can be converted into a practical prescription to obtain explicit memory functions in the equivalent ordered system, the GME treatment of grating observations given in this chapter will be immediately applicable to disordered systems.

Recently work has also been done /3.20/ on the effect of relaxation processes on transient grating observations. The basic idea is to apply to the grating situation a Master equation capable of describing the interplay of relaxation wiht exciton motion. The equation was suggested and analyzed by KENKRE /3.21/ on the basis of the methods of MONTROLL and SHULER /3.22/. We only mention here that when it is used to replace (2.11) and (2.72) for the grating analysis, non-exponential signals of a kind different from (3.12) result. It appears that some of the heat dissipation and relaxation aspects reported by SALCEDO et al. /3.1/ in their experiments can be addressed naturally with the help of this approach.

It may be helpful to comment on the validity of the various limits discussed in Sects. 3.2.1 and 3.2.2. The values $\alpha = 10^{12}\ s^{-1}$, $V = 10^{12}\ s^{-1}$, $\tau = 10^{-8}$ s appear reasonable for some systems. Since $\alpha\tau \gg 1$, no nonexponential signals are expected, at least as a result of coherence. Furthermore, because $4V/\alpha$ is not a small quantity, a and λ will determine whether the completely incoherent case (3.9) is sufficient or whether the analysis in Sect. 3.2.3, incorporating exponential signals in the *coherent* regime, is necessary. In the system of /3.1/, a concentration of 1.6×10^{-3} mol/mol, $a \approx 50$ Å and $\lambda = 5 \times 10^4$ Å would give $16V^2 \sin^2(\eta/2) \ll \alpha$ and one is in the domain of validity not only of (3.9) but even of the continuum limit (3.11). However, if $\alpha = 10^{10}\ s^{-1}$, which might be appropriate for low temperatures, a suitably large value of a/λ would force one to use (3.12) with intermediate coherence. Although these estimates refer strictly to a pure crystal, they provide at least semi-quantitative insights into the behaviour of the mixed crystal of /3.1/.

In closing the discussion on transient grating observations, we reiterate that conceptually they form an exceptionally direct (and theoretically uncomplicated) probe into exciton motion. For pure crystals the theory is complete /2.29,3.15,17/ and awaiting observations. For disordered crystals further work is required. We hope this powerful observational probe will be exploited fully in future experimental work.

3.3 Sensitized Luminescence

Sensitized luminescence is perhaps the oldest experimental method of studying exciton dynamics in molecular crystals and aggregates. The host (in which exciton dynamics is the object of study) is doped with guest or detector molecules. Excitons are created through illumination, an appropriate frequency range being chosen to ensure that only the host (or guest) is excited. The excitons decay through radiative as well as nonradiative processes, and also move within the host. If they arrive within the sphere of influence of the guest molecules, they may be captured. If they are captured, they decay radiatively in a different frequency range. By monitoring the luminescence from the guest or the host, it is thus possible to extract information about exciton motion in the host. The noteworthy characteristic of excitons, which other entities such as electrons do not possess, is that excitons signal their status to the observer through spontaneous emission. It is this characteristic that is the basis of sensitized luminescence studies of exciton motion.

The primary observables in these experiments are the host and guest quantum yields in the case of steady-state observations and the host and guest luminescence intensities in the case of time-dependent observations. The former have been reviewed by WOLF /3.5/ and the latter by POWELL and SOOS /1.20/. Picosecond spectroscopy has brought a resurgence of work in this field, and many old concepts such as time-dependent energy transfer rates are being seriously questioned on experimental /3.6-11,23/, as well as theoretical, grounds. The work on time-dependent observations by SCHMID and collaborators /3.6,9-11/ has been particularly responsible for focusing attention on a number of these issues. In this section we shall analyze only the effect of coherence on these observations and return to other aspects of sensitized luminescence in Chap. 4. The formulation of the problem is given in Sect. 3.3.1 and the unified GME description of the effect of coherence in Sect. 3.3.2. The transport description is completely general in this treatment, but the exciton capture process is first analyzed in terms of a relatively simple model. A more complete analysis of the capture process is given in Sect. 3.3.3, and further calculations are mentioned in Sect. 3.3.4.

3.3.1 Formulation: the Simplest Trapping Model

Our starting point is the GME augmented by the addition of terms describing exciton trapping by guest molecules /3.24/:

$$\frac{dP_m(t)}{dt} + \frac{P_m(t)}{\tau_H} = \int_0^t dt' \sum [\mathcal{W}_{mn}(t-t')\, P_n(t') - \mathcal{W}_{nm}(t-t')\, P_m(t')] - c \sum_r{}' P_m(t)\, \delta_{m,r} \quad . \tag{3.17}$$

Equation (3.17) describes a simple "sink" model of trapping in which the exciton probability or density is taken to be depleted at rate c whenever the exciton finds itself at any of the guest-influenced host sites r. The primed summation is over these sites r. The lifetime in the host is denoted by τ_H. For simplicity, we may assume that a guest-influenced host site is that host site which is closest to a guest site and that the trapping process is nearest-neighbour in range. The solution of (3.17) is given in terms of elementary Green's functions,

$$\widetilde{P}_m(\varepsilon) = \sum_n \widetilde{\psi}_{mn}\, P_n(0) - c \sum_n \widetilde{\psi}_{mn} \sum_r{}' \delta_{n,r}\, \widetilde{P}_r(\varepsilon) \quad , \tag{3.18}$$

where (as elsewhere in this review) ε is the Laplace variable and tildes denote transforms, the first term represents the solution of (3.17) in the absence of guest molecules, and $\psi_m(t)$ is the "propagator", i.e., the solution of (3.17) in the absence of guest molecules and for the initial condition $P_n(0) = \delta_{n,0}$. The "solution" (3.18) is of no real use because the last term which arises from trapping contains the unknown $\tilde{P}_m(\varepsilon)$. The defect technique developed by MONTROLL and collaborators /1.11/ consists of turning (3.18) into a true solution. The technique is useful when the "size" of the defect region (measured by the number of defect or guest sites) is not too large. It consists in writing down particular cases of (3.18) corresponding to m being the various guest sites r, solving the simultaneous equations for probabilities at guest sites, and then using these solutions back in (3.18) to get an explicit usable general solution for all m. It is obvious that this method is particularly useful for small defect regions since the number of simultaneous equations to be solved equals the number of guest sites. However, there are situations, two of which we shall describe below, when this method finds use even for an arbitrary large number of guest sites. We first examine the simple case of a single trap at r, for which we get

$$\tilde{P}_m(\varepsilon) = \tilde{\eta}_m(\varepsilon') - c\,\frac{\tilde{\psi}_{m-r}(\varepsilon')\,\tilde{\eta}_r(\varepsilon')}{1 + c\tilde{\psi}_0(\varepsilon')} \quad . \tag{3.19}$$

The notation $\varepsilon' = \varepsilon + 1/\tau_H$ is used in (3.19) and $\tilde{\eta}_m(\varepsilon)$ is the transform of

$$\eta_m(t) = \sum_n \psi_{m-n}\,P_n(0) \quad , \tag{3.20}$$

which is the solution of (3.17) in the absence of guest sites.

To obtain the host luminescence intensity, we must calculate $n_H(t)$, the total probability that the host is excited:

$$\tilde{n}_H(\varepsilon) \equiv \sum_m \tilde{P}_m(\varepsilon) = \frac{1}{\varepsilon'}\left[1 - \frac{c\tilde{\eta}_r(\varepsilon')}{1 + c\tilde{\psi}_0(\varepsilon')}\right] \quad . \tag{3.21}$$

The luminescence intensity or differential photon count rate is proportional to $n_H(t)/\tau_H$ and requires the inversion of the Laplace transform in (3.21). The host quantum yield ϕ_H, however, defined as

$$\phi_H \equiv \frac{1}{\tau_H}\int_0^\infty dt\; n_H(t) = \lim_{\varepsilon\to 0}\frac{\tilde{n}_H(\varepsilon)}{\tau_H} = \lim_{\varepsilon'\to 1/\tau_H}\frac{\tilde{n}_H(\varepsilon')}{\tau_H} \tag{3.22}$$

requires no inversion. This happy accident, that a popular and accessible observable, the quantum yield, is given directly by Laplace-transformed quantities, simplifies theoretical labor to a great extent. It has been exploited by KENKRE and WONG /3.24/ to carry out exact analytic investigations of the effect of coherence on yield. Equations (3.21,22) give

$$\phi_H = 1 - \frac{\tilde{\eta}_r(1/\tau_H)}{1/c + \tilde{\psi}_0(1/\tau_H)} \quad . \tag{3.23}$$

To obtain guest quantities we note that $n_G(t)$, the probability that the guest is excited, obeys generally

$$\frac{dn_G(t)}{dt} + \frac{n_G(t)}{\tau_G} = c \sum_r{}' P_r(t) \quad , \tag{3.24}$$

where τ_G is the exciton lifetime in the guest. For the single-trap model, we have if the guest is not excited initially,

$$\tilde{n}_G(\varepsilon') = \left[\frac{c}{\varepsilon' + [(1/\tau_G) - (1/\tau_H)]}\right] \cdot \left[\frac{\tilde{n}_r(\varepsilon')}{1 + c\tilde{\psi}_0(\varepsilon')}\right] \quad . \tag{3.25}$$

Again, (3.25) must be Laplace inverted to obtain the guest luminescence intensity, but expressions for the yield require no inversion. The guest yield ϕ_G is defined as

$$\phi_G = \frac{1}{\tau_G} \int_0^\infty dt \; n_G(t) \quad . \tag{3.26}$$

It is given by

$$\phi_G = \frac{\tilde{n}_r(1/\tau_H)}{1/c + \tilde{\psi}_0(1/\tau_H)} \quad . \tag{3.27}$$

We see from (3.23,27) that ϕ_H and ϕ_G add up to 1. This is so because we have assumed that the exciton does not decay nonradiatively through processes such as internal conversion or intersystem crossing. This assumption, while valid for some systems such as anthracene, is rather poor for others such as naphthalene. However, it is trivial to modify the above expressions in the presence of nonradiative decay. Thus, the yields ϕ_H and ϕ_G are merely multiplied by the respective ratios τ_H/τ_H^0 and τ_G/τ_G^0, where τ_H and τ_G are the total lifetimes and τ_H^0 and τ_G^0 the radiative ones. Other expressions are unmodified.

We shall focus attention on the yield expressions (3.23) and (3.27). The effect of initial placement of excitons is present entirely in $\tilde{n}_r(1/\tau_H)$. From (3.20) we see that for uniform initial illumination

$$\tilde{n}_r(1/\tau_H) = \tau_H/N \quad , \tag{3.28}$$

where N is the number of crystal sites, whereas if the guest-influenced host site were alone excited, we would have

$$\tilde{n}_r(1/\tau_H) = \tilde{\psi}_0(1/\tau_H) \quad . \tag{3.29}$$

We shall use (3.28) in the subsequent discussion as it corresponds to the most frequent experimental situation. We then have, from (3.27,28),

$$\phi_G = \frac{\rho}{\frac{1}{c\tau_H} + \frac{1}{\tau_H}\tilde{\psi}_0(1/\tau_H)} \quad , \tag{3.30}$$

the symbol ρ being used here to denote $(1/N)$. It is clearly the guest concentration and should not be confused with the density matrix.

Equation (3.30) for the guest yield ϕ_G, the corresponding expression for the host yield ϕ_H obtained from (3.23), and their counterparts for other illumination conditions contain the effect of transport coherence on yields. We shall use (3.30) to present the general discussion in the next section.

3.3.2 Unified Description of the Effect of Coherence on Yields

Before undertaking the unified description of the effect of exciton transport on quantum yields, we make two remarks about (3.30) which, while not related to coherence, are of some relevance. The yield given by (3.30) is proportional to the concentration ρ. This is a consequence of the fact that the analysis leading to (3.30) considers a single trap. We shall see in Sect. 3.3.4 below that this result is valid for low concentrations but that the ρ dependence is, not unexpectedly, quite different for high-concentration situations. However, in studies of singlet excitons in aromatic hydrocarbon crystals /1.20, 3.5-11/ experimental constraints make the guest concentration always very low, often 10^{-6} or lower, and (3.30) is directly applicable in most cases. The departures from that result will be discussed in Sect. 3.3.4. The second comment concerns the peculiar form of (3.30) as far as the interplay of the capture process and the motion process are concerned. One is reminded of the effective resistance of parallel resistors. In extreme limits, the yield is determined by the slower of the two processes. Thus,

$$\phi_G = \tau_H c \rho \tag{3.31}$$

in the limit that $c \ll [\tilde{\psi}_0(1/\tau_H)]^{-1}$. The physical meaning of this limit is that capture proceeds much more slowly than exciton motion. The quantity $1/\psi_0(1/\tau_H)$ represents a motion rate. In the opposite limit of slow motion, or fast capture, we have

$$\phi_G = \tau_H [\tilde{\psi}_0(1/\tau_H)]^{-1} \rho \quad . \tag{3.32}$$

If the exciton motion is much faster than capture, i.e., if (3.31) is applicable, yield observations will provide absolutely no information concerning exciton motion. This is an important note of warning. Traditional procedures of analyzing yields with energy rates which are proportional to the diffusion constant could be completely invalid in such a situation.

We now return to the unified description of the effect of transport coherence on yields as given in (3.30). Observe that coherence effects are manifest in, and only in, the $\tilde{\psi}_0(1/\tau_H)$ and that (3.30) is valid for *any* degree of coherence. The quantity $\tilde{\psi}_0(1/\tau_H)$ is the Laplace transform of the propagator $\psi_0(t)$ evaluated at the value $\varepsilon = 1/\tau_H$. Stated differently, $(1/\tau_H)\ \tilde{\psi}_0(1/\tau_H)$ is an average weighted value of the propagator, the weighting factor being $\exp(-t/\tau_H)$. The lifetime of the exciton plays the role of a probe time. If it is very small, it picks up only small-t values of the propagator in obtaining $\tilde{\psi}_0(1/\tau_H)$. If it is very large, it samples essentially the entire propagator. We know from earlier discussions in this review that the GME reduces to the (Markoffian) Master equation at long times but possesses coherence behaviour at short times. Therefore, if almost the entire propagator is sampled in $\tilde{\psi}_0(1/\tau_H)$, the result in (3.30) will be essentially that of the incoherent description. However, if only the small-time behaviour of $\psi_0(t)$ is picked up, coherence characteristics will be manifest in yield observations. Thus it is the value of τ_H, the exciton radiative decay time, relative to $1/\alpha$, the decay time of the memory functions in the GME, that determines whether or not coherence effects are observable in steady-state sensitized luminescence experiments.

To make the discussion more concrete, we write from (2.48) or (2.51)

$$\tilde{\psi}_0(1/\tau_H) = \lim_{\varepsilon \to 1/\tau_H} \int dk \, \frac{1}{\varepsilon + \tilde{A}^k(\varepsilon)} \quad . \tag{3.33}$$

If the memory functions $A_{mn}(t)$ are δ-functions, signifying incoherent motion of the exciton, $\tilde{A}^k(\varepsilon)$ is independent of ε. If coherence is present, it will make its presence felt by forcing $A_{mn}(t)$'s to decay in nonzero times and equivalently by making $\tilde{A}^k(\varepsilon)$ depend on ε. The value of $\tilde{A}^k(1/\tau_H)$ appearing in (3.33) will therefore be different from $\tilde{A}^k(0)$. (The latter corresponds to the totally incoherent description.) This affects $\tilde{\psi}_0(1/\tau_H)$ through (3.33) and the observed yields through (3.30). The details of the effect will be clearer in the context of the explicit example to be discussed below.

It is particularly appropriate that the effect will be negligible if $(1/\tau_H)$ is very small. In such a situation, even if $\tilde{A}^k(\varepsilon)$ varies significantly in the ε domain (the memory functions are then significantly non-δ-like in the time domain), no coherence effects will be observed since (3.30) is sensitive only to the value of $\tilde{A}^k(\varepsilon)$ at $\varepsilon = 1/\tau_H$. If $(1/\tau_H)$ is small enough, the value of $\tilde{A}^k(\varepsilon)$ sampled by (3.30) will be essentially that given by the Markoffian approximation, viz. $\tilde{A}^k(\varepsilon) = \tilde{A}^k(0)$, equivalently $A^k(t) = \delta(t) \int_0^\infty dt' \, A^k(t')$. These mathematical details have a simple physical meaning. We have already seen in Chap. 2 that GMEs reduce to the Markoffian Master equation if one waits long enough. Whether one can wait long enough depends on the lifetime of the excitons. To have a very small $(1/\tau_H)$ represents excitons that live long and allow the GME to reduce to the Master equation. Coherence effects are then washed out. However, if excitons decay rapidly, their behaviour is coherent enough during their lifetime to have an effect on the steady-state yield. The coherence phenomenon requires this kind of observation to be fast, specifically to take place earlier than the time in which the memory functions decay to zero. In steady-state luminescence experiments, the experimental setup removes the moving excitons in times of the order of τ_H and collects them in the form of photons in the detector. The GME treatment predicts observable coherent effects if this collection occurs while exciton motion is still coherent.

The above remarks are quite general. If we now describe the situation as in Sect. 2.6.2 and connect the memories for the general and the purely coherent situations through

$$A^k(t) = A^k_{coh}(t)\, e^{-\alpha t} \quad , \tag{3.34}$$

where $A^k_{coh}(t)$ represents the coherent case, we obtain an interesting relation between the propagators in the general and in the purely coherent cases:

$$\tilde{\psi}^k(\varepsilon) = \frac{\tilde{\psi}^k_{coh}(\varepsilon+\alpha)}{1 - \alpha\, \tilde{\psi}^k_{coh}(\varepsilon+\alpha)} \quad . \tag{3.35}$$

Equation (3.35), which may also be written as

$$\tilde{\psi}^k(\varepsilon) = [\tilde{\psi}^k_{coh}(\varepsilon+\alpha)]\;[1 + \alpha\, \tilde{\psi}^k(\varepsilon)] \quad , \tag{3.36}$$

is similar in form to the well-known Lippman - Schwinger equation in perturbation theory. It provides the answer to the general question of how solutions to the GME are related in the general and the purely coherent cases, and is of particular use in sensitized luminescence through (3.30).

To exhibit coherence effects on yield explicitly, we treat the one-dimensional infinite crystal of (2.66) with the memory functions (2.76). We have also analyzed this system in the context of transient grating observations in Sect. 3.2. The propagator corresponding to this case, given in the integral form in /2.29/, has been evaluated explicitly by KENKRE and WONG /3.24/ in terms of elliptic integrals. The integral form is that in (2.80) and the explicitly evaluated expression is

$$\begin{aligned}\psi_0(\varepsilon) = {} & \frac{\alpha}{\left[(\varepsilon^2+2\varepsilon\alpha)(\varepsilon^2+2\varepsilon\alpha+16V^2)\right]^{1/2}} + \frac{(2/\pi)}{\left[(\varepsilon+\alpha)^2+16V^2\right]^{1/2}}\, \mathbb{K}(k) \\ & + \frac{\alpha^2}{\left[(\varepsilon+\alpha)^2+16V^2\right]^{1/2}}\, \frac{(2/\pi)}{\left(\varepsilon^2+2\varepsilon\alpha+16V^2\right)^{1/2}}\, \mathbf{\Pi}(a_1^2,k) \quad ,\end{aligned} \tag{3.37}$$

where $a_1^2 = 16V^2\,(\varepsilon^2+2\varepsilon\alpha+16V^2)^{-1}$, $k = 4V\left|(\varepsilon+\alpha)^2+16V^2\right|^{-1}$ and $\mathbb{K}(k)$ and $\mathbf{\Pi}(a_1^2,k)$ are elliptic integrals of the first and third kinds, respectively. These are defined through

$$\mathbb{K}(k) = \int_0^1 \frac{dx}{(1-x^2)^{1/2}} \left[\frac{1}{(1-k^2x^2)^{1/2}}\right] \quad , \tag{3.38}$$

$$\Pi(a_1^2,k) = \int_0^k \frac{dx}{(1-x^2)^{1/2}} \cdot \frac{1}{(1-k^2x^2)^{1/2}} \cdot \frac{1}{1-a_1^2x^2} \quad . \tag{3.39}$$

In the purely coherent limit $\alpha \to 0$, the first and third terms in (3.37) vanish. We then get, for the key quantity $(1/\tau_H)\ \tilde{\psi}_0(1/\tau_H)$ appearing in the yield expression such as (3.30),

$$(1/\tau_H)\ \tilde{\psi}_0(1/\tau_H) = \frac{(2/\pi)}{(1+16V^2\tau_H^2)^{1/2}}\ \mathbb{K}\left(\frac{4V\tau_H}{(1+16V^2\tau_H^2)^{1/2}}\right) \quad . \tag{3.40}$$

This is the situation when the amplitude (or Schrödinger) equation (2.5) is appropriate. The propagator $\psi_0(t)$ is given by (2.6) to be $J_0^2(2Vt)$ and (3.40) could also be recovered directly from this result.

In the purely incoherent limit, particularly when $\alpha \to \infty$, $V \to \infty$, $2V^2/\alpha \to F$, the second term in (3.37) vanishes and limiting properties of the third elliptic integral give

$$(1/\tau_H)\ \tilde{\psi}_0(1/\tau_H) = (1+4F\tau_H)^{-1/2} \quad . \tag{3.41}$$

This situation corresponds to the Master equation (2.7), the propagator $\psi_0(t)$ being given by (2.8) to be $e^{-2Ft}\ I_0(2Ft)$. Again, (3.41) can be connected independently and trivially to the Laplace transform of the modified Bessel function. Further limits of (3.40,41), when exciton motion is fast or slow with respect to decay, have been explored in /3.24/.

The entire intermediate situation, when excitons are neither too coherent nor too incoherent, is described by (3.37). KENKRE and WONG /3.24/ have provided plots of $(1/\tau_H)\ \psi_0(1/\tau_H)$ for this intermediate region. Figure 3.3 shows one of them. Knowing V from bandwidth calculations or spectral splittings, and τ_H from other ex-

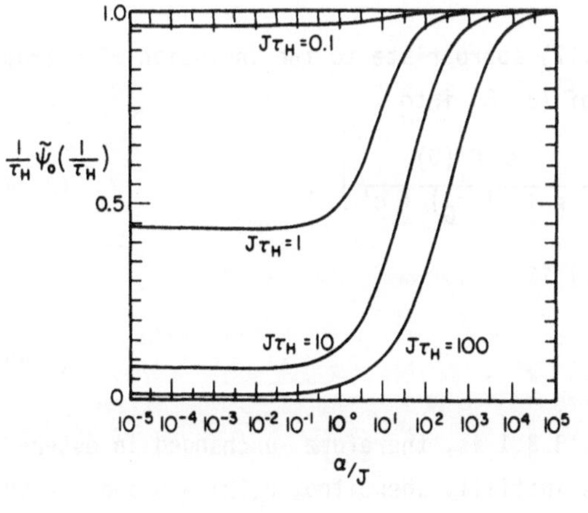

Fig. 3.3. Effect of coherence on steady-state sensitized luminescence observables. The quantity plotted as a function of the incoherence parameter α/J (J is the same as V in the text and is proportional to the exciton bandwidth) is $(1/\tau_H)\ \tilde{\psi}_0(1/\tau_H)$. It equals the ratio of the trap concentration ρ to the guest yield ϕ_G for small concentrations

periments, one selects a particular curve from the figure. In the limit of fast capture, i.e., when (3.32) is applicable, the quantity plotted in Fig. 3.3 is nothing other than the ratio of the concentration ρ to the yield ϕ_G. The ordinate in the plot is therefore directly accessible through experiment. If capture is not too fast, additional information about the capture rate is necessary. In either case we can deduce the value of J/α, i.e., the degree of coherence, directly from the figure.

We have here both a unified description of the effect of coherence on quantum yields (generally on exciton trapping) and a specific prescription, explained in the context of Fig. 3.3, to extract the degree of coherence. It is the generality of the GME that makes possible this unified description. It spans the entire range between the limits represented by the coherent (2.5) and the incoherent (2.7).

3.3.3 More General Trapping Models

Having explained the essence of the GME treatment of coherence effects on sensitized luminescence yields in the context of the simplest trapping model above, we shall now briefly mention more elaborate ways of describing the *capture* process in exciton trapping by guest molecules. The first elaboration is to include detrapping within the model of Sect. 3.3.2. We replace the last term in (3.17) by $-\sum_r' \delta_{m,r}[cP_m(t) - c'P_\theta(t)]$, where $P_\theta(t)$ denotes the probability that the trap is occupied, and write an additional equation for the latter

$$\frac{dP_\theta}{dt} + \frac{P_\theta}{\tau_G} = c\,P_r - c'\,P_\theta \quad . \tag{3.42}$$

While we have assumed here that a guest site communicates with a single host site and no other guest site, this has been done only for simplicity; a generalization is possible in a straightforward manner. The "solution" of (3.42) is

$$\tilde{P}_\theta(\varepsilon) = \frac{c\,\tilde{P}_r(\varepsilon) + P_\theta(0)}{\varepsilon + (1/\tau_G) + c'} \quad . \tag{3.43}$$

When substituted in the extension of (3.17) appropriate to the inclusion of detrapping, it modifies the single-trap case of (3.18) into

$$\tilde{P}_m(\varepsilon) = \tilde{\eta}_m(\varepsilon) - \tilde{\psi}_{m-r}(\varepsilon)\left[\tilde{c}(\varepsilon)\,\tilde{P}_r(\varepsilon) - \frac{c'\,P_\theta(0)}{\varepsilon + (1/\tau_G) + c'}\right] \quad , \tag{3.44}$$

where the generalized capture "rate" $\tilde{c}(\varepsilon)$ is

$$\tilde{c}(\varepsilon) = \frac{c\,[\varepsilon + (1/\tau_G)]}{\varepsilon + (1/\tau_G) + c'} \quad . \tag{3.45}$$

The further treatment described in Sect. 3.3.1 is, therefore, unchanged in essentials. In fact, if the guest molecule is initially unexcited, $P_\theta(0) = 0$ and *all* the equations of Sect. 3.3.1 and 3.3.2 are unmodified in the presence of detrapping, except for the replacement of c by the generalized capture rate $\tilde{c}(\varepsilon)$. Furthermore, the yield expression (3.27) is changed to

$$\phi_G = \frac{\tilde{n}_r(1/\tau_H)}{1/c^{eff} + \tilde{\psi}_0(1/\tau_H)} \quad , \tag{3.46}$$

where the effective capture rate c^{eff} incorporates detrapping effects and is given by

$$c^{eff} = c(1 + c'\tau_G)^{-1} \quad . \tag{3.47}$$

The "sink" model treated so far in Sect. 3.3 has also been analysed, although only in the purely incoherent limit, by a number of authors: HUBER /3.25-27/, PEARLSTEIN /3.28/, LAKATOS-LINDENBERG et al. /2.61,3.29/, and KLAFTER and SILBEY /3.30,31/. The model is simple, tractable and appropriate to physical situations wherein guest molecules are placed interstitially. KENKRE and WONG /3.24/ have presented another model which describes the more commonly occurring case of substitutional traps. The index m now runs over the guest as well as the host sites, the former being denoted by r. This is in contrast to the sink model wherein r denotes guest-influenced *host* sites. There is now no decay of probability out of the totality of sites m through trapping. No capture rates c and c' are introduced, but the memory functions $\mathcal{W}_{mn}(t)$ are generally modified when m or n is a trap site. For simplicity, we assume that the memories or rates from host to guest sites are unmodified but that those from guest to host sites are reduced in strength by the detailed balance factor $\exp(-\beta\Delta)$. The former describe trapping and the latter detrapping. The quantity β equals $1/k_BT$ where T is the temperature and Δ is the amount by which the guest energy is less than the host energy. As a replacement for (3.17) we then have, with asterisks denoting time convolutions,

$$\frac{dP_m}{dt} + \frac{P_m}{\tau_H} = \sum_n (\mathcal{W}_{mn} * P_n - \mathcal{W}_{nm} * P_m) - \sum_r \delta_{m,r}\left[\left(\frac{1}{\tau_G} - \frac{1}{\tau_H}\right) + (e^{-\beta\Delta} - 1)\sum_n \mathcal{W}_{nr} *\right] P_r$$
$$+ (e^{-\beta\Delta} - 1)\,\mathcal{W}_{mr} * P_r \quad . \tag{3.48}$$

It is remarkable that (3.48) can be solved by the defect technique, since the defect size is infinite: the memory (or rate) from the guest site r to *every* host site is reduced. Restricting the analysis again to a single guest site r, one may write a replacement for (3.18) and then, putting m = r, obtain

$$\tilde{P}_r(\varepsilon) = \tilde{n}_r(\varepsilon')\left\{1 + \left(\frac{1}{\tau_G} - \frac{1}{\tau_H}\right)\tilde{\psi}_0(\varepsilon') + (1 - e^{-\beta\Delta})\right.$$
$$\left.\cdot\left[\sum_n \tilde{\psi}_{r-n}(\varepsilon')\,\tilde{\mathcal{W}}_{nr}(\varepsilon') - \tilde{\psi}_0(\varepsilon')\left(\sum_n \tilde{\mathcal{W}}_{nr}(\varepsilon')\right)\right]\right\}^{-1} \quad . \tag{3.49}$$

The infinite size of the defect region is apparent in the infinite summation term that multiplies $(1-e^{-\beta\Delta})$ in (3.49). However, it may be simplified immediately. Translational periodicity of the crystal, which we have continually invoked throughout the analysis above, allows us to switch indices and to write that term as

$$\sum_n \tilde{\mathcal{W}}_{0n}(\varepsilon')\,\tilde{\psi}_n(\varepsilon') - \sum_n \tilde{\mathcal{W}}_{n0}(\varepsilon')\,\tilde{\psi}_0(\varepsilon') = \varepsilon'\,\tilde{\psi}_0(\varepsilon') - 1 \quad . \tag{3.50}$$

The change of form of the term to that in the left-hand side of (3.50) uses the periodicity of the crystal. Recognizing it as the Laplace transform of the right-hand side of the GME (2.11), with m = 0, allows us to write the right-hand side of (3.50). A more detailed and different way of obtaining the result mentioned is given in /3.24/. We thus obtain the exact solution for the probability of the guest site

$$\tilde{P}_r(\varepsilon') = \tilde{\eta}_r(\varepsilon')\left\{e^{-\beta\Delta} + \tilde{\psi}_0(\varepsilon')\left[\varepsilon'(1-e^{-\beta\Delta}) + \left(\frac{1}{\tau_G} - \frac{1}{\tau_H}\right)\right]\right\}^{-1} \quad . \tag{3.51}$$

In this substitutional-trap model, $n_G(t)$ is identical to $P_r(t)$. The guest yield is then immediately given by

$$\phi_G = \frac{\tilde{\eta}_r(1/\tau_H)}{\dfrac{1}{(e^{\beta\Delta}/\tau_G)} + \left[1 - \dfrac{1}{(e^{\beta\Delta}/\tau_G)\tau_H}\right]\tilde{\psi}_0(1/\tau_H)} \quad . \tag{3.52}$$

A comparison of the guest-yield expression (3.52) for the substitutional-trap model and (3.46) for the sink model reveals perhaps unexpected, but convenient, similarities. The quantity c^{eff} defined in (3.47), which equals c in the absence of detrapping, is the effective trapping or capture rate of the sink model. Its counterpart in the substitutional-trap model is $e^{\beta\Delta}/\tau_G$. At low temperatures such that $(\tau_H/\tau_G)\cdot e^{\beta\Delta} \gg 1$, the correspondence between (3.52) and (3.46) is perfect. At higher temperatures there is a slight difference. To appreciate the closeness of the two results one may write the detrapping rate c' in the sink model as $ce^{-\beta\Delta}$. Then, if $ce^{-\beta\Delta}\tau_G \gg 1$ one obtains from (3.47), the result that the effective capture rate c^{eff} *exactly* equals $e^{\beta\Delta}/\tau_G$. These comparisons provide guidelines for deciding when the usual sink model may or may not be used as a representation of the (often) more physical substitutional-trap model.

We now describe the third elaboration in the description of the trapping process: the inclusion of many traps. For reasons of space limitation as well as simplicity, we treat only the *sink* model with no detrapping but with many traps. We return to (3.17,18). An exact solution is impossible unless we know the exact placement of traps. However, if we use (3.18), first by summing over all m to get

$$\tilde{n}_H(\varepsilon) = \frac{1}{\varepsilon'}\left[1 - c\sum_r{}'\,\tilde{P}_r(\varepsilon)\right] \tag{3.53}$$

and then by summing only over the trap sites r or s to get

$$\sum_r{}' \tilde{P}_r(\varepsilon) = \sum_r{}' \tilde{\eta}_r(\varepsilon') - c \sum_s{}' \tilde{\nu}_s(\varepsilon') \tilde{P}_s(\varepsilon) \quad , \tag{3.54}$$

where we introduce the new function $\nu_s(t)$,

$$\nu_s(t) = \sum_r{}' \psi_{r-s}(t) \quad , \tag{3.55}$$

an eminently physical assumption suggests itself. The guest molecules are placed at various locations r , s. Although these locations are not known exactly (only the concentration ρ is available), it is known that observations of sensitized luminescence are reproducible in different samples and therefore insensitive to those exact locations. An ensemble average of trap configurations is therefore appropriate. Unless definite correlations have been deliberately imposed in the doping process, $\nu_s(t)$, the sum of propagators from the trap site s to all trap sites, may be taken to be independent of s. We obtain

$$\sum_r{}' \tilde{P}_r(\varepsilon) = \frac{\sum_r{}' \tilde{\eta}_r(\varepsilon')}{1 + c\tilde{\nu}(\varepsilon')} \quad . \tag{3.56}$$

The immediate consequence of (3.56) is to modify all the observationally relevant results given earlier for the sink model by the simple replacement of $\psi_0(\varepsilon')$ by $\nu(\varepsilon')$. In particular, the guest yield for uniform host illumination is

$$\phi_G = \frac{\rho}{\frac{1}{c\tau_H} + (1/\tau_H)\, \tilde{\nu}(1/\tau_H)} \quad . \tag{3.57}$$

Equation (3.57) is the generalization of (3.30) for many traps. For small concentrations one might argue that (3.30) is recovered because the large distance between trap sites makes (3.55) state the essential equality of $\nu(t)$ and $\psi_0(t)$. In many practical cases this argument is indeed valid. There is, however, the following subtlety. As $t \to \infty$ the equality of $\nu(t)$ and $\psi_0(t)$ breaks down severely. Since, as $t \to \infty$, every propagator tends to 1/N where N is the total number of host crystal sites, $\nu(t)$ tends to the concentration ρ whereas $\psi_0(t)$ tends only to 1/N. There is a difference of a factor of the number of trap sites. These considerations gain in importance for the long-time behaviour of $n_H(t)$ and $n_G(t)$ and, in the context of yields, if τ_H is large enough to allow the difference between $\nu(t)$ and $\psi_0(t)$ to be made apparent in observations.

A general description of the effect of coherence on exciton trapping must therefore use $\tilde{\nu}(1/\tau_H)$ and $\nu(t)$ rather than $\tilde{\psi}_0(1/\tau_H)$ and $\psi_0(t)$ in its discussion. For small enough ρ's and small enough motion rates (if the exciton moves too fast $\psi_0(t)$ decays too rapidly and the onset of the region where it differs from $\nu(t)$ occurs too soon), the one-trap description given in Sect. 3.3.2 is sufficient. However, in other cases a full treatment along the present lines is necessary. SCHER and WU

/3.32/, HUBER et al. /3.25-27/, KLAFTER and SILBEY /3.30,31/ and MONTROLL and WEST /1.11/ have treated the multitrap case for periodic placement of traps, and HUBER et al. /3.25-27/ have also applied an average transfer-matrix approximation technique to random traps. Coherence studies from the $\nu(t)$ function introduced above have been carried out by KENKRE and PARRIS /3.33/. They have approximated $\nu(t)$ in various nonstandard ways for general trap placement, have shown that for a periodic placement of traps in a N-site lattice with a concentration ρ, $\nu(t)$ exactly equals the propagator $\psi_0(t)$ evaluated for a smaller lattice with $1/\rho$ sites, and have calculated $\varepsilon\tilde{\nu}(\varepsilon)$ to have the form

$$\varepsilon\tilde{\nu}(\varepsilon) = \frac{\tanh\,(\xi'/2)}{\tanh\,(\xi'/2\rho)} \tag{3.58}$$

for an infinite one-dimensional crystal with excitons moving incoherently via nearest-neighbour rates F. The definition of ξ' in (3.58) is identical to that in (4.7) of Sect. 4.2.1 below.

Although the case of periodic traps can be solved exactly /3.25-32/, there is the nagging possibility that a "super-lattice" of traps might result in some unphysical predictions not representative of real systems. On the other hand, despite considerable advances by HUBER et al. /3.25-27/ and others, the case of random traps continues to be difficult to analyze. We suggest that a fruitful approach might be to concentrate attention on the function $\nu(t)$ as we have done in the above treatment and to attempt to relate it directly to several experimental observables. As an example we mention that for large times, $\nu(t)$ can be shown to be intimately related to $n_H(t)$, the host luminescence, directly in the time domain. Periodic trap calculations and machine calculations for random traps are being used at present to gain insight into $\nu(t)$. It is to be stressed that a formalism in terms of $\nu(t)$ makes no assumption about any periodic trap placement. The only assumption made is that $\nu_s(t)$ is independent of the trap site s. This assumption is valid exactly for periodic traps. However, it is also very reasonable for random traps, as we have argued above.

3.3.4 Further Calculations

Coherence effects are usually associated with time-dependent observations. However, we have seen that steady-state observables, viz., the quantum yields, also contain coherence information. The reason is that steady-state yield observables do probe exciton motion for a finite time, viz., the exciton lifetime. The yield can be considerably different for the coherent and incoherent cases (Fig. 3.2), and as has been shown /3.24/, neglect of coherence can lead to an underestimation of the exciton diffusion constant interpreted from bulk quenching yield observations.

There are also other observable quantities that can show the effect of coherence.

Thus, KENKRE /3.34/ has shown that the steady-state energy transfer rate, defined and used by WOLF /3.5/ but generally forsaken in recent times in favour of a time-dependent rate, can be a valuable quantity for the analysis of coherence and other effects. Several interesting experimental and theoretical debates in this field will be mentioned in Chap. 4 below. Surely, time-dependent observables can exhibit coherent effects in a more detailed fashion. Yield observables are restricted to the radiative decay time τ_H as the probe time. In many systems this time may be too long to detect coherence. Thus if $\tau_H \approx 10^{-8}$ s, and coherence times are of the order of 10^{-10} s or smaller, coherence effects will be hidden and an incoherent description will suffice. Recent advances in time-dependent spectroscopy allow one to push probe times to the picosecond regime, and it is likely that many exciting coherence effects inaccessible to steady-state experiments are on the verge of being uncovered. The theory for the description of such time-dependent observations is given, in principle, by the Laplace transforms of the luminescence intensities $n_H(t)$ and $n_G(t)$ in equations such as (3.21) and (3.25). PARRIS and KENKRE /3.33/ have recently carried out numerical inversions of these transforms to investigate such effects. Various other illumination conditions are also being studied. They include initial excitation of the guest molecules and effects of crystal size. The time at which the guest luminescence intensity peaks is another observable that has been studied experimentally by SCHMID and collaborators /3.6,9-11/ and theoretically by PARRIS and KENKRE /3.33/.

The unification provided by the GME description of sensitized luminescence is convenient from a practical viewpoint. We are no longer required to do a k-space treatment /3.35-37/ whenever the slightest degree of coherence is suspected, and to return to real space /3.25-32/ when incoherence seems to be the main characteristic of the transport. The nature of the memory functions and the general formalism of the GME indeed permit a description of systems with arbitrary degree of coherence. It should also be noticed that these considerations apply to phosphorescence as well as fluorescence observations. FAYER and HARRIS /3.38/, SHELBY et al. /3.39/, and DLOTT et al. /3.40-42/ have carried out numerous investigations of triplet exciton trapping. The present GME formalism applies to those situations equally easily. Work on these applications is under way. Since the number 10^{-7} s has been often quoted as the coherence time in some of these systems, the results of such work should be particularly interesting. It is also hoped that the GME analysis will assist in the resolution of many important issues about exciton motion that have been recently raised by the elegant experiments of ZEWAIL and collaborators /3.43/. These concern arbitrary concentrations, disordered host lattices (the systems are mixed crystals) and the interplay of traps and supertraps. KOPELMAN /3.44 , 1.3/ has argued that percolation may be present, JORTNER and collaborators /3.45/ that Anderson localization of excitons may be afoot, BLUMEN and SILBEY /3.46/ have shown that a simple kinetic analysis of the process works remarkably well, and BLUMEN and collaborators /3.47-49/ have given a detailed analysis of trapping in disordered structures. In the inorganic realm, ORBACH and collaborators /3.50-53/ have developed a theory of motion in random systems and many new insights have resulted from the experimental work of EL-SAYED and collaborators /3.54/ and of YEN and collaborators /3.55/. These are all intriguing aspects of exciton motion that are being actively explored at the moment. Space limitations and subject restriction force us, however, to refrain from discussing them here.

3.4 Annihilation Observations

When two excitons approach each other, their interaction may lead to their mutual annihilation. The process consists of the deexcitation of one molecule and the excitation to a higher state of the other molecule. The result is a depletion of the number of excitons corresponding to the original electronic state. This phenomenon has been reviewed by AVAKIAN and MERRIFIELD /3.12/ and by GEACINTOV and SWENBERG /3.13/, and used by many experimentalists /3.56-61/ to derive information about exciton motion. The basic experimental procedure consist in creating excitons through optical absorption and observing the luminescence, either in a time-resolved fashion or through steady-state yield observables. The temperature and intensity of illumination are the usual experimental variables. The observed light intensity and yields are used to deduce aspects of exciton motion such as the diffusion constant.

Most interpretations of annihilation observations are based on the assumption that the exciton density is depleted at a rate proportional to the square of the local density. The constant of proportionality is called the annihilation constant γ. Values for this constant γ are deduced and presented for various systems. Usually, the diffusion constant of excitons is extracted from this γ by the application of an expression used by JORTNER et al. /3.62/ and others /3.63,64/ on the basis of the CHANDRASEKHAR-SMOLUCHOWSKI /1.18,19/ analysis of colloidal coagulation:

$$\gamma = 8\pi R D \quad . \qquad (3.59)$$

Here R is a distance characteristic of the annihilation interaction and D is the exciton diffusion constant.

The bilinear rate equation for annihilation in which (3.59) appears is

$$\frac{dn(t)}{dt} + \frac{n(t)}{\tau} = -\gamma n^2(t) \quad , \qquad (3.60)$$

where n(t) is the exciton density, and τ the lifetime of the exciton. A generalization of (3.60),

$$\frac{\partial n(x,t)}{\partial t} + \frac{n(x,t)}{\tau} = -\gamma n^2(x,t) + D\nabla^2 n(x,t) \quad , \qquad (3.61)$$

is also often used to analyze situations in which the exciton density $n(x,t)$ is inhomogeneous, i.e., is a function of the location x in the molecular crystal or aggregate.

The following questions are of importance in exciton annihilation. Is a bilinear rate equation of the form (3.60) or (3.61) always valid, and if it is not, what is its range of validity? When it is not valid, what constitutes an alternative description of exciton annihilation? When it is valid, how may we calculate the annihilation constant γ from microscopic inputs? In particular, when is (3.59) applicable and is there any effect of dimensionality on the expression for γ? Finally, what is the effect of coherence on γ, or more generally on luminescence observables in the presence of annihilation? We shall show below how we may answer all these questions with the help of the GME approach.

3.4.1 Technique of Solution

Our analysis /3.65/ will apply, without approximation, to a system of a single pair of excitons moving in a crystal of N sites with an arbitrary degree of coherence in their motion. For clarity, let us, however, begin with a one-dimensional crystal with incoherent motion via nearest-neighbour transition rates. If the excitons were not to decay or annihilate, (2.7) would therefore describe their motion. Let us first assume that the annihilation interaction has an extremely short range: the excitons annihilate each other at a rate 2b when they occupy the same site. This short-range assumption can be relaxed immediately but is being made for simplicity.

We construct a space of twice as many dimensions as the crystal, two in the present case, and study the evolution of $P_{m,n}(t)$, the probability that the first exciton is at site m and the second at site n. We follow the motion of the system point in this higher space and take its projections on the first and second subspace to obtain the motion of the individual excitons. The remarkable outcome of this construction is that the exciton annihilation problem is reduced to a trapping problem in the higher space. In the one-dimensional case it is clear that annihilation occurs when the system point is on the line of slope 1 through the origin in the higher (two-dimensional) space. Occupation of points along that line by the system point means that $m = n$, i.e., the two excitons occupy the same site. The evolution equation for $P_{m,n}(t)$ corresponding to (2.7) is

$$\frac{dP_{m,n}}{dt} = F\,(P_{m+1,n} + P_{m-1,n} + P_{m,n+1} + P_{m,n-1} - 4P_{m,n}) - \delta_{m,n}\, 2b\, P_{m,m} \quad . \tag{3.62}$$

The solution of this equation may be approached via the trapping analysis of Sect. 3.2. Thus

$$\widetilde{P}_{m,n}(\varepsilon) = \widetilde{n}_{m,n}(\varepsilon) - 2b \sum_r \widetilde{\Psi}_{m-r,n-r}(\varepsilon)\, \widetilde{P}_{r,r}(\varepsilon) \quad . \tag{3.63}$$

If we put m = n in (3.63) and define

$$\tilde{P}^k(\varepsilon) = \sum_m \tilde{P}_{m,m}(\varepsilon)\, e^{ikm} \quad , \tag{3.64}$$

we obtain from (3.63)

$$\tilde{P}^k(\varepsilon) = \frac{\tilde{\eta}^k(\varepsilon)}{1 + 2b\,\tilde{\Psi}^k(\varepsilon)} \quad . \tag{3.65}$$

Equation (3.64) has no counterpart in the trapping analysis. Its pictorial meaning is that, having converted the annihilation problem into a higher-dimensional trapping problem, we now reconvert the latter to what is, more or less, a trapping problem in the original space. In the present case the "trapping line" has been reduced to a trapping point. The Ψ and η in (3.65) and (3.63) are quantities similar to those introduced in Sect. 3.3. For instance,

$$\eta_{m,n}(t) = \sum_{m',n'} \Psi_{m-m',n-n'}(t)\, P_{m',n'}(0) \tag{3.66}$$

and $\Psi_{m,n}(t)$ is the propagator for the homogeneous part (without annihilation terms) of (3.62). This propagator is the product of the propagators $\psi_m(t)$:

$$\Psi_{m,n}(t) = \psi_m(t)\, \psi_n(t) \quad . \tag{3.67}$$

As in Sect. 3.3, $\psi(t)$'s refer to the single-exciton equation, in the present case (2.7).

The exact solution for $P_{m,n}(t)$ in (3.63) is available on substituting the *Fourier* inverse of (3.65) in (3.63):

$$\tilde{P}_{m,n}(\varepsilon) = \tilde{\eta}_{m,n}(\varepsilon) - \frac{2b}{N} \sum_r \sum_k \left[\tilde{\Psi}_{m-r,n-r}(\varepsilon)\right] \left[\frac{e^{-ikr}\, \tilde{\eta}^k(\varepsilon)}{1 + 2b\,\tilde{\Psi}^k(\varepsilon)}\right] \quad . \tag{3.68}$$

We show below that all the physically relevant quantities, such as the quantum yield and luminescence intensities, can be directly obtained from (3.68). The single-exciton motion equation such as (2.7) provides us with the single-exciton propagators $\psi_m(t)$. Their products in the time domain give us the two-exciton propagators through (3.67). The initial condition of illumination yields the η's through (3.66). Equation (3.68) then contains all information about annihilation up to quadratures. Calculations of yield are easier than those of time-dependent exciton number of luminescence intensity because the former require no Laplace inversions. However, the latter too can be obtained through analytic asymptotic techniques for extreme time limits or through numerical methods for all times.

Equation (3.68) is valid in the form given, for arbitrary dimensions and arbitrary degree of coherence. We have shown the analysis for a one-dimensional crystal with (2.7) for the evolution, but there is no difference in the final result, whatever the complexity of the full GME for the single-exciton motion and of the lattice structure of the crystal. The indices m , n are generally vectors and coherence information enters through the actual form of the propagators ψ_m and therefore through $\Psi_{m,n}$.

It is also straightforward to include long-range annihilation. Thus, if the last term in (3.62) is replaced by $-2\sum_\ell \delta_{m,n+\ell} b_\ell P_{m,n}(t)$, where $2b_\ell$ is the annihilation rate when the excitons are separated by ℓ (ℓ is a vector of the dimensions of the crystal), the result (3.63) is generalized to

$$\tilde{P}_{m,m+\ell}(\varepsilon) = \tilde{\eta}_{m,m+\ell}(\varepsilon) - \sum_{r,\ell'} 2b_{\ell'}\, \tilde{\Psi}_{m-r,m-r+\ell-\ell'}(\varepsilon)\, \tilde{P}_{r,r+\ell'}(\varepsilon) \quad . \tag{3.69}$$

Replacing (3.64) by the more general

$$\tilde{P}^k_\ell(\varepsilon) = \sum_m \tilde{P}_{m,m+\ell}(\varepsilon)\, e^{ikm} \quad , \tag{3.70}$$

one obtains

$$\tilde{P}^k_\ell(\varepsilon) = \tilde{\eta}^k_\ell(\varepsilon) - 2\sum_{\ell'} b_{\ell'}\, \tilde{\Psi}^k_{\ell-\ell'}(\varepsilon)\, \tilde{P}^k_{\ell'}(\varepsilon) \quad , \tag{3.71}$$

which has (3.65) as a particular case for $b_\ell = b\delta_{\ell,0}$. The details of how observables may be calculated from (3.71) for long-range annihilation are available in /3.65/.

In demonstrating the technique of solution we have started with (3.62), which does not include exciton decay. When it is introduced by adding a decay term, viz., $-P_m(t)/\tau$, in the one-exciton equation such as (2.7), the term $-2P_{m,n}(t)/\tau$ appears in the two-exciton equation such as (3.62): the probability $P_{m,n}$ can be depleted through the decay of either particle. Furthermore, it is necessary now to introduce $p_m(t)$, the probability that there is *only a single exciton* in the system and it is at m. Without radiative (or nonradiative) decay, excitons are depleted only through annihilation and their number is either 2 or 0. When decay is introduced, $P_{m,n}$ cannot provide a complete description. If we also introduce q(t), the probability that no exciton exists in the system, we have

$$\sum_m p_m + \frac{1}{2}\sum_{m,n} (P_{m,n} + P_{n,m}) + q = 1 \tag{3.72}$$

as the statement of conservation of probability. The probable number of excitons which, except for a volume factor (in three dimensions, or an area in two dimensions, or length in one dimension), is the n(x,t) mentioned in (3.61), is denoted here by $f_m(t)$. It is obviously given by

$$f_m(t) = p_m(t) + \sum_n [P_{m,n}(t) + P_{n,m}(t)] \quad . \tag{3.73}$$

3.4.2 Expressions for Observable Quantities

We can now state the most general form of the starting point of this analysis. In a two-exciton system with arbitrary crystal structure, dimensionality, and degree of coherence, we use $P_{m,n}(t)$, $p_m(t)$, and $q(t)$, respectively, in the two-exciton, one-exciton, and no-exciton space as the primary quantities. Their evolution obeys

$$\frac{dP_{m,n}}{dt} + \frac{2P_{m,n}}{\tau} = \sum_{m',n'} \left[\mathcal{W}_{mm',nn'} * P_{m',n'} - \mathcal{W}_{nn',mm'} * P_{m,n} \right] - \sum_{\ell} \delta_{m,n+\ell}\, 2b_{\ell} * P_{m,n} \quad , \tag{3.74}$$

$$\frac{dp_m}{dt} + \frac{p_m}{\tau} = \sum_n \left[\mathcal{W}_{mn} * p_n - \mathcal{W}_{nm} * p_m \right] + \frac{1}{\tau} \sum_n \left[P_{m,n} + P_{n,m} \right] \quad , \tag{3.75}$$

$$\frac{dq}{dt} = \frac{1}{\tau} \sum_m p_m + 2 \sum_{m,\ell} b_{\ell} * P_{m,m+\ell} \quad . \tag{3.76}$$

The last terms in (3.74,76) have been written as being non-Markoffian and spatially long-range for generality. In the rest of the discussion we shall take them to be local in time and space.

The experimentally relevant differential photon count rate, or the monitored luminescence intensity $I_L(t)$, is given by

$$I_L(t) = - \sum_m \left(\frac{df_m}{dt} \right)_{\text{radiative}} = \frac{1}{\tau} \sum_m p_m + \frac{2}{\tau} \sum_{m,n} P_{m,n} \tag{3.77}$$

since it is these terms that transfer excitons from (3.74,75) into (3.76). Appropriate modifications of τ to represent the fact that the radiative lifetime τ_0 is not the total lifetime can be made, if required, in the manner explained in Sect. 3.3. The quantum yield is given by

$$\phi = \frac{1}{2} \int_0^{\infty} dt\, I_L(t) = \frac{1}{2} \int_0^{\infty} dt \sum_m \left(\frac{df_m(t)}{dt} \right)_{\text{radiative}} \quad , \tag{3.78}$$

where the normalization $\frac{1}{2}$ corresponds to the number of excitons being initially 2.

Observations and usual interpretations in annihilation experiments refer to the exciton number $f_m(t)$ given by (3.73) and the yield ϕ given by (3.78). We compute these quantities from (3.74-76) through (3.68), keeping in mind that the latter has been written for $\tau \to \infty$. It can be shown /3.65/ that

$$I_L(t) = \frac{2}{\tau} e^{-t/\tau} \left[e^{-t/\tau} Q(t) + \frac{1}{\tau} \int_0^t dt'\, e^{-t'/\tau} Q(t') \right] \quad , \tag{3.79}$$

$$\phi = \frac{2}{\tau} \tilde{Q}(2/\tau) \quad , \tag{3.80}$$

thus necessitating the evaluation of the single quantity

$$Q(t) = \sum_{m,n} P_{m,n}(t) \quad , \tag{3.81}$$

where the right-hand side is to be computed for the case $\tau \to \infty$, i.e., from (3.68). Summation of (3.68) over m and n give the simple result

$$\tilde{Q}(\varepsilon) = \frac{1}{\varepsilon}\left[1 - \frac{(2b)\tilde{\eta}^0(\varepsilon)}{1+(2b)\tilde{\Psi}^0(\varepsilon)}\right] \quad . \tag{3.82}$$

The observables, viz., the luminescence intensity $I_L(t)$ and the yield ϕ, are thus determined entirely by $\tilde{\Psi}^0(\varepsilon)$ and $\tilde{\eta}^0(\varepsilon)$ as far as the contribution of exciton motion is concerned. Furthermore, these two-exciton quantities can be connected to single-exciton ones very simply. The result

$$\psi_\ell(t_1+t_2) = \sum_m \psi_{\ell-m}(t_1)\,\psi_m(t_2) \quad , \tag{3.83}$$

relating the (one-exciton) propagators for arbitrary time intervals t_1 and t_2, is a special case of the general chain condition obeyed by propagators no matter what the complexity of the $W_{mn}(t)$'s is. Symmetry and the particular case $t_1 = t_2$ reduce (3.83) to

$$\sum_m \psi_{m+\ell}(t)\,\psi_m(t) = \psi_\ell(2t) \quad . \tag{3.84}$$

Therefore we have

$$\tilde{\Psi}^0(\varepsilon) = \frac{1}{2}\,\tilde{\psi}_0(\varepsilon/2) \quad , \tag{3.85}$$

$$\tilde{\eta}^0(\varepsilon) = \frac{1}{2}\sum_{m',n'} P_{m',n'}(0)\,\tilde{\psi}_{m'-n'}(\varepsilon/2) \quad . \tag{3.86}$$

The substitution of (3.85,86) into (3.82) and the use of (3.79,80) gives the final results for annihilation observables in terms of *single-exciton quantities only*. For instance, for the completely delocalized initial condition when $P_{m',n'}(0) = (1/N^2)$,

$$\phi = 1 - \frac{(b\tau/N)}{1 + b\tilde{\psi}_0(1/\tau)} \tag{3.87}$$

For the localized initial condition when the excitons are initially ℓ apart,

$$\phi = 1 - \frac{b\tilde{\psi}_\ell(1/\tau)}{1 + b\tilde{\psi}_0(1/\tau)} \quad , \tag{3.88}$$

with the particular case, when they occupy the same site,

$$\phi = \frac{1}{1 + b\tilde{\psi}_0(1/\tau)} \quad . \tag{3.89}$$

The simplicity and remarkable similarity of these results to the trapping expressions [compare, e.g., (3.86,28)] should be noted.

The effect of coherence, dimensionality, lattice structure, and other transport characteristics enters expressions (3.86-88) through the *single-exciton* propagators $\widetilde{\psi}_0(1/\tau)$ just as in the trapping case. There should be no need to repeat either the evaluation of propagators which we have done in Sect. 3.3, or the discussion of the effects of coherence and other features. Further details, particular cases for one-, two- and three-dimensional crystals, and for arbitrary degree of coherence as expressed in the GME (2.66) with memories (2.76), may be found in /3.65/. Only three points need be mentioned here. The factor 1/N appearing in the above expressions for the yield ϕ and the luminescence intensity $I_L(t)$ is to be replaced in large systems simply by $\rho/2$, where ρ is the initial concentration of excitons. This is valid if ρ is sufficiently small. The situation is similar to the single-trap discussion in Sect. 3.3. The analysis presented here, while quite general in other respects, is therefore restricted to initial exciton concentrations that are small. For illumination intensities which are very high, a significant modification of the calculations is necessary. The second point is that although the final results for the observables display striking similarities with the trapping problem, subtle differences which limit the usefulness of the analogy do exist. In Sect. 3.3.4 we have presented a discussion of trapping in terms of the $\nu_s(t)$ function and have described the case of periodic traps to gain useful insights into the multitrap problem. Those studies have no direct analogy in the annihilation problem. Periodic trapping corresponds to the totally unphysical case of annihilation interactions that might exhibit oscillatory behaviour with inter-exciton distance. Relations such as (3.83) have helped the analysis of annihilation through considerations of trapping in higher spaces. However, it appears that the many-exciton annihilation problem must be approached in an entirely different way. Some efforts in this direction are described in Sect. 3.4.4.

The third point that needs to be stressed here is that in presenting expressions for experimentally accessible variables, viz. the yield and the luminescence intensity, and in analyzing the effect of exciton transport characteristics on them, we have not made use of the traditional bilinear annihilation equation (3.60). This is a definite advantage of this treatment because the range of validity of (3.60) is *not* yet understood properly. We emphasize that (3.60) is not required to analyze experiments at all. However, because it is a familiar equation in the literature, we study aspects of its validity in the next subsection /3.66/.

3.4.3 Validity of the Bilinear Equation and Expressions for γ

While (3.60,61) describe the evolution of n(x,t), the proper quantity to analyze in a discrete crystal is clearly the exciton number $f_m(t)$, which reduces to n(x,t) in the continuum limit. The discrete counterpart of (3.61) is, if we also include coherence in the analysis,

$$\frac{df_m}{dt} + \frac{f_m}{\tau} = \sum_n (W_{mn} * f_n - W_{nm} * f_m) - \gamma' f_m^2 \quad , \tag{3.90}$$

where γ' is related to the annihilation constant through

$$\gamma' = \frac{1}{v}\gamma \quad . \tag{3.91}$$

Here v is a unit-cell volume, area, or length in a three-, two-, or one-dimensional crystal, respectively. Relation (3.73) and the analysis presented in Sect. 3.4.2 allow us to write, at once, the exact equation

$$\frac{df_m}{dt} + \frac{f_m}{\tau} = \sum_n (W_{mn} * f_n - W_{nm} * f_m) - \int_0^t dt' \sum_n \lambda_{mn}(t-t')\, P_n^2(t') \quad . \tag{3.92}$$

Although we have used asterisks to denote time convolutions in (3.90) and in part of (3.92), we have retained the full form of the last term in (3.92) to emphasize its nature.

The study of the range of validity of (3.90) is therefore based on the comparison of (3.90) and (3.92), and on the study of how and under what conditions the last term in (3.90) approximates the last term in (3.92). This latter term contains the quantity $P_m(t)$, which is the solution of the exciton motion problem in the absence of annihilation but in the presence of the given initial condition,and the kernel $\lambda_{mn}(t)$ given by

$$\lambda_{mn}(t) = \int d\varepsilon\, e^{\varepsilon t} \frac{b}{N} \sum_k e^{-ik(m-n)} \left[1 + 2b\, \tilde{\Psi}^k\left(\varepsilon + \frac{2}{\tau}\right)\right]^{-1} \quad . \tag{3.93}$$

The ε-integration is on the Bromwich contour.

The non-Markoffian nature of the last term in (3.92) does not require coherence in transport and arises even from a Master equation for $P_m(t)$. The term is also nonlocal in space. The differences between the exact (3.92) and the traditional approximation (3.90) are therefore three: non-Markoffian nature, nonlocal nature, and the fact that the exact term contains the exciton number in the absence of annihilation.

The replacement of $P_m(t)$ by a constant times $f_m(t)$, which is inherent in (3.90), is difficult to justify. Its only advantage is that it makes (3.90) *closed* in the quantities $f_m(t)$. There is no reason to make it, since exact solutions can be obtained for $\lambda_{mn}(t)$ as well as $P_m(t)$ from the analysis in Sect. 3.4.2. The problem is similar to the justification of the Boltzmann equation /3.67/ which replaces higher members of the BBGKY hierarchy /3.67/ by first-order distribution functions. Let us make that approximation and investigate here the validity of the replacement

$$\lambda_{mn}(t) = \gamma' \, \delta_{m,n} \, \delta(t) \quad , \tag{3.94}$$

which allows (3.92) to be reduced to (3.90).

The replacement (3.94) involves the Markoffian approximation, whereby $\tilde{\lambda}_{mn}(0)$ is used in place of $\tilde{\lambda}_{mn}(\varepsilon)$, and the local approximation, whereby $\lambda^k(t)$ is replaced by $\lambda^0(t)$. The former assumes that $\lambda_{mn}(t)$ decays rapidly in time. While coherence has little to do with this decay directly, a rapid decay corresponds to fast motion of the exciton, as is clear from (3.93). This is so because the "memory function" here is basically the Laplace inverse of $[1+2b\,\tilde{\Psi}^k(\varepsilon+\frac{2}{\tau})]^{-1}$ and is controlled by exciton motion. The replacement of $\lambda_{mn}(t)$ by $\delta(t)\,[\int_0^\infty dt'\,\lambda_{mn}(t')]$ thus corresponds to fast exciton motion.

The local approximation could be rather objectionable in general, although its applicability is also decided by how fast the excitons move. The replacement of Ψ^k by Ψ^0 corresponds to the assumption that the one-exciton propagator ψ_m is strongly peaked in m space. But this is true at short times. As time goes on, all ψ_m's begin to be identical to one another. The local approximation is therefore good at short times.

One thus discovers /3.66/ the local and the Markoffian elements of the approximation (3.94) to be in conflict with each other. If characteristic exciton motion times are of the order of picoseconds, one would conclude that the Markoffian approximation is good in most experiments but that the local approximation should never be used. What is more important, the two approximations should not be used together.

It would appear that the nonoverlapping nature of the two approximations in (3.94) make the recovery of the traditional (3.60) impossible. This is, perhaps fortunately, incorrect. For homogeneous initial conditions, f_m is independent of m [i.e., n(x,t) is n(t)] and may be taken out of the summation. Equation (3.60) is recovered when the Markoffian approximation is made.

The analysis means that (3.61) has doubtful validity but (3.60) may be used for *homogeneous* illumination (such as in two-photon absorption) and for times large with respect to decay times of $\lambda_{mn}(t)$.

KENKRE /3.66/ has given several alternatives to (3.60,61) to be used when they are not valid. When the bilinear equation (3.60) is applicable, expressions for the annihilation constant γ, equivalently the rate γ' [see (3.90) for the relation], must be computed. These are obtained /3.66/ through (3.93,94). Generally,

$$\gamma = \frac{v}{\frac{1}{b} + \tilde{\psi}_0(1/\tau)} \tag{3.95}$$

holds. If annihilation occurs much more slowly than trapping, γ is given by bv and is insensitive to motion characteristics. If, however, the motion process is the slower one /3.14/, then

$$\gamma = v\,[\tilde{\psi}_0(1/\tau)]^{-1} \quad . \tag{3.96}$$

The effect of coherence, dimensionality, and other transport features on the annihilation constant are immediately obtained through (3.96). Charts of the effect of dimensionality have been given by SUNA /3.14/ and by KENKRE /3.66/. The latter reference also contains the analysis in the limits of slow, as well as fast, exciton motion with respect to decay. Specifically, it has been shown /3.66/ that for incoherent exciton motion with nearest-neighbour transition rates F and rapid annihilation (large b), the annihilation rate γ' is given by

$$\gamma' = (1+4F\tau)^{1/2}\,(1/\tau) \quad , \tag{3.97}$$

$$\gamma' = \frac{1+4F\tau}{(2\tau/\pi)}\left[\mathbb{K}\left(\frac{4F\tau}{1+4F\tau}\right)\right]^{-1} \quad , \tag{3.98}$$

$$\gamma' = 2F\left[I(0,0,0\,;1\,;1+\frac{1}{6F\tau})\right]^{-1} \quad . \tag{3.99}$$

in one, two, and three dimensions, respectively (simple square and simple cubic structures being assumed in the latter 2 cases). Here $\mathbb{K}$ is the elliptical integral of the first kind, and

$$I(a,b,c\,;\alpha\,;\beta) = \int_0^\infty dt\; e^{-(2+\alpha)\beta t}\; I_a(t)\; I_b(t)\; I_c(\alpha t) \quad . \tag{3.100}$$

The function in (3.100) has been defined and tabulated by MARADUDIN et al. /3.68/ in the context of lattice dynamics. KENKRE /3.66/ has further shown that the proportionality of the annihilation constant and the diffusion constant, (3.59), holds only for three-dimensional crystals and only in the limit of motion which is fast with respect to decay. Since exciton motion in many aromatic hydrocarbon crystals is often said to be effectively two-dimensional, (3.59) must be applied with caution.

The effect of coherence on the annihilation constant can be examined by substituting the propagator (3.37) in the expression (3.96) for the annihilation constant. Thus, for instance, in one-dimensional crystals, the purely coherent limit is

$$\gamma = V\left[\frac{(1+16V^2\tau^2)^{1/2}}{(2\tau/\pi)}\right]\left[\mathbb{K}\left(\frac{4V\tau}{(1+16V^2\tau^2)^{1/2}}\right)\right]^{-1} \tag{3.101}$$

and the totally incoherent limit is (3.97).

3.4.4 Further Calculations

The various questions posed at the beginning of Sect. 3.4 have been answered within the framework of the GME theory /3.65,66/. The unified description it provides of the effects of a variety of transport features, its analysis of discrete crystals rather than of continua, and its ability to handle any initial conditions including nonsymmetric ones, are the advantages it possesses over earlier theories /3.14/. However, extensions in a number of directions are required. Magnetic field effects /3.12,13/ including recent observations of quantum beats /3.69/ must be incorporated perhaps along the lines laid out in /3.14/. Some striking effects of crystal size have been recently found in theoretical investigations in terms of the analysis presented here. They appear counterintuitive: under certain conditions, restraining excitons within finite boundaries results, not in enhanced annihilation (expected as a consequence of more frequent encounters) but in reduction of annihilation. These effects are being actively pursued at the moment. One of the most important and difficult extensions of this theory is in the direction of higher exciton concentrations. Surely, this is not peculiar to this theory. No exact high-concentration theory of exciton annihilation exists in the literature. It appears that radically new approaches must be sought since high exciton densities (certainly attainable experimentally) present a true many-body problem. Recently, KENKRE and VAN HORN /3.70/ have given an analysis of pairwise annihilations of *stationary* particles on a lattice. The methods used are quite different from those explained above. Master equations in configuration (i.e., many-body) space are used and the n-particle annihilation problem is solved exactly in one dimension. That theory is directed at the description of pycnonuclear reactions in the interior of stars and is *not* restricted to low concentrations. But it allows no motion of the particles. On the other hand, exciton motion is appreciable and of definite importance in molecular crystals. Efforts are being made to combine the methods of the two theories to provide a description which applies to finite motion and high concentration.

4. Other Issues and Other Master Equation Techniques

4.1 Introduction

Coherence is no doubt a central issue in exciton physics. But there are also a number of other problems of equal importance. Some of them have been mentioned in Sect. 1.2. We discuss them in this chapter. In keeping with the theme of the article, the analysis used employs master equations. These, however, are Markoffian in character, i.e., involve no (or delta-function) memories. They are of three types. The first operate in real space and are identical to the GME's discussed in Chaps. 2 and 3 except for the absence of memories. They are used in Sect. 4.2 to study surface quenching and in Sect. 4.3 to analyze questions concerning the so-called energy transfer rate. The second kind operate in momentum space and are similar to linearized Boltzmann equations used for electron transport in metals. They are used in Sect. 4.4 in the context of exciton trapping and sensitized luminescence. The third kind use a space which is a mixture of real space and an energy space. They describe the interplay of vibrational relaxation and exciton motion and are given in Sect. 4.5. The respective discussions contain an account of experimental observations as well as their theoretical analyses.

4.2 Master Equations for Surface Quenching

One of the earliest measurements of the diffusion constants of excitons in aromatic crystals was by SIMPSON /4.1/. He illuminated one side of an anthracene crystal and detected the excitons thus created with the help of a coating of guest molecules of tetracene (sometimes called naphthacene) placed at the other end. A measurement of the exciton flux at the detector end, or more specifically of the quantum yield of the detector, was used to deduce the diffusion length and hence the diffusion constant. Similar experiments were carried out by KURIK /4.2/, by TOMURA and TAKAHASHI /4.3/, and by GALLUS and WOLF /4.4/. Conceptually, this kind of experiment is very direct since it essentially involves measuring the time taken by the excitons to cross from one end to the other, the specific observed quantity being the fraction of excitons that survive (against decay) during the end-to-end migration. In contrast to the case of bulk-quenching experiments discussed in Sect. 3.2, there is little uncertainty here about the distance the exciton must travel to get captured. There is, however, considerable discrepancy in the reported values of the diffusion constant D as has been discussed by POWELL and SOOS /1.20/. TOMURA and TAKAHASHI /4.3/ suggested that the source of the discrepancy could lie in the detectors not being perfect absorbers. This has also been pointed out by POWELL and SOOS /1.20/.

The discrepancy is, however, rather large (several orders of magnitude), and it was not clear that it could be quantitatively explained by the suggestion of imperfect absorption. In fact, calculations based on continuum diffusion equations such as the one used by SIMPSON /4.1/ lead one to believe that only small correction factors of the order of 4 can result from these considerations. It has been recently shown by KENKRE and WONG /4.5/, however, that a Master-equation treatment does indeed result in a possible quantitative explanation of the large span of reported values of D. We begin with that treatment in Sect. 4.2.1.

4.2.1 Effect of Imperfect Absorption by End Detectors

Since the experimental setup in /4.1-4/ suggests the consideration of a chain of N host lattice sites with a guest site appended at one end, the appropriate Master equation is

$$\frac{dP_1}{dt} + \frac{P_1}{\tau_H} = F\,(P_2 - P_1) \quad , \tag{4.1}$$

$$\frac{dP_N}{dt} + \frac{P_N}{\tau_H} = F\,P_{N-1} + f\,P_G - (F + \mathcal{F})P_N \quad , \tag{4.2}$$

$$\frac{dP_m}{dt} + \frac{P_m}{\tau_H} = F\,(P_{m+1} + P_{m-1} - 2P_m) \qquad N-1 \geqslant m \geqslant 2 \quad , \tag{4.3}$$

$$\frac{dP_G}{dt} + \frac{P_G}{\tau_G} = \mathcal{F}\,P_N - f\,P_G \quad . \tag{4.4}$$

For simplicity, nearest-neighbour transition rates have been assumed, F denoting the host-host motion rate, $\mathcal{F}$ the host-guest trapping rate, and f the guest-host detrapping rate. The guest quantum yield is

$$\phi_G = \int_0^\infty dt\,\frac{P_G(t)}{\tau_G} \tag{4.5}$$

if the total lifetime equals the radiative lifetime. In the following expressions this assumption will be made, but it may be immediately relaxed, if necessary, by multiplying the right-hand side of (4.5) by the ratio of the lifetimes as in Sect. 3.3.

The solution of (4.1-4) also employs the defect technique /1.11/ but cannot be based on the propagators used in Sect. 3.2 and 3.3 because they are not appropriate to a finite chain with ends. Discrete Fourier transforms cannot be used profitably for our present system because translational symmetry is broken by the ends. LAKATOS-LINDENBERG et al. /3.29/ have, however, given the appropriate propagator. We call it $\psi_{mn}(t)$. It is not a function merely of $m-n$ and is not therefore a single-site-index quantity. It is the solution $P_m(t)$ of equations (4.1-3) with $(1/\tau_H) = f = F = 0$ and with $P_\ell(0) = \delta_{\ell,n}$. It thus corresponds to the present system in the absence of radiative decay and of the detector. Its Laplace transform is given /4.5/ as

$$\tilde{\psi}_{mn}(\varepsilon) = \left[F(\sinh \xi')(\sinh N\xi')\right]^{-1} \times \left\{\cosh\left[\left(\frac{\xi'}{2}\right)(2N - |m+n-1| - |m-n|)\right]\right\} \left\{\cosh\left[\left(\frac{\xi'}{2}\right)(|m+n-1| - |m-n|)\right]\right\} , \tag{4.6}$$

$$\cosh \xi' = 1 + \frac{\varepsilon}{2F} . \tag{4.7}$$

Application of the defect technique to (4.1-4) on the other hand gives /4.5/

$$\phi_G = \left[\frac{F}{1 + f\tau_G + F\,\tilde{\psi}_{NN}(1/\tau_H)}\right] \left[\sum_{n=1}^{N} \tilde{\psi}_{Nn}(1/\tau_H)\; P_n(0)\right] . \tag{4.8}$$

The second factor in the right-hand side of (4.8) contains the initial condition $P_n(0)$ which corresponds to the detail of initial illumination. If the absorption coefficient of the host crystal is k, and if we define its discrete version κ by

$$\kappa = ka \tag{4.9}$$

where a is the lattice constant of the host crystal, the initial $P_n(0)$ may be written as

$$P_n(0) = e^{-\kappa(m-1)} \left[\sum_{m'=1}^{N} e^{-\kappa(m'-1)}\right]^{-1} = e^{-\kappa(m-1)}\,(1 - e^{-\kappa})(1 - e^{-\kappa N})^{-1} . \tag{4.10}$$

Substitution of (4.10) and of the propagator expression (4.6) into (4.8) finally gives the central result of this analysis:

$$\phi_G = \left|\frac{(F/F)}{(F/F) + c}\right| \left\{\mathrm{sech}[\xi(N - \tfrac{1}{2})]\right\} I' . \tag{4.11}$$

The reason for displaying ϕ_G in the particular form of (4.11) will be apparent in the following. The quantity I' in (4.11) is decided primarily by the details of the initial exciton distribution (4.10) and is given by

$$I' = \frac{1}{2}\left[\frac{1 - e^{-\kappa}}{1 - e^{-\kappa N}}\right] \left[\frac{e^{\xi/2}\,(1 - e^{-N(\kappa-\xi)})}{1 - e^{-(\kappa-\xi)}} + \frac{e^{-\xi/2}\,(1 - e^{-N(\kappa+\xi)})}{1 - e^{-(\kappa+\xi)}}\right] . \tag{4.12}$$

Its extreme limits are

$$I' = (1/N)(\sqrt{F\tau_H})\;[\sinh(N\xi)] \tag{4.13}$$

for $\kappa \to 0$, signifying spatially uniform initial illumination, which would hold reasonably well for a relatively thin sample, and

$$I' = \cosh(\xi/2) \tag{4.14}$$

for $\kappa \to \infty$, which describes excitons initially placed completely at one end of the sample. The quantity c in the general expression (4.11) for the yield is given by

$$c = \left(\frac{1 + f\tau_G}{\sqrt{F\tau_H}}\right) \left[\sinh(N\xi)\right] \left\{\operatorname{sech}[\xi(N - \tfrac{1}{2})]\right\} \quad . \tag{4.15}$$

Note that ξ, which appears in (4.11-15), is given by

$$\cosh \xi = 1 + \frac{1}{2F\tau_H} \quad , \tag{4.16}$$

i.e., is obtained from (4.7) by the substitution of ε by $1/\tau_H$.

Although the primary contribution of the result (4.11) comes from the discrete Master-equation treatment underlying it, it is instructive to study how it reduces to the earlier continuum analysis. The continuum limit is $a \to 0$. However, it simultaneously involves

$$\lim_{a\to 0} Fa^2 = D \quad , \tag{4.17}$$

$$\lim_{a\to 0} \mathcal{F}a = \mathcal{D} \quad , \tag{4.18}$$

$$\lim_{a\to 0} \kappa/a = k \quad , \tag{4.19}$$

$$\lim_{a\to 0} P_n(t)/a = p(x,t) \quad , \tag{4.20}$$

$$\lim_{a\to 0} Na = L \quad , \tag{4.21}$$

all the quantities on the right-hand sides being finite. Here L is the length of the crystal, D the exciton diffusion constant, and $p(x,t)$ the exciton probability density. It is obvious, of course, that in the continuum limit, the number of sites N tends to infinity and κ and P_n tend to 0. However, the behaviour of the transition rates F, $\mathcal{F}$, and f deserves comment. The latter is unaffected in the limit, while the former two go to infinity as $1/a^2$ and $1/a$, respectively. This is so because F controls a second derivative in space, $\mathcal{F}$ describes the loss of excitons from 1/N of the host crystal, i.e., from a single site, and f describes such a loss from the entire detector. With (4.17-21) and with the limiting value

$$\xi = 1/\sqrt{F\tau_H} \tag{4.22}$$

of the quantity ξ, obtained by expanding $\cosh \xi$ as $1 + \xi^2/2$, the continuum limit of (4.11) can be shown /4.5/ to be

$$\phi_G = \left\{ \frac{1+f\tau_G}{\mathcal{D}\tau_H} + \left[\ell_D \tanh(L/\ell_D)\right]^{-1} \right\}^{-1} \left\{ k(1-e^{-kL})^{-1} \left[\ell_D \sinh(L/\ell_D)\right]^{-1} \ell_D^2 \right.$$
$$\left. \times (k^2-\ell_D^2)^{-1} \left[k - e^{-kL} \left(k\cosh(L/\ell_D) + (1/\ell_D)\sinh(L/\ell_D) \right) \right] \right\} , \tag{4.23}$$

where ℓ_D is the diffusion length $\sqrt{D\tau_H}$ and $\mathcal{D}$, given by (4.18), is a trapping parameter of the dimensions of centimeter/second. If the further approximation

$$1 + f\tau_G \ll \mathcal{D}\tau_H \tag{4.24}$$

is made, (4.23) reduces to

$$\phi_G = \frac{k^2\ell_D^2}{k^2\ell_D^2 - 1} \left[\operatorname{sech}(L/\ell_D) - e^{-kL} \left(1 + \frac{\tanh(L/\ell_D)}{k\ell_D} \right) \right] \frac{1}{1-e^{-kL}} . \tag{4.25}$$

When the factor $(1-e^{-kL})^{-1}$, which appears in (4.25) from normalization, is identified with the intensity I, and the yield ϕ_G with the "exciton flux" in earlier expressions /4.1/, the recovery of those expressions from (4.11) is seen to be complete. It must be emphasized that earlier interpretations neglected the term $(1+f\tau_G)/\mathcal{D}\tau_H$ in addition to making the continuum approximation.

Having thus obtained the SIMPSON analysis /4.1/ as an approximate case of the Master-equation result (4.11), we now return to a possible explanation of the spread in observed values of D. This explanation is not restricted to small factors. For simplicity, consider the case of a thin sample or equivalently of uniform initial illumination across the length of the crystal. Equation (4.13) holds and (4.11) gives

$$\phi_G = \phi_G^s \left[\frac{(F/\mathcal{F})}{(F/\mathcal{F}) + c} \right] . \tag{4.26}$$

The quantity c, given by (4.15), and ϕ_G^s, given by

$$\phi_G^s = \frac{\sqrt{F\tau_H}}{N} \left[\sinh(N\xi)\right] \left\{ \operatorname{sech}[\xi(N-1/2)] \right\} = (1/N) \left(\frac{F\tau_H}{1+f\tau_G} \right) c , \tag{4.27}$$

are both independent of the trapping rate $\mathcal{F}$. The yield ϕ_G of (4.26), when plotted as a function of the relative detrapping parameter $F/\mathcal{F}$, saturates at the value ϕ_G^s (Fig. 4.1). This saturation value corresponds to a perfect absorber being used as the detector and therefore to the earlier analyses. The analysis of KENKRE and WONG /4.5/ shows, however, that for any realistic value of $F/\mathcal{F}$, the yield will be lower than in the earlier analyses. Conversely, the interpreted value of the diffusion constant, equivalently of F, will be larger in the exact analysis. Thus, the perfect absorber assumption always results in an underestimation of D. The yield ϕ_G has been plotted in Fig. 4.1 for N = 1000 and for three different values of D (equi-

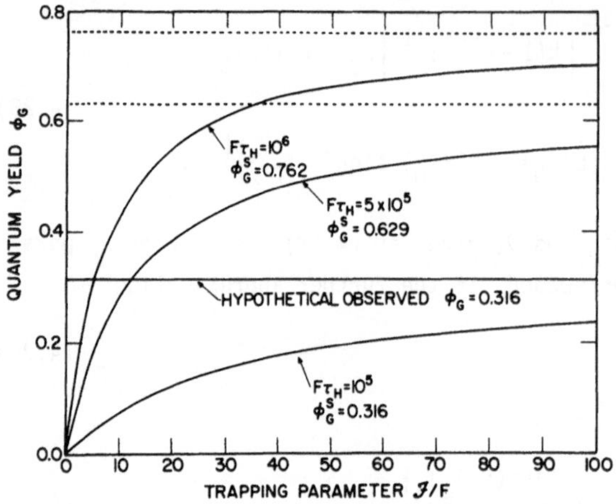

Fig. 4.1. Possible explanation of disparity in reported values of the singlet exciton diffusion constant D. The detector quantum yield ϕ_G or flux is plotted as a function of the ratio of the trapping rate F to the motion rate F (proportional to D) for three values of D. The dotted lines represent ϕ_G for perfect absorbers. The solid straight line shows a hypothetical observed ϕ_G = 0.316 and the assumed value of $f\tau_G$ is 10^4

valently of F). For $\tau_H = 10^{-8}$ s, and a = 10 Å, they are 1, 0.5, and 1/10 cm^2/s. A hypothetical observed ϕ_G, also shown in the figure, would be interpreted through the earlier analyses as being indicative of D = 0.1 cm^2/s. However, if the detrapping rate F is of the order of F, say 5F, the figure shows that the true D equals 1 cm^2/s. Various detectors for the same host crystal would correspond to various points on the F/F axis and a disparity in reported values of D could thus result.

We now comment on the *magnitude* of the correction factors provided by the above theory. That magnitude is crucially dependent on how rapidly the yield curves for various F's rise to their saturation values. If the rise is always rapid, ϕ_G would equal ϕ_G^S (the perfect absorber value) for all realistic values of F/F. The above "explanation" would then be of little use. Equation (4.26) shows that the quantity c is a measure of how rapidly the ϕ_G curves rise: when c equals the trapping parameter F/F, ϕ_G equals half the saturation value. For very thick samples we have

$$c = \frac{1 + f\tau_G}{\sqrt{F\tau_H}} \quad , \tag{4.28}$$

whereas for very thin samples we have

$$c = \left(\frac{1 + f\tau_G}{F\tau_H}\right) N \quad , \tag{4.29}$$

the sample length being compared to the diffusion length here. A typical value of N might be 10^3 if the sample length is half a micron and the lattice constant is 5 Å. Values of $F\tau_H$ might lie anywhere between 10^2 and 10^6 if reported values of D are used for estimates. Equations (4.28,29) then show clearly that the $f\tau_G$ can be crucial to the above argument. It has never been considered in earlier interpretations based on continuum equations. However, it can have a large value even when thermal factors (detailed balance) suggest that f should be neglected in most expressions: although the ratio f/F might be negligible, the ratio $f/(1/\tau_G)$ may be quite large. For large $f\tau_G$ and consequently large c, the ϕ_G curves rise slowly and considerable difference can exist between ϕ_G^s and ϕ_G for a reasonable value of $\mathcal{F}/F$.

It appears therefore that the theory given by KENKRE and WONG /4.5/ on the basis of a development of the ideas of TOMURA and TAKAHASHI /4.3/, may indeed provide an explanation of the spread in the reported values of D. These considerations show that if no other effects interfere, the largest reported value of D is closest to its actual value.

4.2.2 Effect of Variation in Penetration Length

The discussion in Sect. 4.2.1 might lead one to believe that the extraction of the transition rate F or of the diffusion constant D from observations of surface quenching necessitates prior knowledge of the trapping and detrapping rates (F and f). Such knowledge is indeed required in interpretations of end-to-end migration observations of the kind made in /4.1-4/. However, on the basis of a special feature of the general result (4.11), KENKRE /4.6/ has recently shown how the need for information about F and f may be bypassed in a different kind of surface quenching observations.

The proposal is based on three peculiarities of the result (4.11) for the yield ϕ_G:

(a) the dependence of ϕ_G on κ, (or on the absorption coefficient $k = \kappa/a$) is present only in the factor I',

(b) the factor I' is independent of the trapping rate F and the detrapping rate f, and

(c) the factor I' does depend on the motion rate $\mathcal{F}$.

Evidently F (equivalently the diffusion constant D) can be deduced from observations of the variation of the yield with variation in $\kappa = ka$. The absorption coefficient k, or equivalently the penetration length $\ell_p = 1/k$, can be varied by changing the wavelength of excitation. The $\phi_G - \ell_p$ curve does not depend on F or f but does depend on $\mathcal{F}$. A fit to the observed $\phi_G - \ell_p$ curve will thus yield the value of the diffusion constant without any prior knowledge of trapping interactions. Figure 4.2 shows predicted curves for three $F\tau_H$'s, corresponding to a sample with N = 1000 and

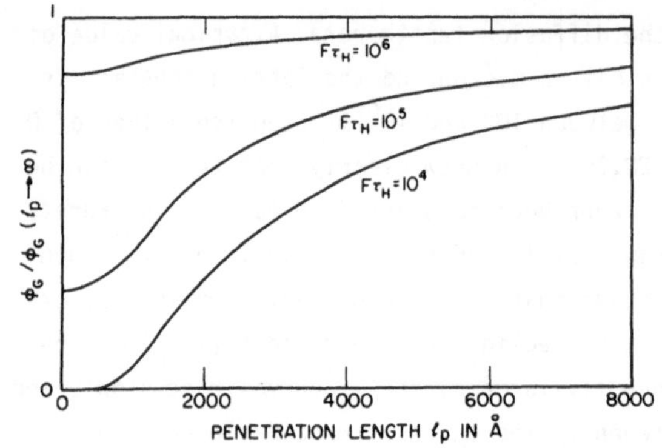

Fig. 4.2. Proposal to measure the exciton diffusion constant D without trapping information. The detector yield ϕ_G is plotted as a function of the penetration length ℓ_p for three values of $F\tau_H$, equivalently of D. The yield is normalized to its limiting value for uniform illumination. Sample length L = 5000 Å and lattice distance a = 5 Å

a = 5 Å, i.e., with sample length L = 0.5 microns.

While it is remarkable that trapping experiments capable of yielding exciton motion characteristics without the need for exciton trapping information exist, the pecularities (a - c) stated above correspond to a simple physical argument. The yield is insensitive to the penetration length if the excitons move so fast that they arrive at the detector rapidly (and thus make a contribution to the yield), irrespective of where they are placed initially in relation to the detector end. The top curve in Fig. 4.2 corresponds to this situation. On the other hand, if exciton motion is slow, the initial placement, consequently the penetration length, can be critical. If ℓ_p is too small, excitons may never be able to arrive at the detector end within their lifetime. This situation corresponds to the bottom curve in Fig. 4.2. What makes the proposal in /4.6/ possible is that the magnitudes are probably just right to enable the effect to resolve the discrepancy. As shown in Fig. 4.2, the effect is appreciable in the debated range of D. Incidentally, the proposed experiment /4.6/ is quite different from those of MULDER /4.7/ and HAARER and CASTRO /4.8/ who create and detect excitons on the same side of the crystal. The analysis is different in that case and the treatment of /4.7,8/ is quite adequate.

4.2.3 Further Developments in Surface Quenching

There is much room for further use of Master-equation techniques in the context of surface quenching observations. The quantitative application of the results of the first part of Sect. 4.2.1 to the observations in /4.1-4/ requires knowledge of f and F. One way of obtaining these is through the application of the spectral prescription of FÖRSTER /1.15/ and DEXTER /1.16/. However, there are problems to be faced, not the least of them being that the FÖRSTER-DEXTER prescription is not necessarily appropriate in molecular crystals. Correlations, whose neglect may be justifiable in solution systems, can introduce significant errors in crystals, as has been pointed out by MIYAKAWA and DEXTER /4.9/ and by SOULES and DUKE /2.43/. Attempts to obtain those corrections also from spectra, but of a different kind, are being made. The analysis of DUKE and MEYER /4.10/ is pertinent to that work.

The effect of coherence on surface quenching observations has not been studied properly because there still exist severe theoretical problems in the derivation of memory functions for systems without translational invariance. Neither the method of KENKRE /2.29,32,34/ nor of KÜHNE and REINEKER /2.35/ is directly useful for the computation of $\mathcal{W}_{mn}(t)$'s in systems with ends. An approximate treatment of coherence effects on surface quenching observations has been given by KENKRE and WONG /4.5/ who showed that like the perfect-absorber assumption, the assumption of incoherent transport leads to an understimation of the diffusion constant. However, that treatment is valid only for the case of a low degree of coherence.

A promising line of research is provided by the Master-equation analysis of *time-resolved* observations of *surface* quenching. It appears that such observations have not been made, although they have been proposed /3.33/. It can be shown /4.5/ that $\tilde{n}_H(\varepsilon)$, the Laplace transform of the host excitation probability in time-resolved observations with experimental setups as in /4.1-4/, is given by

$$\tilde{n}_H(\varepsilon) = \frac{1}{\varepsilon} - \left(\frac{\varepsilon + (1/\tau_G) - (1/\tau_H)}{\varepsilon}\right) \times \left(\frac{F}{\varepsilon + f + (1/\tau_G) - (1/\tau_H) + [\varepsilon + (1/\tau_G) - (1/\tau_H)]\, F\tilde{\psi}_{NN}(\varepsilon)}\right) \sum_{n=1}^{N} \tilde{\psi}_{Nn}(\varepsilon)\, P_n(0) \ . \qquad (4.30)$$

The work in /3.33/ consists of predictions of time-resolved observations on the basis of numerical Laplace inverses of expressions such as (4.30).

By extending the one-dimensional analysis of Sect. 4.2.1 to calculations on realistic three-dimensional lattices, it is possible to study the anisotropic character of the diffusion constant. On the basis of such a theory,observations have been proposed /4.11/ in experimental arrangements wherein the detector coating is placed on other parts of the host crystal as well as the end opposite to the illuminated end. These are only a few of the many possibilities in this field.

4.3 Master Equation Analysis of the Energy Transfer Rate

The energy transfer rate is one of the most discussed quantities in exciton dynamics. It was first introduced /3.5/ as a time-independent quantity k appearing in the simplest description of sensitized luminescence:

$$\frac{dn_H}{dt} + \frac{n_H}{\tau_H} = -kn_H \quad , \tag{4.31}$$

$$\frac{dn_G}{dt} + \frac{n_G}{\tau_G} = kn_H \quad . \tag{4.32}$$

If the above description, which we shall call the two-state description, is valid, the guest intensity n_G rises and decays as a sum of two exponentials, and the host luminescence decays exponentially with a decay constant which is a sum of $1/\tau_H$ and the "energy transfer rate" k. Time-resolved spectroscopy, introduced into this field by POWELL and KEPLER /4.12/ and discussed in detail in the review by POWELL and SOOS /1.20/, resulted in observations which could not be explained by the two-state description of (4.31,32). A time-dependent energy transfer rate k(t) was therefore introduced through the definition

$$k(t) = -\frac{1}{n_H(t)}\frac{dn_H(t)}{dt} - \frac{1}{\tau_H} \quad . \tag{4.33}$$

Clearly, if the two-state description applies, (4.33) makes k(t) reduce to k. The introduction of k(t) brought with it the CHANDRASEKHAR-SMOLUCHOWSKI /1.18,19/ expression

$$k(t) = A + Bt^{-1/2} \quad , \tag{4.34}$$

where A and B are constants. As the review of POWELL and SOOS /1.20/ shows, many attempts were made to fit experimental observations to (4.34) and to deduce, from A and B thus obtained, values for the exciton diffusion constant. There have been statements in the literature that a time-independent k(t) (i.e. the validity of the two-state description) means that the exciton diffusion model is appropriate and that a time-dependence signals the breakdown of exciton diffusion. There have been attempts to borrow an expression such as (4.34) from FÖRSTER's analysis of *random* structures and to use them for periodic crystals. Time-dependent k(t)'s have been challenged on experimental grounds /3.7,8/. The particular form (4.34) has been challenged on the basis of some theoretical considerations /3.25-27,35-37 , 4.13/ but defended on the basis of others /3.30/. Perhaps the most interesting development has been the recent experimental observations /3.6,23/ that $n_H(t)$'s turn out to be exponential on closer scrutiny and that k(t)'s do not really have any t-dependence.

4.3.1 Time-Dependent Energy Transfer Rates

We leave many of these exciting topics untouched in this review and select for our discussion two questions in this vast subject of the energy transfer rate. What is the real meaning of time-dependent and time-independent rates (in particular of the recently observed /3.6,23/ lack of time dependence)? And is it possible to extract information about exciton motion from the steady-state energy transfer rate defined by WOLF /3.5/ even when k(t) is time-dependent?

To address the first question we point out that the two-state description of (4.31,32) is excessively simplified. In fact, if we were to divide the crystal into a region near the trap with the excitation probability n_H^n and a region distant from the trap with n_H^d, we could write a three-state model

$$\frac{dn_H^d}{dt} + \frac{n_H^d}{\tau_H} = M(n_H^n - n_H^d) \quad , \tag{4.35}$$

$$\frac{dn_H^n}{dt} + \frac{n_H^n}{\tau_H} = M(n_H^d - n_H^n) - Cn_H^n \quad , \tag{4.36}$$

$$\frac{dn_G}{dt} + \frac{n_G}{\tau_G} = Cn_H^n \quad . \tag{4.37}$$

We represent the motion process, which makes the exciton move between the near and the distant region, by the motion rate M, and the capture process, which transfers excitation from the near region to the guest molecules, by C. These equations can be solved trivially /3.34/ to bring out many facets of energy transfer. The energy transfer rate k(t) turns out to be time-dependent even in this simplest extension of the two-state description. But this outcome should be obvious. Transfer from the host to the guest occurs from certain host locations and there is change of the excitation probability among these locations as a result of exciton motion. The "internal structure" of the host, i.e., the fact that it comprises many states (sites) rather than a single one as in (4.31,32), not all of which are equally connected to the guest, makes it inevitable that $n_H(t)$ should be nonexponential and that k(t) should develop a time dependence. This has nothing whatsoever to do with validity of exciton diffusion. Exciton diffusion does result generally in a time-dependent k(t). So does almost any other kind of exciton propagation. The only way one could have a truly constant k(t) is if one had really just two sites in the crystal, or if every host site was occupied by a guest, or if every host site delivered excitation equally to the guest site irrespective of their mutual distance.

To state this in terms of a Master-equation treatment, one need only examine (3.21) and observe that a time-independent k(t) would correspond to an exponential $n_H(t)$. Casting (3.53,56) for uniform illumination ($\tilde{\eta}_r(\varepsilon') = \rho/\varepsilon'$) as

$$\tilde{n}_H(\varepsilon) = \frac{1}{\varepsilon' + \tilde{k}_m(\varepsilon')} \ , \tag{4.38}$$

we see that a t-independent $k(t)$ requires an ε-independent $\tilde{k}_m(\varepsilon)$. But the "energy transfer memory" $k_m(t)$ is given through its transform by

$$\tilde{k}_m(\varepsilon) = \frac{\rho}{\frac{1}{c} + \frac{1}{\varepsilon}\left[\varepsilon\tilde{\nu}(\varepsilon) - \rho\right]} \ . \tag{4.39}$$

That the energy transfer memory does not require coherence should go without saying. It is always possible to cast n_H in the form (4.38). Since $k_m(t)$ is never really a δ-function except in the two-state system or in trivial limits in other systems, $k(t)$ is always t-dependent, even in the "exciton diffusion model".

How then should we interpret those observations /3.6-8,23/ in which $k(t)$ has been found to be t-independent? The discussion on coherence and GME's given in Chaps. 2 and 3 is of indirect help in answering this question. We know that for no real system do memory functions $\mathcal{W}_{mn}(t)$ really have a δ-function behaviour. However, the incoherent Master equation is perfectly adequate for many systems. They are the ones in which experimental probe times are much larger than characteristic times for the decay of $\mathcal{W}_{mn}(t)$'s. Although coherence effects are not under discussion, this is precisely what happens in the present context. If probe times are much larger than the decay times of $k_m(t)$ defined in (4.39), the time dependence in $k(t)$ will not be apparent since $n_H(t)$ will have exponential character under the experimental resolution. The decay of the energy transfer memory $k_m(t)$ is fast if exciton motion is fast. The recent observations of t-independent $k(t)$'s mean, therefore, that exciton motion is occurring faster than probe times in those systems. Some of those experiments /3.6/ employ picosecond spectroscopy. The conclusion to be drawn is that characteristic times for exciton motion in those systems are smaller than 10^{-12} s. There is no doubt that $\tilde{k}_m(\varepsilon)$ is generally ε-dependent and that the energy transfer rate is generally t-dependent. To observe the time dependence, faster spectroscopy is required. We conjecture that experimental efforts in the subpicosecond regime will uncover the time dependence of $k(t)$ in the near future.

The above comments can be illustrated with specific results from the Master-equation treatment, in particular, from the exact analytic solutions that have been obtained /3.33/ for the one-dimensional crystal as well from the numerical determination of $n_H(t)$, $n_G(t)$, and $k(t)$. The essence of the argument can be appreciated, however, from the three-state model of (4.35-37). A straightforward calculation /3.34/ shows that $k(t)$ is given by

$$k(t) = A_1 - \Omega\left[\frac{\sinh \Omega t + \left(\frac{A_1 - Cf_1}{\Omega}\right)\cosh \Omega t}{\cosh \Omega t + \left(\frac{A_1 - Cf_1}{\Omega}\right)\sinh \Omega t}\right] \ , \tag{4.40}$$

where $A_1 = M + (C/2)$, $\Omega = [M^2 + (C^2/4)]^{1/2}$, and f_1 is the fraction $[n_H^n(0)/n_H(0)]$ which can be made to represent the trap concentration. Equation (4.40) shows that the energy transfer rate starts at the value Cf_1 and tends to $A_1 - \Omega$ at long times. The asymptotic value of $A_1 - \Omega$ equals the smaller of the rates M or C/2 in extreme limits. It is usually assumed that the capture rate is much larger than the motion rate. Such an assumption is implicit in expressions such as $4\pi RD$ which are often used for k /1.18-20 , 3.6,23/. If we make that assumption, (4.40) shows that k(t) will indeed appear time-independent for times much larger than motion times 1/M. Incidentally, the energy transfer memory $k_m(t)$ for the three-state model for $f_1 = 0$ is extremely simple and physically transparent:

$$k_m(t) = MC\, e^{-t(2M+C)} \quad . \tag{4.41}$$

Notice that it decays with a rate which is a sum of the motion and capture rates and that, if the latter differ widely, its Markoffian limit is M or C/2 (whichever is smaller).

Expressions identical to the above ones emerge from the analysis of a system in which the propagator sum $\nu(t)$ is approximated

$$\nu(t) = \rho + (1-\rho)\, e^{-2Mt} \quad . \tag{4.42}$$

Equation (4.42) represents the basic behaviour of $\nu(t)$ (discussed in Sect. 3.3) through an exponential approximation. Expressions which are very similar to (4.40, 41) also appear in the exact analysis of one-dimensional crystals. These details may be found in /3.33/.

4.3.2 Steady-State Energy Transfer Rates

We now turn /3.34/ to the time-independent energy transfer rate studied by WOLF /3.5/:

$$k^S = \frac{1}{\tau_H}\frac{\phi_G}{\phi_H} \quad . \tag{4.43}$$

The reason for defining it in this form is that it equals k in the two-state model and is always accessible from simple steady-state experimental observables. It should be clear from the discussion in Sect. 3.3 that the effects of coherence as well as of other transport features are contained in (4.43). We can calculate k^S generally from (3.57):

$$k^S = \frac{\rho}{\frac{1}{c} + \tau\left[(1/\tau)\,\tilde{\nu}(1/\tau) - \rho\right]} \quad . \tag{4.44}$$

For small concentrations and small radiative lifetimes, $(1/\tau)\,\tilde{\nu}(1/\tau)$ may be approximated by $(1/\tau)\,\tilde{\psi}_0(1/\tau)$. We then have

$$k^{s} = \frac{\rho}{\frac{1}{c} + \tau\left[(1/\tau)\,\tilde{\psi}_0(1/\tau) - \rho\right]} \quad . \tag{4.45}$$

However, as $\rho \to 1$, we know that ν tends to its infinite-t value which, from an Abelian theorem, equals the limit, as $\tau \to \infty$, of $(1/\tau)\,\tilde{\nu}(1/\tau)$. We then have

$$k^{s} = c\rho = c \quad . \tag{4.46}$$

The general concentration dependence of k^s, which has been observed to have some peculiar behaviour, can be described by (4.45). The effects of motion characteristics enter into k^s through $\tilde{\psi}_0$ or $\tilde{\nu}$. The most noteworthy feature of (4.45) is its similarity to (4.39). Indeed, we observe that except for a factor of $\delta(t)$, the WOLF rate (4.43) is *the Markoffian limit of the energy transfer memory* $k_m(t)$. The relation of $k_m(t)$ and k^s is thus identical to the relation of $\mathcal{W}_{mn}(t)$ and F_{mn} of Chaps 2 and 3.

Energy transfer in sensitized luminescence may be analyzed in terms of $k(t)$, $k_m(t)$, or k^s. We have seen that k^s equals $\int_0^\infty dt\, k_m(t)$. The recent experiments that have uncovered no t-dependence to $k(t)$ are thus probing k^s, the Markoffian approximation to $k_m(t)$. Further elaboration, including the ρ-dependence of k^s may be found in /3.34/. On the basis of the present theory, KENKRE and SCHMID /4.14/ have recently suggested a resolution of an apparent paradox in the temperature dependence of the annihilation rate on one hand and of the energy transfer rate on the other. The traditional expressions $8\pi RD$ for the annihilation constant and $4\pi RD$ for the energy transfer rate make it difficult to understand how γ and k^s can have different temperature dependences. A number of observations /3.5,10,11/ have, however, shown that γ is temperature-independent but k^s does depend on temperature. The resolution /4.14/ lies in the comparison of the expressions (4.45) for the energy transfer rate k^s and (3.95) for the annihilation constant γ. If we assume that $b \gg [\tilde{\psi}_0(1/\tau)]^{-1}$ but that a similar extreme relation does not hold between c and $\tilde{\psi}_0(1/\tau)$, the resolution is immediate. It is possible that the temperature dependence of $\tilde{\psi}_0(1/\tau)$ is weak but that of c is not. It is thus suggested that the slow motion limit might be valid in the annihilation case but not necessarily in the capture case.

We believe the time-dependent energy transfer rate $k(t)$ has little role to play in sensitized luminescence in *ordered* molecular crystals. It is not a direct experimental observable and it is seldom a direct outcome of theory. The procedure followed by experimentalists consists in measuring the host luminescence $n_H(t)$ and then in *computing* $k(t)$ numerically (and laboriously) through (4.33). This has been done in the past to fit expressions, such as (4.34), believed to have an established theoretical basis. The extensive debates in the literature concerning (4.34) show that this is not true. Because Laplace-transform techniques are natural to this subject, it would appear that the energy transfer memory $k_m(t)$ is the natural quantity to focus attention on. It is given by (4.38,39) and is a δ-function if $k(t)$ is t-independent. If we find $k_m(t)$ to be an unfamiliar concept, we need only observe that its Markoffian limit yields the well known WOLF rate k^s /3.5/ which is directly obtained from the yield ratio. Direct experimental studies of $k_m(t)$ may also be possible through modulation spectroscopy.

4.4 Momentum-Space Master Equations for Trapping

Unlike Master equations employed in Sect. 4.2 and 4.3, those used in the present section are not Markoffian versions of the GME's discussed in Chaps. 2 and 3. Instead, they bear kinship to the linearized Boltzmann equation used in the theory of electrons in metals /3.67/. If exciton motion is largely coherent, it might be appropriate to use momentum space (k-space) for its study. A theory for sensitized luminescence observations has been constructed along these lines by KENKRE /3.35,36/ and extended by WONG and KENKRE /3.37/. Although the relevant experimental setup is the same as in Sect. 3.2, the Master equations are quite different. No translational invariance applies in k-space, Fourier-transform techniques are therefore of no use and relaxation-time procedures have to be developed.

4.4.1 k-Space Equations and Trapping Rates

The phase space used in this treatment is that suggested by FAYER and HARRIS /3.38/. If $f_k(t)$ is the probability that the Frenkel exciton is in state k at time t, and if $f_\theta(t)$ denotes the corresponding probability that it occupies the trap state, we have the coupled equations

$$\frac{df_k}{dt} + (\frac{1}{\tau_H} + \alpha_k)f_k = \alpha_k^\dagger f_\theta + \sum_{k'} (Q_{kk'} f_{k'} - Q_{k'k} f_k) \quad , \tag{4.47}$$

$$\frac{df_\theta}{dt} + \left(\frac{1}{\tau_G} + \sum_k \alpha_k^\dagger\right) f_\theta = \sum_k \alpha_k f_k \quad . \tag{4.48}$$

Here $Q_{kk'}$ is the scattering rate from state k' to state k arising from exciton-phonon or exciton-imperfection interactions, α_k is the trapping rate from state k, and $\alpha_k^\dagger$ the detrapping rate to state k from the trap. No trap-to-trap interactions are described. The k dependence of the host lifetime is suppressed.

The observables are the host excitation probability $n_H(t)$ given by

$$n_H(t) = \sum_k f_k(t) \quad , \tag{4.49}$$

the guest or trap excitation probability $n_G(t)$ which is essentially $f_\theta(t)$, and the quantum yields ϕ_H and ϕ_G given by (3.22,26). Their calculation is carried out /3.35 -37/ in two steps: model evaluation of the trapping rates α_k and detrapping rates

$\alpha_k^\dagger$, and solution of (4.47,48) followed by computation of the observables as in (4.49). The scattering rates $Q_{kk'}$ may be obtained in principle from observations such as those by BURLAND et al. /4-15/ or by SCHMIDT et al. /4.16/, but exact solution after they are substituted in (4.47) is impossible. This is a standard state of affairs in electron transport where extensive use is made of the relaxation-time assumption /3.67/. However, its use for trapping requires care. KENKRE /4.17/ has suggested an extended relaxation-time approximation and used it /3.35/ in solving (4.47). The procedure consists in replacing the second term in the right-hand side of (4.47) through

$$\sum_{k'} [Q_{kk'} f_{k'}(t) - Q_{k'k} f_k(t)] \approx \Gamma^k [f_k^{th} n_H(t) - f_k(t)] \quad , \tag{4.50}$$

where $1/\Gamma^k$ is the relaxation time and f_k^{th} is the thermalized distribution function. The usual relaxation-time approximation /3.67/ corresponds to the absence of true sinks of the probability and would replace the host excitation probability $n_H(t)$ in (4.50) by 1. Resultant absurdities are eliminated /4.17/ by the use of the above prescription (4.50).

The model evaluation of trapping rates α_k may be carried out by trivial application of the Golden Rule. Denoting by V_m the matrix element for the trapping interaction between the trap state θ and the mth site state (i.e., a localized, e.g. Wannier, state), we get /3.36/

$$\alpha_k = \text{const.} \sum_{m,n} V_m V_n^* \, e^{-ik(m-n)} \quad . \tag{4.51}$$

This result is independent of dimensionality. The constant in (4.51) includes the density of states and other proportionality factors and differs by the Boltzmann factor $\exp(-\Delta/k_BT)$ in the detrapping case (in the calculation of $\alpha_k^\dagger$) from the trapping one. Here Δ is the energy difference between the host k state and the guest θ state. We emphasize that the relevant density of states to be included in the constant in (4.51) is *not* the one obtained from the k derivative of the band energy and other parameters since our interest lies in the transition rates from (and to) *individual* k states and not from (and to) the entire band. The relevant density of state refers to states of the bath such as phonons which must assist in the transition.

If the host-guest interaction V_m is sufficiently short-ranged so that only the terms V_0, V_1 and V_{-1} may be kept, and if we restrict ourselves to one dimension,

$$\alpha_k = \text{const.} \, [(V_0^2 + V_1^2 + V_{-1}^2) + 2V_0(V_1 + V_{-1}) \cos k + 2V_1V_{-1} \cos 2k] \quad . \tag{4.52}$$

Furthermore, if the guest molecule is placed substitutionally, or generally in such a way that $V_1 = V_{-1}$, then

$$\alpha_k = (V_0 + 2V_1 \cos k)^2/\gamma_0 \quad , \tag{4.53}$$

where we have called the constant γ_0. If the guest communicates with a single host site, the above expressions show that α_k (and $\alpha_k^\dagger$, given by multiplying α_k by the detailed balance factor) is constant over the host band. In the general case the rate has its maximum at the k = 0 edge of the band. The value at the other band edge ($k = \pm\pi$) is a minimum if $V_0 \geqslant 2V_1$ but is larger than the minimum if $2V_1 > V_0$. For $V_0 = 0$, which signifies interaction with only the two side states, the value of the rate at $k = \pm\pi$ is a maximum and equals the value at k = 0. These considerations have been used in /3.36/ to show that one can conclude from available experiments that trapping rates are *not* proportional to group velocities in the exciton band.

4.4.2 Expressions for Luminescence Observables

Using the extended relaxation-time approximation /4.17/ given in (4.50), and neglecting the detrapping rates for simplicity, we can get the host excitation probability $n_H(t)$ from (4.47):

$$\tilde{n}_H(\varepsilon) = \left\{1 - \sum_k [\varepsilon + (1/\tau_H) + \alpha_k + \Gamma_k]^{-1} \Gamma_k f_k^{th}\right\}^{-1}$$
$$\times \left\{\sum_k [\varepsilon + (1/\tau_H) + \alpha_k + \Gamma_k]^{-1} f_k(0)\right\} \quad . \tag{4.54}$$

The contents of the second set of braces in (4.54) describe the effect of the initial distribution of the exciton in the band determined by the method of illumination. A detailed analysis of the effect of these initial conditions is available in /3.37/. Here we shall assume that $f_k(0) = \delta_{k,0}$, which would correspond to optical selection rules, and make the further simplification whereby Γ_k is replaced by an average Γ over the band. The latter simplification, widely used in electron transport/3.67/, describes the situation exactly if $Q_{kk'}$ is independent of k and k'.

KENKRE /3.35/ replaced the trapping rates α_k by two rates, α_0 and α, the first corresponding to the initially populated band state (k = 0) and the second to an average over the band, and obtained the observables in the form

$$n_H(t) = \left[\frac{e^{-t/\tau_H}}{\Gamma + (\alpha_0 - \alpha)}\right]\left[\Gamma\, e^{-t\alpha} + (\alpha_0 - \alpha)e^{-t(\alpha_0+\Gamma)}\right] \tag{4.55}$$

$$k(t) = \frac{\Gamma\alpha\, e^{-t\alpha} + (\alpha_0 - \alpha)(\alpha_0 + \Gamma)e^{-t(\alpha_0+\Gamma)}}{\Gamma\, e^{-t\alpha} + (\alpha_0 - \alpha)e^{-t(\alpha_0+\Gamma)}} \quad , \tag{4.56}$$

and an expression involving a sum of three exponentials for the guest luminescence

$n_G(t)$. The host yield expression is

$$\phi_H = \left[\frac{1}{1+\alpha_0\tau_H}\right]\cdot\left[\frac{1+(\Gamma\tau_H)/(1+\alpha\tau_H)}{1+(\Gamma\tau_H)/(1+\alpha_0\tau_H)}\right] = \left[\frac{1}{1+\alpha\tau_H}\right]\cdot\left[\frac{1+(\alpha\tau_H)/(1+\Gamma\tau_H)}{1+(\alpha_0\tau_H)/(1+\Gamma\tau_H)}\right] \quad . \tag{4.57}$$

Equation (4.57) has been written in two ways to make clear the various limits. For very coherent motion the scattering rate Γ is small, (4.57) shows that the yield is determined primarily by trapping from the initially occupied k state, (4.55) shows a similar result, and a single decaying exponential with exponent $1/\tau_H+\alpha_0$ describes $n_H(t)$ completely. The energy transfer rate k(t) is seen from (4.56) to be merely α_0, the trapping rate from the k = 0 state. In the opposite limit of highly incoherent exciton motion, represented in the present momentum-space description by large scattering, i.e., large Γ, the exciton is distributed over the entire band rapidly. The yield in (4.57) is then $1/(1+\alpha\tau_H)$, i.e., determined by the average trapping rate α, the host luminescence (4.55) is again a single decaying exponential but with exponent $1/\tau_H+\alpha$, and the energy transfer rate is now equal to the average trapping rate α. The intermediate coherence situation results in a k(t) which varies from α_0 to α in time.

WONG and KENKRE /3.37/ did not make the approximation of replacing α_k's by the two trapping rates but employed (4.53) for the rates. Their analysis shows that, surprisingly, the results (4.55-57) are quite adequate for the description of the problem in the largely coherent and the largely incoherent limits, and that the prescription given in /3.35/ from the calculation of the average of α_k over the band is exact. This prescription is based on an arithmetic rather than harmonic average,

$$\alpha = (1/N)\sum_k \alpha_k \quad , \tag{4.58}$$

where N is the number of states. The meaning of the prescription may be discussed /3.35/ in terms of memory functions.

As an example of the correction terms that can be provided /3.37/ for the expressions (4.55-57) we show

$$n_H(t) = n_H^0(t) + e^{-t/\tau_H}\int_0^t ds \int_0^s du\,[h(s,u)]\,\exp\Big\{-[(\Gamma+\alpha_0)(t-s) + (\Gamma+\alpha)(s-u)]\Big\} \quad , \tag{4.59}$$

where n_H^0 is the right-hand side of (4.55), h(s,u) is given by

$$h(s,u) = \Gamma(\alpha_0-\alpha)\,e^{-\alpha u}\,u\,(s^2-u^2)^{-1/2}\,I_1\left[(\alpha_0-\alpha)(s^2-u^2)^{1/2}\right] \quad , \tag{4.60}$$

and I_1 is the modified Bessel function. This result applies for $V_0 \gg V_1$ so that (4.52) may be written as

$$\alpha_k = (V_0^2 + 4V_0V_1 \cos k)/\gamma_0 \quad . \tag{4.61}$$

The correction term in (4.59) can be shown to vanish both for large and small scattering rates /3.37/.

Many further calculations of sensitized luminescence observables may be found in /3.37/. They include the effects of detrapping from guest states back into the host band, details of initial illumination effects, alternatives to (4.61) and asymptotic analysis of k(t) for large times. No evidence has been found of any $t^{-1/2}$ behaviour in the rates. The k-space analysis lends no support to (4.34).

The momentum-space theory discussed above was constructed to exploit the translational invariance of the crystal from the very beginning. It is meant to be applied particularly to situations wherein there is substantial coherence. It complements the real-space treatment described in Sects. 3.3, 4.2, and 4.3. Analyses of this kind (but without trapping terms) will be especially useful in studying recent experiments such as those of SCHMIDT and collaborators /4.16/ in which direct observation of the time evolution of the band state populations $f_k(t)$ has been reported. Combining the k-space treatment and the real-space treatment into a single unified theory constitutes an interesting theoretical project. Some work towards this end has been carried out along the lines of the analysis of RAHMAN et al. /2.60/.

4.5 Master Equations for the Interplay of Vibrational Relaxation with Exciton Motion

One of the questions posed in Chap. 1 concerning exciton motion was whether exciton motion is hot or cold, i.e., whether it occurs before, during, or after vibrational relaxation. The traditional theories of exciton transport such as those of FÖRSTER /1.15,17/, DEXTER /1.16/, and TRLIFAJ /4.18/ assume, implicitly or explicitly, that the exciton undergoes complete vibrational relaxation before each hop. DEXTER and FOWLER /4.19,20/ pointed out that such an assumption could be quite inappropriate for certain systems and DEXTER /4.21/ provided an analysis in the limit of no relaxation. The effects of vibrational relaxation on exciton motion over the complete (intermediate) range have been studied by HIZHNYAKOV and TEHVER /4.22-24/ and by KENKRE /3.21/. The latter treatment is based on Master equations and is described below.

4.5.1 Formalism and Technique of Solution

The explicit thermalization assumption of the traditional theory /1.15-17/ of excitation transfer makes it inapplicable to situations wherein transfer (exciton motion) times are comparable to vibrational relaxation times. Values usually assigned to the latter are 10^{-12} s or less in the condensed phase, and they are generally believed to be smaller by two orders of magnitude or more than motion times. If this extremely fast relaxation were a universal feature, the above-mentioned inapplicability of the traditional theory would not be a practical limitation. However, many recent observations /4.25-27/ of dramatically slow relaxation have been made in the solid and liquid phases. There is no doubt at all that systems exist in which exciton motion and vibrational relaxation have comparable characteristic times.

Our analysis /3.21/ is directed at the quantity $P_{m,\mu}$, the probability that the exciton resides on the mth site and that the system is in the μth vibrational state. The label μ generally represents the vibrational state not of a single molecule, but of the entire system. The Master equation

$$\frac{dP_{m,\mu}(t)}{dt} = \sum_{n,\eta} [R_{m,\mu\ n,\eta}\ P_{n,\eta}(t) - R_{n,\eta\ m,\mu}\ P_{m,\mu}(t)] \qquad (4.62)$$

describes the relaxation and the motion processes with the help of the combined transition rates $R_{m,\mu\ n,\eta}$. If we drop the subscripts μ,η, etc. and write $R_{mn} = F_{mn}$, we obtain (2.27). On the other hand, if we drop m,n etc., we can obtain the description of pure relaxation without motion. On the basis of a detailed analysis of the mixed site-and-energy space, it can be shown /3.21/ that a physically appropriate form of the rates R is

$$R_{m,\mu\ n,\eta} = \gamma^{v}_{\mu\eta}\ \delta_{m,n} + F_{mn}(\mu)\ \delta_{\mu,\eta} \quad , \qquad (4.63)$$

where $\gamma^{v}_{\mu\eta}$ is the vibrational relaxation rate from state η to μ, the superscript v being used to distinguish this vibrational γ from other γ's used in this review, and $F_{mn}(\mu)$ is the rate for exciton motion from site n to site m in the system vibrational state μ. The relaxation rates γ^{v} have been assumed independent of the site. For unlike molecules this assumption can be suitably modified. Motion rates F have been taken to depend on the vibrational state μ, but the change of the vibrational state during motion has been suppressed for simplicity. The dependence of F_{mn} on μ is the primary characteristic of this analysis.

The motion rates $F_{mn}(\mu)$ are essentially FÖRSTER-DEXTER rates /1.15-17/ but modified through Franck-Condon factors to describe the dependence on μ. Machine calculations have shown that if we assume the molecules to be harmonic oscillators (the excited-state oscillator being displaced relative to the ground-state one), a reasonably good representation of the dependence on μ is /3.21/

$$F_{mn}(\mu) = B^{v} + \mu C^{v} \quad . \qquad (4.64)$$

Here B^{v} and C^{v} are constants and μ specifically represents the vibrational level of the oscillator. It is therefore proportional to the vibrational energy of the oscillator.

The vibrational relaxation rates γ^{v} are taken to be given by the LANDAU-TELLER /4.28/ prescription

$$\gamma^{V}_{\mu\eta} = k^{V}\left[(\mu+1)\ \delta_{\eta,\mu+1} + \mu\ e^{-\beta^{V}}\ \delta_{\eta,\mu-1}\right] , \tag{4.65}$$

which assumes that the relaxation of the harmonic oscillator (of frequency ω) representing the molecule occurs through an interaction with the bath proportional to the displacement of the oscillator. The delta functions in (4.65) correspond to selection rules for matrix elements of the displacement in a harmonic oscillator, k^V is a parameter describing the strength of the bath-oscillator interaction, and β^V is the *dimensionless* inverse temperature ($\hbar\omega/k_BT$). The substitution of the rates (4.63 -65) in (4.62) allows an immediate solution of the Master equation through a combination of several known techniques. Let us consider a crystal, i.e., a system with translational periodicity. If we define $A_{mn}(\mu) = -F_{mn}(\mu)$ for $m \neq n$ and $A_{mm}(\mu) = \sum_n F_{nm}(\mu)$, and use Fourier transforms $A^k(\mu)$ as in (2.50),

$$A^k(\mu) = \sum_{m-n} A_{mn}(\mu)\ e^{ik(m-n)} , \tag{4.66}$$

we obtain /3.21/

$$\frac{dP^k_\mu}{dt} + A^k(\mu)\ P^k_\mu = \sum_n \left(\gamma^V_{\mu\eta}P^k_\eta - \gamma^V_{\eta\mu}P^k_\mu\right) . \tag{4.67}$$

As a result of translational periodicity of the crystal, (4.67) contains no cross-k terms and represents the problem of the vibrational relaxation of a fictitious k-oscillator in the presence of "sink" processes causing true depletion of the P^k_μ's through the "sink rates" $A^k(\mu)$. This observation shows that the present problem is isomorphic to that of the interplay of vibrational relaxation with a true sink process such as luminescence. But the latter problem has been analyzed by SESHADRI and KENKRE /4.29-32/ in great detail. That analysis may therefore be used immediately to solve for P^k_μ's. Space reasons do not allow the demonstration of the details of the technique here. Suffice it to say that the definition of a generating function

$$G^k(z,t) = \sum_{\mu=0}^{\infty} z^\mu\ P^k_\mu(t) , \tag{4.68}$$

where μ is not a superscript but the power to which z is raised, allows us to obtain a first-order partial differential equation for G(z,t). Its solution is obtained through the standard method of characteristics. Exact solutions for $P^k_\mu(t)$ are provided for arbitrary initial conditions. SESHADRI and KENKRE /4.29-32/ have treated four initial distributions in the vibrational space of the μ's: δ-functions, Poisson distributions, Laguerre polymonials, and Boltzmann distributions. The first three correspond /4.31/ to the usual methods of excitation. Laser excitation corresponds to δ-function excitation, broad-band excitation at zero temperature to Poisson distributions (as a result of Franck-Condon factors), and broad-band excitation at nonzero temperatures to Laguerre polynomials. The Boltzmann initial distribution at nonenvironmental temperatures corresponds to the situation wherein a rapid quasi-equilibrium followed by a slower thermalization occurs. The Boltzmann distribution also introduces considerable algebraic simplicity into the problem and by a suitable choice of temperature may be profitably used to represent the actual distribution. We shall show some of the results for the Boltzmann initial distribution in Sect. 4.5.2.

There are two ways to use the exact solutions for $P_\mu^k(t)$ obtained through the methods of /4.29-32/, for the purposes of the investigation of the interplay of vibrational relaxation and exciton motion. We may use $P_\mu^k(t)$ directly or invert the Fourier transform to give $P_{m,\mu}(t)$. The former use is illustrated by calculations of transient grating signals /3.1-4/ in the presence of relaxation. It is clear from (3.2,3) that these signals are related more closely to the Fourier transforms P_μ^k than to $P_{m,\mu}$. We point out that k corresponds here to $(4\pi a/\lambda)\sin(\theta/2)$ of (3.1) and that such an analysis results in nonexponential grating signals which have nothing to do with the coherence effects shown in Fig. 3.2.

The more straightforward way of obtaining the desired observables from $P_\mu^k(t)$ proceeds through the Fourier inversion

$$P_{m,\mu}(t) = \frac{1}{N}\sum_k P_\mu^k(t)\, e^{-ikm} \quad . \tag{4.69}$$

4.5.2 Explicit Expressions and Time-Dependent Transition Rates for Exciton Motion

We illustrate the method by showing calculations for the simple case of a dimer /3.21/. The original equations for quantities such as $P_{1,\mu}$, the probability that the exciton is on molecule 1 and the vibrational state is μ, are transformed into

$$\frac{dP_\mu^+}{dt} = \sum_\eta (\gamma_{\mu\eta}^v P_\eta^+ - \gamma_{\eta\mu}^v P_\mu^+) \tag{4.70}$$

$$\frac{dP_\mu^-}{dt} + 2F(\mu)P_\mu^- = \sum_\eta (\gamma_{\mu\eta}^v P_\eta^- - \gamma_{\eta\mu}^v P_\mu^-) \tag{4.71}$$

for the probabilities of transformed "oscillators" + and - defined via

$$P_\mu^\pm(t) = P_\mu^1(t) \pm P_\mu^2(t) \quad . \tag{4.72}$$

The solution of (4.70,71) is obtained through the methods outlined above. Although detailed information of the occupation of individual vibrational states is available in those solutions /3.21/, we may sum over μ to obtain $P_1(t)$ or $P_2(t)$, the probability that the exciton is on the first or second molecule. KENKRE /3.21/ has shown that for an initial condition in which the molecule 1 alone is occupied and the vibrational state is Boltzmann with a nonenvironment temperature T_0, one obtains

$$P_1(t) = \frac{1}{2}\left[1 + \left(\frac{1-e^{-\beta_0}}{1-e^{-\beta(t)}}\right)\left(\frac{1-\Gamma^- e^{-\beta(t)}}{1-\Gamma^- e^{-\beta_0}}\right)\right]\exp\left\{-t\left[2B^V + k^V e^{-\beta}(1-\Gamma^-)\right]\right\} \qquad (4.73)$$

$$P_2(t) = \frac{1}{2}\left[1 - \left(\frac{1-e^{-\beta_0}}{1-e^{-\beta(t)}}\right)\left(\frac{1-\Gamma^- e^{-\beta(t)}}{1-\Gamma^- e^{-\beta_0}}\right)\right]\exp\left\{-t\left[2B^V + k^V e^{-\beta}(1-\Gamma^-)\right]\right\} \quad . \qquad (4.74)$$

Here β_0 equals $\hbar\omega/k_BT_0$, the time-dependent dimensionless inverse temperature $\beta(t)$ is given by

$$\beta(t) = \ln\frac{\Gamma^+(1-e^{-\beta_0}\Gamma^-) - \Gamma^-(1-e^{-\beta_0}\Gamma^+)\exp[-tk^V e^{-\beta^V}(\Gamma^+-\Gamma^-)]}{(1-e^{-\beta_0}\Gamma^-) - (1-e^{-\beta_0}\Gamma^+)\exp[-tk^V e^{-\beta^V}(\Gamma^+-\Gamma^-)]} \quad , \qquad (4.75)$$

and the quantities $\Gamma^\pm$ are given by

$$\Gamma^\pm = \frac{1}{2}(e^{\beta^V}+1+\delta) \pm \left[\frac{1}{4}(e^{\beta^V}+1+\delta)^2 - e^{\beta^V}\right]^{1/2} \quad , \qquad (4.76)$$

$$\delta = 2C^V e^{\beta^V}/k^V \quad . \qquad (4.77)$$

Although these expressions have a somewhat overbearing appearance, interesting physics resides in them. Equations (4.73,74) should be compared to

$$P_1(t) = \frac{1}{2}\left(1 + \exp\left\{-t[2B^V + 2C^V(e^{\beta^V}-1)^{-1}]\right\}\right) \quad , \qquad (4.78)$$

$$P_2(t) = \frac{1}{2}\left(1 - \exp\left\{-t[2B^V + 2C^V(e^{\beta^V}-1)^{-1}]\right\}\right) \quad , \qquad (4.79)$$

which would result from the traditional FÖRSTER-DEXTER theory using thermalized rates F^{th} for exciton motion:

$$F^{th} \equiv \sum F(\mu)\, e^{-\mu\beta^V}(1-e^{-\beta^V}) = B^V + C^V(e^{\beta^V}-1)^{-1} \quad . \qquad (4.80)$$

We see at once that the interplay of vibrational relaxation with motion makes the transition rates for motion time-dependent. Equations (4.73,74) correspond to

$$\frac{dP_1(t)}{dt} = F(t)\,[P_2(t) - P_1(t)] \quad , \qquad (4.81)$$

$$\frac{dP_2(t)}{dt} = F(t)\,[P_1(t) - P_2(t)] \quad . \qquad (4.82)$$

Whereas F(t) would be taken to have the time-independent thermalized value in (4.80) in the traditional theory, our analysis /3.21/ gives

$$F(t) = B^V + C^V(e^{\beta(t)} - 1)^{-1} \quad . \tag{4.83}$$

The time-dependent transition rate F(t) has the initial value $F(0) = B^V + C^V(e^{\beta_0} - 1)^{-1}$ corresponding to the initial vibrational distribution, and it changes in time as a consequence of the relaxation process. If C^V is positive, the exciton moves faster if it is more excited vibrationally. Figure 4.3 shows a plot of F(t) and uncovers a striking result: at large times the rate *does not tend* to the thermalized value given by (4.80). Instead, it tends to a smaller value

$$\lim_{t\to\infty} F(t) = B^V + C^V(e^{\beta(\infty)} - 1)^{-1} = B^V + C^V(\Gamma^+ - 1)^{-1} \quad . \tag{4.84}$$

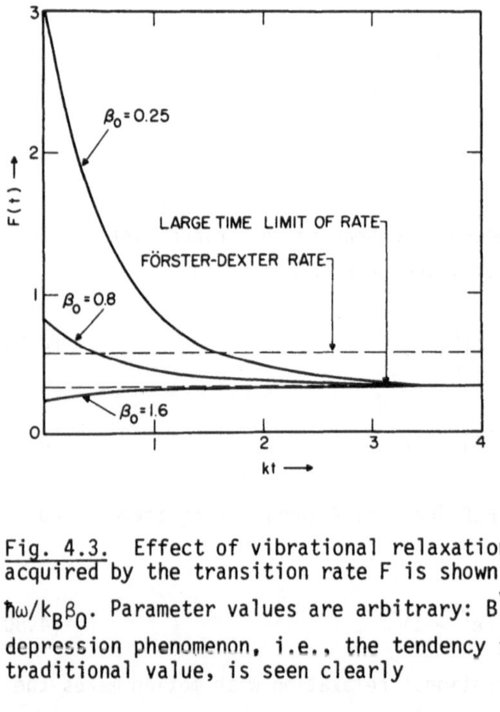

Fig. 4.3. Effect of vibrational relaxation on exciton motion. The time dependence acquired by the transition rate F is shown for various initial temperatures $T_0 = \hbar\omega/k_B\beta_0$. Parameter values are arbitrary: $B^V = 0$, $C^V = 1$, $\hbar\omega/k_BT = 1$. The rate-depression phenomenon, i.e., the tendency for F(t) to tend to a value below the traditional value, is seen clearly

This behaviour of the rate is intimately related to an interesting cooling effect described in /4.29-32/ which arises from the interplay of luminescence and vibrational relaxation.

Whether the initial value F(0) of the rate is lower or higher than the thermalized or FÖRSTER-DEXTER rate, F^{th} depends on the conditions of excitation. As Fig.

4.3 shows, if the rate F(t) is initially high, it slows down in time, crosses the value F^{th}, and ends up lower than the traditional value. If the temperature of the initial distribution is low, the rate rises to $F(\infty)$ and never reaches the FÖRSTER-DEXTER value. Effective rates may therefore have values profoundly different from F^{th}. This rate depression phenomenon is dependent for its magnitude on $C^V/k^V e^{-\beta^V}$: it is small or large according as this quantity is small or large. It should be observable for slow relaxation and if transfer rates $F_{mn}(\mu)$ depend strongly on the vibrational state μ, i.e., if C^V is large. The phenomenon is general and by no means restricted to initial Boltzmann distributions. It has been explicitly proved /4.32/ that $F(\infty)$ is independent of the initial distribution and always lower than F^{th} if C^V is positive.

Whether the *average* rate of transfer is slower or faster than the thermalized rate F^{th} will depend on the relative values of δ of (4.77), and of the initial and environment temperatures. WEBER /4.33/ and others /4.34-36/ have observed a dependence of the efficiency of excitation transfer on the wavelength of excitation. KENKRE /3.21/ has shown how this theory of the effect of vibrational relaxation on transfer could explain that phenomenon. He has also extended the theory to calculate quantum yields and luminescence intensities in the presence of relaxation, and has given several generalizations to treat more complex situations including coherence and nonperiodic arrays.

The emergence of time-dependent transition rates, the rate-depression effect, the possible explanation of the dependence of transfer efficiency on excitation wavelength, and the prediction of nonexponential transient grating signals constitute the primary outcomes of this theory /3.21/. It is hoped that some of these effects will be measured experimentally.

5. Conclusion

5.1 Summary of the Master Equation Approach

The approach of generalized master equations, set out in the first part of this review, may be summarized as follows. The coherence question, specifically the problem of a unified description of coherent, wavelike, band, or flowing motion on one hand and of incoherent, diffusive, random-walk-like or hopping motion on the other, leads to the introduction of the GME. The special characteristic of the GME, which it derives from its non-Markoffian (i.e., nonlocal in time) kernel, is that it describes coherent motion at short times and incoherent motion at long times. The transition time (and there may exist several such times in a given system), is essentially the decay time of the memory functions and may be called the coherence time . We have often referred to it as $1/\alpha$. The GME reduces to the traditional

(Pauli) Master equation for times large compared to $1/\alpha$. What this means is that traditional results are recovered not only in a parameter limit (when $\alpha \to \infty$) but also, for any given α, for times large with respect to $1/\alpha$. The GME prediction is, therefore, that experiments with resolution times that are smaller than $1/\alpha$ can probe directly into coherence. The primary advantages of the GME approach are that the GME is a formally exact consequence of microscopic dynamics, that it has built-in potentialities to treat coherence in a unified manner, that its memory functions can be obtained in well-defined ways, that its application to experiments is straightforward, and that for large times it reduces to the familiar Master equation, the basis of the traditional theory of exciton transport. Let us briefly examine these features one by one.

Under a large class of initial conditions of experimental interest we have seen that the GME is exact. They include delocalized, as well as localized, initial placement of the exciton in the molecular crystal or aggregate. Furthermore, when exciton wavepackets are created, necessitating appending the initial term of (2.29) to the GME, it is known how to treat its effect: one analyzes the driven GME (2.18). The built-in potentiality of the GME to unify coherent and incoherent behaviour, discussed, for instance, in Sect. 2.5, stems from the fact that any natural memory function is intermediate between a constant (or a purely oscillatory function) and a δ-function. Every real system is coherent at short times and (unless pathological) is incoherent at long times. The GME approach offers us the memory function as a natural tool for coherence investigations. We have seen in Sect. 2.6 that these memory functions may be obtained in some cases from exact calculations, in others from perturbation methods, and in yet others directly from observed spectra. The importance of the latter method should not be forgotten. Assuming that one is within the range of validity of the spectral prescription of Sect. 2.6.4, one obtains from it detailed information about exciton dynamics without having to make a large number of model assumptions. Furthermore, the traditional theory which relates rates to spectra /1.15,16/ is obtained as a special case. It is hoped that this review will have removed occasional misrepresentations appearing in the literature that the GME is only a Born-approximated entity, or one that neglects off-diagonal elements of the density matrix, or one that is too formal for experimental investigations. Indeed, for the analysis of energy transfer, annihilation, relaxation, diffusion constants, transient grating observations, scattering, and other such direct migration experiments, the GME is the most direct transport instrument available. It cannot be used directly for the study of optical line shapes, magnetic-resonance experiments and other observations requiring explicitly the off-diagonal elements of the density matrix. For these the SLE, discussed in the next article in this book, is more appropriate. The GME treatment of direct migration experiments proceeds through the exciton propagator as explained in Chap. 3 where explicit coherence effects on observables have been shown. If coherence is experimentally unimportant, the Master equation is obtained as a special case and one returns to the familiar and traditional concepts.

Since the first modern treatment of coupled coherent and incoherent exciton motion, given by HAKEN and collaborators, was in terms of the SLE, the relations of the GME and the SLE acquire special significance. We have seen in Sect. 2.7.2 that when we cast the SLE into the form of the GME, a memory function which has two widely different time constants emerges. The SLE analyses coupled coherent and incoherent motion as the competition of two additive parts in the motion. The emphasis of the SLE is on two coexisting mechanisms for motion, one termed coherent and the other incoherent. Figure 5.1 makes this clear in the context of the Hamiltonian of Sect. 2.6.3. The polaronic transformation pioneered in this field by SILBEY and collaborators transforms the bare picture in (a) of Fig. 5.1 into (b). A phonon-assisted motion interaction appears as a result of this dressing, and the phonon-independent part (horizontal in the figure) is reduced from J to $\tilde{J}$. This is the familiar phenomenon of polaronic bandwidth reduction discussed by HOLSTEIN, EMIN, and others /5.1-3/. The Fourier transform of the memory function is shown in (c): the δ-function corresponds to the reduced bandwidth $\tilde{J}$ and the broad band to the phonon-assisted (nonhorizontal) transitions in (b). They respectively result in the constant com-

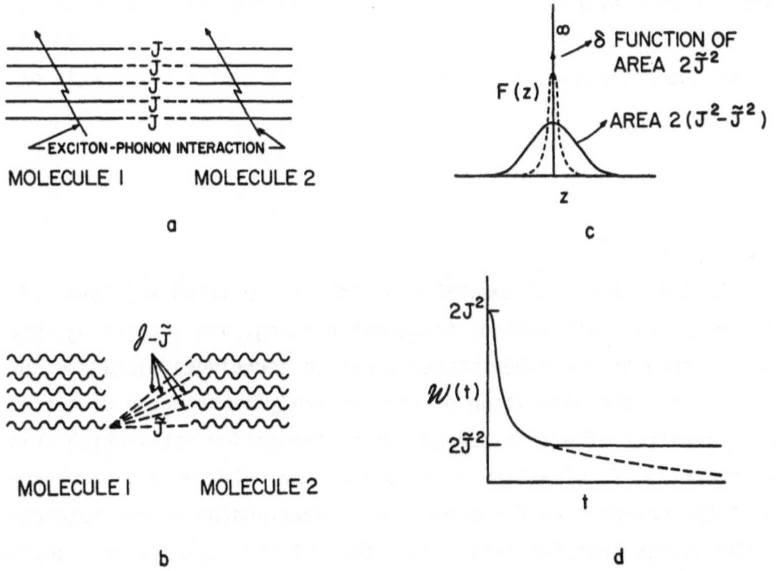

Fig. 5.1a-d. A pictorial representation of exciton-phonon interactions and exciton transport: (a) is the bare picture, (b) is the dressed or polaronic picture with the reduced polaronic band width and phonon-assisted transitions, (c) shows the Fourier transforms of the memory functions, and (d) depicts them in the time domain, clarifying their multiple-time nature. The symbol V in Chap. 2 and J here are interchangeable

ponent and the decaying component of the memory in (d) in the time domain. The dotted lines in (c) and (d) describe further broadening as discussed in Sect. 2.6.3.

Although the SLE and the GME are equivalent in some ways, subtle differences exist in their approach to coherence. This difference is hopefully clarified by Fig. 5.1. If the polaronic bandwidth $\tilde{J}$ in Fig. 5.1 were negligible, the SLE would reduce immediately to the Master equation appropriate to incoherent motion. The GME, however, might still discuss coherence as relating to the decay of the other part of the $W(t)$ in (d). It may be appropriate to state that the GME aims at the *unification* of coherent and incoherent behaviour, whereas the SLE aims at the description of the *coexistence* of two channels of motion which may be termed phonon-hindered and phonon-assisted. These considerations are useful also to the study of polaronic transport /5.1-12,2.24/ in charge carrier motion in organic crystals. Perhaps the most general transport instrument in exciton dynamics is a combination /2.57/ of the GME and the SLE, i.e., an SLE with memory functions /2.57-60/.

The Master-equation techniques developed in Chap. 4 for the analysis of issues other than coherence require little comment in summary. Those used for sensitized luminescence are Markoffian versions of the GME. Although GME's could have been used in momentum space, there appears to be little reason to do so as no similar coherence issues appear in k-space. In momentum space, coherence comes to be associated with the lifetime of the k-states rather than with oscillatory motion in that space. The Master equations for the interplay of motion and vibrational relaxation are composites in real and energy space and as we have seen, they result in new predictions for transfer rates and the dependence of transfer efficiency on the wavelength of initial excitation.

5.2 Future Work

The experimentally relevant outcome of the GME approach is a unified treatment of coherence in energy transfer, annihilation, transient grating, and scattering observations. The similar outcome of the other master equation techniques discussed in Chap. 4 consists of the possible resolution of the disparity in reported values of the singlet diffusion constant, the time dependence of transition rates which arises as a result of the interplay of relaxation with motion, the rate-depression phenomenon, the concept of the energy transfer memory, an understanding of the temperature dependence of the energy transfer rate versus that of the annihilation constant, the study of the range of validity of the bilinear annihilation equation, effects of transport features including dimensionality on the annihilation constant, and some understanding of the trapping dynamics of excitons in momentum space. These analyses suggest a number of new directions for further research in this field, theoretical as well as experimental. We mention some of the latter below.

Many further picosecond and subpicosecond observations are required, especially in sensitized luminescence and grating experiments. These could unravel details of exciton transport including coherence. The older Ronchi grating technique, perfected by ERN and collaborators /5.13,14/, still has considerable potential and should be pursued. Effects of size on trapping and annihilation should be studied since theory appears to predict novel phenomena. Size effects on trapping are also related to caging effects which could arise on creating barriers within a crystal through deuteration /5.15/. Variation of exciton flux with penetration length, if observed, could give us a fairly good idea of the magnitude of exciton diffusion constants. Observations of trap effects of saturation are also called for. High-concentration trap phenomena /3.43,44,5.16/ are certain to provide new transport information. Measurement of the scattering function $S(q,\omega)$ of excitons through interaction with probe beams such as photons or neutrons might be interesting in the light of the theory of Sect. 2.7.5. Time resolved studies of end-detected migration and the investigation of anisotropy by placing detector layers not at the end but elsewhere in the crystal sides are other possibilities. Observations of evolution in k-space /4.16,5.17/ would also be very useful, if made over a wide temperature range.

In regard to coherence it might be useful to mention that two types of experiments are possible: those that probe exciton motion on a definite time scale and those that measure global features (over all time) of the dynamics. An example of the former is the nonexponential transient grating signals predicted in Sect. 3.2.2 and that of the latter is the behaviour of the exponential signals in Sect. 3.2.3. The former are crucially dependent on whether the time resolution of the experiment is smaller or larger than the coherence time $1/\alpha$. If smaller, the experiment is able to measure coherence. If much larger, coherence features are washed out. A number of experiments belong to this class. The other kind, illustrated by the discussion in Sect. 3.2.3, is concerned with the value of $1/\alpha$ relative, not to the probe time τ_p, but to the oscillation time $1/V$. This is a nontrivial difference which can be understood with the example of the damped harmonic oscillator. We could ask whether our probe time is sufficiently small to observe the oscillations in an underdamped oscillator. This question is related to the first kind of experiment mentioned above. But we could also ask whether the oscillator's own parameters are such that it is underdamped or overdamped. This question has little to do with the probe time and it represents the second kind of experiment mentioned above. The first kind of coherence experiment accepts that there is always coherence (roughly, oscillations) at short enough times, and attempts to catch the exciton when it is young and coherent. The second kind of coherence experiment asks the global question whether V is larger or smaller than α, and is similar to the cyclotron resonance question whether or not an electron completes a revolution before being scattered. Many experimental definitions of coherence are given in the literature /1.5 , 5.18/ in terms of oscillations, lifetimes, and similar quantities. They all possess very similar meanings and there is little point in analyzing them further. However, we believe it is important to distinguish between the above two kinds of coherence measurement from the experimental point of view.

The subject of exciton dynamics is intensely alive and full of controversy and activity. There is no doubt that many exciting developments will follow in the near future.

Acknowledgments. The work reviewed in this article is the product of a close interaction among the general ideas of nonequilibrium statistical mechanics, the specific techniques of extracting physics from transport equations, and the interesting features and idiosyncrasies of the quasiparticle we call an exciton. It is a pleasure for me to acknowledge my debt to Prof. M. Dresden, Prof. E.W. Montroll, and Prof. R.S. Knox in these three respective areas. I owe much to my students, past and present, particularly Dr. V. Seshadri and Dr. Y.M. Wong; and to many other colleagues, particularly Prof. R.J. Silbey, Prof. Z. Soos, and Prof. D. Schmid.

I thank Prof. H. Haken for inviting me to write this review, and him as well as Prof. G. Höhler for displaying infinite patience during many delays. A part of this review was completed in Ulm and Stuttgart where I enjoyed the excellent hospitality of Prof. P. Reineker and Prof. H. Haken. I especially wish to thank my friends A. Scheuing and J. Domay, for making my stay in Germany a memorable one. I acknowledge the award of a visiting professorship in 1979 by the Deutsche Forschungsgemeinschaft and support by the National Science Foundation under grant no. DMR 7919539.

This article is dedicated to the memory of Prof. David L. Dexter. The world of exciton physics will miss him a great deal. I shall miss him no less as an exceptional person and irreplaceable friend.

References

1.1 R. Silbey: Ann. Rev. Phys. Chem. *27*, 203 (1976)
1.2 Y. Toyozawa: in *Proceedings of the 1975 International Conference on Luminescence*, ed. by S. Shionaya, S. Nagakura, S. Sugano (North-Holland, Amsterdam 1976)
1.3 R. Kopelman: in *Radiationless Processes in Molecules and Condensed Phases*, ed. by F.K. Fong, Topics in Applied Physics, Vol. 15, (Springer, Berlin, Heidelberg, New York 1976)
1.4 C.B. Harris, D.A. Zwemer: Ann. Rev. Phys. Chem. *29*, 473 (1978)
1.5 D. Burland, A. Zewail: Adv. Chem. Phys. *50*, 385 (1980)
1.6 R.S. Knox: in *Primary Processes of Photosynthesis*, ed. by J. Barber (North-Holland, Amsterdam 1977) p. 55
1.7 I. Oppenheim, K.E. Shuler, G.H. Weiss: *Stochastic Processes in Chemical Physics: The Master Equation* (MIT Press, Cambridge, Mass. 1977)
1.8 N. Van Kampen: Adv. Chem. Phys. *15*, 65 (1965)
1.9 F. Haake: "Statistical Treatment of Open Systems by Generalized Master Equations", in Springer Tracts in Modern Physics, Vol. 66 (Springer, Berlin, Heidelberg, New York 1973) p. 98-168
1.10 G.S. Agarwal: "Quantum Optics", in Springer Tracts in Modern Physics, Vol. 70 (Springer, Berlin, Heidelberg, New York 1974)
1.11 E.W. Montroll, B. West: in *Fluctuation Phenomena*, ed. by E.W. Montroll, J.L. Lebowitz (North-Holland, Amsterdam 1979)
1.12 R.S. Knox: *Theory of Excitons*, Solid State Physics, Supplement 5, ed. by F. Seitz, D. Turnbull (Academic, New York 1963)
1.13 D.L. Dexter, R.S. Knox: *Excitons* (Interscience, New York 1965)
1.14 A.S. Davydov: *Theory of Molecular Excitons* (Plenum, New York 1979)
1.15 Th. Förster: Ann. Phys. (Leipzig) (b) *2*, 55 (1948)

1.16 D.L. Dexter: J. Chem. Phys. *21*, 836 (1953)
1.17 Th. Förster: in *Comparative Effects of Radiation*, ed. by J.S. Kirby-Smith, J.L. Magee (Wiley, New York 1960) Chap. 13
1.18 S. Chandrasekhar: Rev. Mod. Phys. *15*, 1 (1943)
1.19 M. Smoluchowski, Ann. Phys. (Leipzig) *48*, 1103 (1915); Z. Phys. Chem. (Leipzig) *92*, 192 (1917)
1.20 R. Powell, Z. Soos: J. Lumin. *11*, 1 (1975)

2.1 W.T. Simpson, D.L. Peterson: J. Chem. Phys. *26*, 588 (1957)
2.2 G.W. Robinson, R.P. Frosch: J. Chem. Phys. *37*, 1962 (1962); *38*, 1187 (1963)
2.3 A.S. Davydov: Phys. Status Solidi *30*, 357 (1969)
2.4 D.L. Dexter, Th. Förster, R.S. Knox: Phys. Status Solidi *34*, K159 (1969)
2.5 L. Van Hove: Physica *21*, 517, 901 (1955); *22*, 343 (1956); *23*, 441 (1957)
2.6 E.W. Montroll: in *Fundamental Problems in Statistical Mechanics*, ed. by E.G.D. Cohen (North-Holland, Amsterdam 1962)
2.7 N. van Kampen: in Ref. 2.6; Physica *20*, 603 (1954)
2.8 I. Prigogine, P. Resibois: Physica *24*, 795 (1958); *27*, 629 (1961)
2.9 R.W. Zwanzig: in "*Lectures in Theor. Phys.*", ed. by W. Downs, J. Downs (Gordon and Breach, Boulder, Colorado 1961) Vol. III; Physica *30*, 1109 (1964)
2.10 G.G. Emch: Helv. Phys. Acta *38*, 164 (1965); *37*, 532 (1964)
2.11 G.V. Chester: Rep. Prog. Theor. Phys. *26*, 410 (1963)
2.12 I. Oppenheim, K.E. Shuler: Phys. Rev. *138*, B1007 (1965)
2.13 P.M. Morse, H. Feshbach: *Methods of Theoretical Physics* (McGraw-Hill, New York 1953) Vol. I, p. 865
2.14 V.M. Kenkre: Phys. Rev. B*11*, 3406 (1975)
2.15 V.M. Kenkre: in *Statistical Mechanics and Statistical Methods in Theory and Application*, ed. by U. Landman (Plenum, New York 1977)
2.16 V.M. Kenkre, R.S. Knox: Phys. Rev. B*9*, 5279 (1974)
2.17 V.M. Kenkre, T.S. Rahman: Phys. Lett. *50A*, 170 (1974)
2.18 V.M. Kenkre: Phys. Rev. B*12*, 2150 (1975)
2.19 W. Pauli: in *Festschrift zum 60. Geburtstag A. Sommerfeld* (Hirzel, Leipzig 1928)
2.20 V.M. Kenkre: J. Stat. Phys. *19*, 333 (1978)
2.21 C. Aslangul, Ph. Kottis: Phys. Rev. B*13*, 5544 (1976)
2.22 V.M. Kenkre, Y.M. Wong: to be published
2.23 M.K. Grover, R. Silbey: J. Chem. Phys. *54*, 4843 (1971)
2.24 R. Silbey, R. Munn: J. Chem. Phys. *72*, 2763 (1980); R. Munn, R. Silbey: Mol. Cryst. Liq. Cryst. *57*, 131 (1980)
2.25 D. Yarkony, R. Silbey: J. Chem. Phys. *67*, 5818 (1977) S. Rackovsky, R. Silbey: Mol. Phys. *25*, 61 (1973)
2.26 Th. Förster: in *Modern Quantum Chemistry*, Part III, ed. by O. Sinanoglu (Academic Press, New York 1965) p. 93
2.27 V.M. Kenkre, R.S. Knox: Phys. Rev. Lett. *33*, 803 (1974)
2.28 V.M. Kenkre: Phys. Lett. *47A*, 119 (1974)
2.29 V.M. Kenkre: Phys. Rev. B*18*, 4064 (1978)
2.30 K.D. Philipson, K. Sauer: Biochem. *11*, 1880 (1972)
2.31 R. Kühne, M. Morsch, P. Reineker: Z. Phys. B*34*, 297 (1979)
2.32 V.M. Kenkre: Phys. Lett. *63A*, 367 (1977)
2.33 F.F. Sokolov: Phys. Status Solidi (b) *76*, K131 (1976)
2.34 V.M. Kenkre: Phys. Lett. *65A*, 391 (1978)
2.35 R. Kühne, P. Reineker: Solid State Comm. *29*, 279 (1979)
2.36 V.M. Kenkre, R. Silbey: unpublished
2.37 G. Wannier: *Elements of Solid State Theory* (Cambridge Univ., Cambridge 1959)
2.38 J. Hajdu: in *Electronic Structures in Solids*, ed. by E.D. Haidemenakis (Plenum, New York 1969) p. 305
2.39 P. Avakian, V. Ern, R.E. Merrifield, A. Suna: Phys. Rev. *165*, 974 (1968)
2.40 P. Reineker, R. Kühne: Phys. Rev. B *21*, 2448 (1980)
2.41 V. Subrta, V. Capek: Phys. Status Solidi (b) *90*, 295 (1978)
2.42 C.B. Duke, T. Soules: Phys. Lett. A*29*, 117 (1969)
2.43 T. Soules, C.B. Duke: Phys. Rev. B*3*, 262 (1971)

2.44 H. Haken: *Quantum Theory of Solids: An Introduction* (North-Holland, Amsterdam 1976)
2.45 V. Capek, I. Rips: Phys. Status Solidi (b) *97*, K93 (1980)
2.46 M. Lax: J. Chem. Phys. *20*, 1752 (1952)
2.47 A. Maradudin: in *Solid State Phys.* ed. by F. Seitz, D. Turnbull (Academic, New York 1966) Vol. 18
2.48 I. Abram, R. Silbey: J. Chem. Phys. *63*, 2317 (1975)
2.49 V.M. Kenkre, R.S. Knox: J. Lumin. *12*, 187 (1976)
2.50 R. Hochstrasser, P. Prasad: J. Chem. Phys. *56*, 2814 (1972)
2.51 A. Muriel, M. Dresden: Physica *43*, 424, 449 (1969)
2.52 R. Balescu: Physica *27*, 693 (1961)
R. Swenson: Physica *29*, 1174 (1963)
2.53 H. Haken, G. Strobl: in *The Triplet State*, ed. by A.B. Zahlan (Cambridge University, Cambridge 1967)
2.54 H. Haken, G. Strobl: Z. Phys. *262*, 135 (1973)
2.55 H. Haken, P. Reineker: Z. Phys. *249*, 253 (1972)
2.56 R. Kubo: in *Fluctuation, Relaxation and Resonance in Magnetic Systems*, ed. by D. ter Haar (Oliver and Boyd, Edinburg 1962)
2.57 V.M. Kenkre: Phys. Rev. *11*, 1741 (1975)
2.58 H. Sumi: J. Chem. Phys. *67*, 2943 (1977)
2.59 R. Kühne, P. Reineker, V.M. Kenkre: Z. Phys. B *41*, 181 (1981)
2.60 T.S. Rahman, R.S. Knox, V.M. Kenkre: Chem. Phys. *44*, 197 (1979)
2.61 R.P. Hemenger, K. Lakatos-Lindenberg, R.M. Pearlstein: J. Chem. Phys. *60*, 3271 (1974)
2.62 E.W. Montroll, G.H. Weiss: J. Math. Phys. *6*, 167 (1965)
2.63 H. Scher, M. Lax: Phys. Rev. B*7*, 4491 (1973)
2.64 H. Scher, E.W. Montroll: Phys. Rev. B*12*, 2455 (1975)
2.65 G. Pfister, H. Scher: Adv. in Phys. *27*, 747 (1978)
2.66 V.M. Kenkre, E.W. Montroll, M.F. Shlesinger: J. Stat. Phys. *9*, 45 (1973)
2.67 W.J. Shugard, H. Reiss: J. Chem. Phys. *65*, 2827 (1976)
2.68 U. Landman, E.W. Montroll, M. Shlesinger: Proc. Nat. Acad. Sci. (USA) *74*, 430 (1977)
2.69 J. Klafter, R. Silbey: Phys. Rev. Lett. *44*, 55 (1980)
2.70 R. Kubo: J. Phys. Soc. Japan *12*, 570 (1957)
2.71 V.M. Kenkre, R. Kühne, P. Reineker: Z. Phys. B *41*, 177 (1981)
2.72 P. Reineker, R. Kühne: Phys. Lett. *70A*, 187 (1979)
2.73 P. Reineker: Z. Phys. *261*, 187 (1973)
2.74 K. Sköld: in *Hydrogen in Metals I*, ed. by G. Alefeld, J. Volkl, Topics in Applied Physics, Vol. 28 (Springer, Berlin, Heidelberg, New York 1978)
2.75 V.M. Kenkre: to be published; D.W. Brown, V.M. Kenkre: in Proceedings of the International Symposium on Hydrogen in Metals (Richmond, USA March 1982), Plenum, to be published

3.1 J.R. Salcedo, A.E. Siegman, D.D. Dlott, M.D. Fayer: Phys. Rev. Lett. *41*, 131 (1978)
3.2 G.S. Hamilton, D. Herman, J. Feinberg, R.W. Hellworth: Opt. Lett. *4*, 124 (1979)
3.3 P.F. Liao, L.M. Humphrey, D.M. Bloom, S. Geschwind: Phys. Rev. B *20*, 4145 (1979)
3.4 J.J. Eichler: Opt, Acta *24*, 631 (1977)
3.5 H.S. Wolf: in *Adv. in Atomic and Molecular Physics*, Vol. 3, ed. by D.R. Bates, I. Estermann (Academic Press, New York 1967)
3.6 D. Schmid: in Proceedings of the IX International Molecular Crystals Symposium (Kleinwalsertal, Austria, 1980) p. 234
3.7 A.J. Campillo, S.L. Shapiro, C.E. Swenberg: Chem. Phys. Lett. *52*, 11 (1977)
3.8 S.I. Golubov, Yu.V. Konobeev: Phys. Status Solidi *79*, 79 (1977)
3.9 H. Auweter, A. Braun, U. Mayer, D. Schmid: Z. Naturforsch. *34A*, 761 (1979)
3.10 A. Auweter, U. Mayer, D. Schmid: Z. Naturforsch. *33A*, 651 (1978)
3.11 A. Braun, H. Pfisterer, D. Schmid: J. Lumin. *17*, 15 (1978)
3.12 P. Avakian, R. Merrifield: Mol. Cryst. *5*, 37 (1968)
3.13 N.E. Geacintov, C.E. Swenberg: in *Organic Molecular Photophysics*, ed. by J.B. Birks (Wiley, New York 1973) Vol. I
3.14 A. Suna: Phys. Rev. B*1*, 1716 (1970)
3.15 V.M. Kenkre: Phys. Lett. *82A*, 100 (1981)

3.16 L. van Hove: Phys. Rev. *95*, 1374 (1954)
3.17 Y. Wong, V.M. Kenkre: Phys. Rev. B *22*, 3072 (1980)
3.18 S. Haan, R.W. Zwanzig: J. Chem. Phys. *68*, 1874 (1978)
3.19 C.R. Gochanour, H.C. Anderson, M.D. Fayer: J. Chem. Phys. *70*, 4254 (1979)
3.20 V.M. Kenkre, E.W. Montroll: unpublished
3.21 V.M. Kenkre: Phys. Rev. A *16*, 766 (1977)
3.22 E.W. Montroll, K.E. Shuler: J. Chem. Phys. *27*, 454 (1957)
3.23 J.O. Williams, A.C. Jones, K. Janecka-Styrcz, D.A. Elliott: in Proceedings of the IX International Molecular Crystals Symposium (Kleinwalsertal, Austria, 1980) p. 298
3.24 V.M. Kenkre, Y.M. Wong: Phys. Rev. B *23*, 3748 (1981)
3.25 D.L. Huber: Phys. Rev. B *20*, 2307 (1979); *20*, 5333 (1979)
3.26 K.K. Ghosh, D.L. Huber: J. Lumin. *21*, 225 (1980)
3.27 D.L. Huber: J. Chem. Phys. *75*, 4749 (1981)
3.28 R.M. Pearlstein: J. Chem. Phys. *56*, 2431 (1972)
3.29 K. Lakatos-Lindenberg, R.P. Hemenger, R.M. Pearlstein: J. Chem. Phys. *56*, 4852 (1972)
3.30 J. Klafter, R. Silbey: J. Chem. Phys. *74*, 3510 (1981)
3.31 J. Klafter, R. Silbey: J. Chem. Phys. *72*, 843 (1980)
3.32 H. Scher, C.H. Wu: Proc. Nat. Acad. Sc. (USA) *78*, 22 (1981)
3.33 V.M. Kenkre, P. Parris: Phys. Rev. B, to be published
3.34 V.M. Kenkre: Chem. Phys., to be published
3.35 V.M. Kenkre: Chem. Phys. *36*, 377 (1979)
3.36 V.M. Kenkre: Phys. Status Solidi B *89*, 651 (1978)
3.37 Y. Wong, V.M. Kenkre: Phys. Rev. B *20*, 2438 (1979)
3.38 M.D. Fayer, C.B. Harris: Phys. Rev. B *9*, 748 (1974)
3.39 R.M. Shelby, A.H. Zewail, C.B. Harris: J. Chem. Phys. *64*, 3192 (1976)
3.40 D.D. Dlott, M.D. Fayer, R.D. Wieting: J. Chem. Phys. *69*, 2752 (1978)
3.41 R.D. Weiting, M.D. Fayer, D.D. Dlott: J. Chem. Phys. *69*, 1996 (1978)
3.42 D.D. Dlott, M.D. Fayer, R.D. Wieting: J. Chem. Phys. *67*, 3808 (1977)
3.43 D.D. Smith, D.P. Millar, A.H. Zewail: J. Chem. Phys. *72*, 1187 (1980)
D.D. Smith, R.C. Powell, A.H. Zewail: Chem. Phys. Lett. *68*, 309 (1979)
3.44 A.H. Francis, R. Kopelman: in *Excitation Dynamics in Molecular Solids*, Topics in Applied Physics ed. by W.M. Yen and P.M. Selzer (Springer, Berlin, Heidelberg, New York 1981) p. 214
3.45 K. Godzik, J. Jortner: Chem. Phys. Lett. *63*, 428 (1979);
J. Klafter, J. Jortner: Chem. Phys. Lett. *60*, 5 (1978)
3.46 A. Blumen, R. Silbey: J. Chem. Phys. *69*, 3589 (1978)
3.47 A. Blumen: J. Chem. Phys. *4*, 2632 (1980)
3.48 K. Allinger, A. Blumen: J. Chem. Phys. *8*, 4608 (1980);
A. Blumen, J. Manz: J. Chem. Phys. *71*, 4694 (1979)
3.49 A. Blumen, J. Klafter, R. Silbey: J. Chem. Phys. *10*, 5320 (1980)
3.50 J. Bernasconi, S. Alexander, R. Orbach: Phys. Rev. Lett. *41*, 185 (1978)
3.51 S. Alexander, J. Bernasconi, W.R. Schneider, R. Orbach: in *Physics in One Dimensional*, ed. by J. Bernasconi, T. Schneider: Springer Series in Solid State Sciences (Springer, Heidelberg, New York 1981) p. 277-288
3.52 S. Alexander, J. Bernasconi, W.R. Schneider, R. Orbach: Rev. Mod. Phys. *53*, 175 (1981)
3.53 T. Holstein, S.K. Lyo, R. Orbach: Phys. Rev. Lett. *36*, 891 (1976)
3.54 P. Avouris, A. Campion, M. El-Sayed: Chem. Phys. Lett. *50*, 9 (1977);
J. Morgan, M.A. El-Sayed: J. Phys. Chem. (to be published) and references therein
3.55 P.M. Selzer, D.L. Huber, B. Barnett, W.M. Yen: Phys. Rev. B*17*, 4979 (1978)
J. Hegarty, D.L. Huber, W.M. Yen: Phys. Rev. B*23*, 6271 (1981)
3.56 F. Heisel, J.A. Miehe, B. Sipp, M. Schott: Chem. Phys. Lett. *56*, 178 (1978)
3.57 S. Fujiwara, T. Nakayama, N. Itoh: Phys. Status Solidi B*78*, 519 (1976)
3.58 S.D. Babenko, V.A. Benderskii, V.I. Goldanskii, A.G. Laorushko, V.P. Tychinskii: Phys. Status Solidi B*45*, 91 (1971)
3.59 G. Paillotin, C.E. Swenberg, J. Breton, N.E. Geacintov: Biophys. J. *25*, 513 (1979)
3.60 G. Paillotin, C.E. Swenberg: Ciba Found. Symp. *61*, 201 (1979)
3.61 R.R. Alfano, S.L. Shapiro, M. Pope: Opt. Commun. *9*, 388 (1973)

3.62 J. Jortner, S.I. Choi, J.L. Katz, S.A. Rice: Phys. Rev. Lett. *11*, 323 (1963); J. Chem. Phys. *46*, 4252 (1963)
3.63 A. Bergman, M. Levine, J. Jortner: Phys. Rev. Lett. *18*, 593 (1967)
3.64 A. Bergman, D. Bergman, J. Jortner: Israel J. Chem. *10*, 471 (1972)
3.65 V.M. Kenkre: Phys. Rev. B*22*, 2089 (1980)
3.66 V.M. Kenkre: Z. Phys. B*43*, 221 (1981)
3.67 M. Dresden: Rev. Mod. Phys. *33*, 265 (1961)
3.68 A. Maradudin, E.W. Montroll, G.H. Weiss, R. Herman, H.W. Milnes: Acad. R. Belg. XIV, 1709 (1960)
3.69 M. Chabr, U.P. Wild, J. Funfschilling, I. Zschokke-Granacher: Proceedings of the IX International Molecular Crystals Symposium, Kleinwalsertal, Austria, 1980, p. 65;
M. Chabr, D.F. Williams: Phys. Rev. B*16*, 1685 (1977)
3.70 V.M. Kenkre, H.M. Van Horn: Phys. Rev. A*23*, 3200 (1981)

4.1 O. Simpson: Proc. Royal Soc. London A *238*, 402 (1956)
4.2 M.V. Kurik: Fiz. Tverd. Tela *13*, 2877 (1971) [English transl.: Sov. Phys.-Solid State *13*, 2421 (1971)]
4.3 M. Tomura, Y. Takahashi: J. Phys. Soc. Japan *25*, 647 (1968); *26*, 1325 (1969); Y. Takahashi, M. Tomura: J. Phys. Soc. Japan *27*, 1369 (1969); *29*, 525 (1970); *31*, 1100 (1971)
4.4 G. Gallus, H.C. Wolf: Z. Naturforsch. *23*, 1333 (1968); Phys. Status Solidi b *16*, 277 (1966)
4.5 V.M. Kenkre, Y.M. Wong: Phys. Rev. B*22*, 5716 (1980)
4.6 V.M. Kenkre: Chem. Phys. Lett. *82*, 301 (1981)
4.7 B.J. Mulder: Phillips Res. Rep. Suppl. *4* (1968)
4.8 D. Haarer, G. Castro: J. Lumin. *12*, 233 (1976)
4.9 T. Miyakawa, D.L. Dexter: Phys. Rev. B*1*, 2961 (1970)
4.10 C.B. Duke, R. Meyer: Phys. Rev. B*23*, 2111 (1981)
4.11 V.M. Kenkre: To be published
4.12 R.C. Powell, R.G. Kepler: Phys. Rev. Lett. *22*, 636, 1232 (1969); J. Lumin. *1*, 254 (1970)
4.13 B. Hughes, V.M. Kenkre: To be published
4.14 V.M. Kenkre, D. Schmid: Chem. Phys. Lett., to be published
4.15 D.M. Burland, U. Konzelmann, R.M. McFarlane: J. Chem. Phys. *67*, 1926 (1977)
4.16 A.J. Van Strien, J.F.C. Van Kooten, J. Schmidt: Chem. Phys. Lett. *76*, 7 (1980)
4.17 V.M. Kenkre: Phys. Letters *70A*, 381 (1979)
4.18 M. Trlifaj: Czech. J. Phys. *8*, 510 (1958)
4.19 W.B. Fowler, D.L. Dexter: Phys. Rev. *128*, 2154 (1962)
4.20 D.L. Dexter, W.B. Fowler: J. Chem. Phys. *47*, 1379 (1967)
4.21 D.L. Dexter: Phys. Status Solidi (b) *51*, 571 (1972)
4.22 V.V. Hizhnyakov, I.J. Tehver: Zh. Eksp. Teor. Fiz. *69*, 599 (1975) [English transl.: Sov. Phys.-JETP *42*, 305 (1975)]; Phys. Status Solidi *39*, 67 (1970)
4.23 I.J. Tehver, V.V. Hizhnyakov: Zh. Eksp. Teor. Fiz. Pis'ma Red. *19*, 338 (1974) [English transl.: JETP Lett. *19*, 191 (1974)]
4.24 K.K. Rebane, I.J. Tehver, V.V. Hizhnyakov: In *Proceedings of the First Joint U.S.A. - U.S.S.R. Symposium on the Theory of Light Scattering in Condensed Matter*, ed. by B. Bendow, J.L. Birman, V.M. Agranovich (Plenum, New York 1976) p. 393
4.25 H. Dubost, L. Abouaf-Marguin, F. Legay: Phys. Rev. Lett. *29*, 145 (1972)
4.26 K. Dressler, O. Ochler, D.A. Smith: Phys. Rev. Lett. *34*, 1364 (1975)
4.27 H. Dubost, R. Charnean: Chem. Phys. *12*, 407 (1976)
4.28 L. Landau, E. Teller: Phys. Z. Sowjetunion *10*, 34 (1936)
4.29 V. Seshadri, V.M. Kenkre: Phys. Lett. *56A*, 75 (1976)
4.30 V.M. Kenkre, V. Seshadri: Phys. Rev. A*15*, 197 (1977)
4.31 V. Seshadri, V.M. Kenkre: Phys. Rev. A*17*, 223 (1978)
4.32 V. Seshadri, V.M. Kenkre: Z. Physik B *33*, 289 (1979)
4.33 G. Weber: Biochem. J. *75*, 335 (1960);
G. Weber, M. Shinitzky: Proc. Natl. Acad. Sci. U.S.A. *65*, 823 (1970)
4.34 K. Itoh, T. Axumi: J. Chem. Phys. *62*, 3431 (1975)

4.35 P. Saari, K. Rebane: Solid State Comm. *7*, 887 (1969)
4.36 R.W. Anderson, R.M. Hochstrasser, H. Lutz, G.W. Scott: J. Chem. Phys. *61*, 2500 (1974)

5.1 T. Holstein: Ann. Phys. (N.Y.) *8*, 325, 343 (1959)
5.2 D. Emin: Adv. Phys. *22*, 57 (1973); *24*, 305 (1975)
5.3 D. Emin: Phys. Rev. Lett. *25*, 1751 (1970); Phys. Rev. B*3*, 1321 (1971); Phys. Rev. B*4*, 3639 (1971)
5.4 A. Madhukar, W. Post: Phys. Rev. Lett. *39*, 1424 (1977)
5.5 G.G. Roberts, N. Apshey, R.W. Munn: Phys. Rep. *60*, 59 (1980)
5.6 C.B. Duke, L.B. Schein: Physics Today *33* (2), 42 (1980)
5.7 L.B. Schein, A.R. McGhie: Chem. Phys. Lett. *62*, 356 (1979)
5.8 Y. Toyozawa: In *Proceedings of the 4th International Conference on Vacuum Ultraviolet Radiation Physics*, ed. by E. Koch, R. Haensel, C. Kunz (Pergamon, New York, 1974); J. Lumin. *12/13*, 13 (1976)
5.9 Y. Toyozawa: In *Relaxation of Elementary Excitations*, ed. by R. Kubo, E. Hanamura, Springer Series in Solid State Sciences, Vol. 18 (Springer, Berlin, Heidelberg, New York 1980) pp. 3-18
5.10 H. Sumi: Solid State Comm. *28*, 309 (1978); *29*, 495 (1979) J. Chem. Phys. *70*, 3775 (1979); *71*, 3403 (1979)
5.11 S. Efrima, H. Metiu: Chem. Phys. Lett. *60*, 226 (1979)
5.12 P. Reineker, R. Kühne, V.M. Kenkre: Phys. Lett. *84*A, 294 (1981)
5.13 V. Ern, M. Schott: in *Localization and Delocalization in Quantum Chemistry*, ed. by O. Chalvet (D. Reidel Publishers, Dordrecht-Holland, 1976) Vol. II, p. 249
5.14 V. Ern, A. Suna, Y. Tomkiewiecz, P. Avakian, R.P. Groff: Phys. Rev. B*5*, 3222 (1972)
5.15 H. Port, D. Vogel, H.C. Wolf: Chem. Phys. Lett. *34*, 23 (1975)
5.16 H.C. Wolf, H. Port: J. Lumin. *12*, 33 (1976)
5.17 D.D. Smith, A.H. Zewail: J. Chem. Phys. *71*, 3533 (1979)
5.18 D.M. Hanson: CRC Crit. Rev. Solid State Sci. *3*, 243 (1973)

Stochastic Liouville Equation Approach: Coupled Coherent and Incoherent Motion, Optical Line Shapes, Magnetic Resonance Phenomena

By P. Reineker

1. Introduction

Since the introduction of the concept of excitons by FRENKEL /1/ in 1931, these quasi-particles have played an outstanding role in the description of electronic excitations both in inorganic and organic materials. A summary of the theoretical description of excitons in inorganic materials is given by HAKEN /2/ and KNOX /3/. The state of research into high densities of excitons generated in semiconductors may be taken from /4,5/.

On account of the weak van der Waals interaction between the molecules in organic molecular crystals, the constituents keep their identity in a first approximation. The basic concepts of excitons in molecular crystals are represented in the monographs by DAVYDOV /6,7/ and by CRAIG and WALMSLEY /8/. The present knowledge about the properties of electronic states in molecular crystals has been summarized recently by SILINSH /9/. Following SOOS /10/, we may classify organic crystals according to the charge transfer in the ground and excited states.

In the first class of molecular crystals the molecules remain neutral when the crystal is excited, e.g. by irradiation of light. In this class the excited electrons remain with "their" molecules and form Frenkel excitons. Standard examples are crystals of naphthalene and anthracene /11-14/, but also 1,2,4,5-tetrachlorobenzene /15/ and 1,4-dibromonaphthalene /16/, in which the electronic excitations move predominantly in one dimension.

Crystals of the second class contain at least two molecules in the unit cell, one of which is a weak donor, the other one a weak acceptor. The crystal is formed by alternating stacks of donor and acceptor molecules. In the ground state of the crystal both molecules are neutral; during the excitation, however, an electron is transferred from the donor to the acceptor forming a charge-transfer state. Examples are biphenyl/tetracyanobenzene (TCNB) /17/, naphthalene/TCNB /18/, anthracene/TCNB /19,20/, and anthracene-pyromellitic dianhydride /21-23/. Recent reviews are given by MÖHWALD /24/ and HAARER /25/.

In charge-transfer crystals formed by phane molecules, donors and acceptors are bound together chemically, and the charge transfer occurs within such a molecule pair. The interaction of π electrons in such materials, embedded also in glass matrices /26/, has been discussed recently by SCHWEITZER /27/.

In the last group of this classification scheme, the crystals are composed of strong donor and acceptor molecules, which are also ionized in the ground state, forming in this way organic salts /10/. Examples for donors are tetrathiofulvalene (TTF) and NNN'N'-tetramethyl-p-phenylenediamine (TMPD); as acceptors tetracyano-quinodimethane (TCNQ) and chloranil are used. Sometimes the molecules are arranged in alternating stacks; segregated stacks, however, are more usual.

In contrast to the materials discussed above, which are insulators or semiconductors /28/, the conductivity of the organic salts is sometimes comparable to that of metals. Therefore, in recent years, these materials have been the subject of extremely active experimental and theoretical research /29,30/.

In this article we shall mainly consider electronic excitations in crystals of the first class, i.e. Frenkel excitons. If a molecule in such a crystal is excited, on account of the interaction between the molecules this localized excitation is not a stationary state of the crystal but will be transferred to other crystal sites. The most challenging problem in this connection for experimentalists as well as for theorists are the following questions. Is this transfer coherent or incoherent, or in other words, does the excitation propagate like a wave packet, whose time development is determined by the Schrödinger equation /31/, or does it move via a hopping process /32/, which may be described by a master equation /33/ ?

The answer to these questions - as we shall see later on - is that these two kinds of exciton transport are limiting cases of a more general kind of exciton dynamics called coupled coherent and incoherent exciton motion. The characteristic feature of this unified description of exciton dynamics is that phase relations, characteristic for a wave-like motion, decay in the course of time on account of interactions with lattice vibrations and structural perturbations. If this decay occurs rapidly, the exciton motion can be described by a hopping process, in the opposite limit the wave picture applies.

The first theoretical approach to the coupled coherent and incoherent motion of excitons is based on the stochastic model of HAKEN and STROBL /34/ and results in the stochastic Liouville equation (SLE) description of exciton dynamics according to HAKEN, REINEKER, and STROBL /34-36/. The coherent part of the dynamics is described by a Hamiltonian containing the excitation energy of the molecules at the various lattice sites and transfer matrix elements between them, being composed of Coulomb and exchange interaction integrals. The phase of this coherent, wave-like motion decays on account of the interaction with intra- and intermolecular vibrations. These vibrations give rise to fluctuations of the excitation energies and of the transfer matrix elements and are taken into account in the Hamiltonian by an additional, stochastically time-dependent contribution. Averaging the equation of motion of the density operator over the fluctuations /36-40/ results in the SLE /41/ for the coupled coherent and incoherent exciton motion. This model has been applied to a variety of experimental situations and allows for their consistent description /42,43/.

In the second approach to exciton dynamics the electronic degrees of freedom as well as those of the phonons are treated quantum mechanically. Considering the phonons as a heat bath /44-46/, which is eliminated from the density operator equation of the coupled system, one arrives at an equation of motion for the reduced density operator of excitons. In the procedure introduced by GROVER and SILBEY /47,48/, the main part of the exciton-phonon interaction is taken into account by a canonical transformation in a way similar to the treatment of the electron-phonon interaction in the small polaron model /49,50/; the remaining part is then treated perturbationally. This method has also been applied to perturbed systems /51,52/ and extended by a variational procedure /53,54/. Within the context of the microscopic treatment of exciton-phonon interaction, quadratic coupling to the phonons has also been considered /55,56/. The electronic energy transfer description based on these methods and their relations to those described above and below in this section have been reviewed by SILBEY /57/. Recent developments in this field are discussed in /58,59/.

In the third method /32,33,60/ the evolution in time of the probability of finding an exciton at a particular lattice site is calculated from a Markovian master equation /61,62/. As a generalization of this procedure, KENKRE and KNOX /63,64/ have recently used the non-Markovian generalized master equation (GME) of NAKAJIMA /65/ and ZWANZIG /66/ for the calculation of the excitonic occupation probability. In contrast to the ordinary master equation, the GME with its non-Markovian memory function allows one to take care of the coupled coherent and incoherent nature of exciton motion. KENKRE and KNOX have evaluated the kernel of the GME in the Born approximation and could in this way relate the kernel to optical absorption and emission spectra /67/. Calculations of the memory functions of the GME, starting from a microscopic description of the exciton-phonon interaction, have been carried out by KENKRE and RAHMAN /68/ and by SMITH /69/ within the Born approximation. Starting from the same microscopic picture, SOKOLOV and HIZHNYAKOV /70,71/ have succeeded in obtaining an approximate expression for the memory function without using the Born approximation. The influence of a non-thermal phonon distribution on the memory function has been investigated by CAPEK and RIPS /72/. The relations of the GME treatment to those of HAKEN, REINEKER, and STROBL /34-36/ and of GROVER and SILBEY /47,48/ have been discussed by KENKRE /73,74/, and recently the connection between the SLE and the GME has been investigated by KENKRE /75-77/ and by KÖHNE and REINEKER /78-81/. A more detailed representation of these relations is given in the article by KENKRE in this volume.

In the following, some experiments will be referred to from which information about the coupled coherent and incoherent dynamics of excitons can be drawn /15,82/. In the context of this paper the list can naturally not be complete.

The most direct information on the dynamics of excitons is obtained from their diffusion tensor. In anthracene crystals the components of the diffusion tensor of triplet excitons in the *ab* plane have been determined either by observing the delayed fluorescene /13,83-89/ or indirectly by a spectroscopic approach /89,90/. Theoretically, the diffusion tensor is obtained from the long-time limit of the mean-square displacement and has been determined in all three approaches to exciton dynamics mentioned above /35,37,48,63,91-95/. From the comparison between theory and experiment the numerical value of the effective hopping rate for triplet excitons in anthracene, which contains contributions from the incoherent and the coherent part of the motion, is derived as $\approx 10^{10}\ s^{-1}$. The exciton motion on finite molecular chains has been investigated by ASLANGUL and KOTTIS /96/. The diffusion of singlet excitons is reviewed in /97/.

From optical investigations, information on electronic interaction matrix elements (Davydov splittings) and on the interaction between electronic degrees of freedom and phonons (line shape and width) may be derived. Thus from isotopically mixed naphthalene crystals, naphthalene-h_8 as guest in naphthalene-d_8, the interaction within guest dimers /98/ and higher aggregates /99/ is obtained. In pure naphthalene crystals the singlet-triplet transition has been investigated in /100-102/ and singlet-singlet spectra are represented in /103/. Triplet spectra in anthracene crystals are reported in /89,90, 104-107/ and their relation to the naphthalene spectra is discussed by PORT and RUND /108/. Optical investigations of singlet excitations are represented in /109-112/. Line shapes of triplet excitations in pyrene crystals are investigated in /113,114/ and a comparison with triplet spectra in anthracene is given in /115/. BURLAND et al. /116/ have investigated the spectral line shapes of triplet excitons in the quasi-one-dimensional crystal 1,2,4,5-tetrachlorobenzene. Optical investigations of 1,4-dibromonaphthalene /117/ have shown that also in this system the motion of electronic excitations is mainly one-dimensional. From the optical line shape of triplet excitons, information on their dephasing has been derived /118/. Finally 1,4-dibromonaphthalene has also been investigated by Stark spectroscopy /119/.

Theoretically, optical line shapes for dimers of two differently oriented molecules have been calculated by HAKEN and REINEKER /35/ and by ASLANGUL and KOTTIS /39/. Line shapes for finite molecular aggregates are derived in /40,96,120/. For molecular crystals with two differently oriented molecules in the unit cell, optical line shapes have been obtained by SCHWARZER and HAKEN /121/ and by REINEKER /122/. The homogeneous optical line shapes of the calculation just mentioned are Lorentzian, because the fluctuations are assumed to be δ correlated in the original Haken-Strobl model. This assumption corresponds to the fast modulation limit in line shape theory /123/. Generalizing the model by taking into account exponentially decaying energy fluctuations, SUMI /124/ and BLUMEN and SILBEY /125/ succeeded in deriving approximate expressions for the optical line shape, which allow the Lorentzian line shape to be recovered in the fast modulation limit but result in a Gaussian shape for slow modulation. Higher-order correction terms to the line shape formula of /125/ have recently been published by RIPS and CAPEK /126/. Absorption band profiles for singlet transitions in molecular crystals have been calculated by DISSADO /127,128/ and by CRAIG and DISSADO /129/, and a theory of depolarization of fluorescence in molecular pairs on account of energy transfer has been published recently by RAHMAN et al. /130/.

Another class of experiments, from which information on exciton motion can be drawn, represents the investigation of electron-spin resonance. The simplest systems, at least from the theoretical point of view, are molecular pairs, with the excitation being able to move between their constituents. Pairs of translationally equivalent molecules of 1,4-dibromonaphthalene-h_6 in the perdeutero host have been investigated with optical detection of magnetic resonance (ODMR) by HOCHSTRASSER and ZEWAIL /131/, and pairs of 1,2,4,5-tetrachlorobenzene-h_2 in the perdeutero host by ZEWAIL and HARRIS /132,133/ at a temperature of about 1.5 K. Recently, in crystals of naphthalene-d_8 doped with naphthalene-h_8, triplet excitons in pairs of translationally equivalent molecules of naphthalene-h_8 (AA pairs) have been observed /134/. In the same system, triplet excitons in pairs of inequivalent naphthalene-h_8 molecules (AB pairs), formed by the two differently oriented molecules in the unit cell of the crystal, have been detected by SCHWOERER and WOLF /135,136/ by electron-spin resonance (ESR), soon after the identification of the long-lived excited state of single naphthalene molecules in a durene matrix as a triplet state by HUTCHISON and MANGUM /137/. These pair measurements were first interpreted /138/ in the hopping model of exciton motion. From optical measurements /98,139/ of such doped naphthalene crystals it has been postulated /140,141/ that excitons in the AB pair should possibly be described within the coherent model. The final confirmation for this conjecture was derived by BOTTER et al. /142/ from ODMR and spin-echo measurements using a model of VAN'T HOF and SCHMIDT /143/ for the interpretation of their results. Recent ESR measurements of HINKEL /144/ of the same system have been discussed /145/ in the coherent model of the exciton motion, taking into account an inhomogeneous distribution of the energy levels of the pair /146/.

In crystals of 1,4-dibromonaphthalene, as mentioned above, excitons move mainly along linear chains of equivalently oriented molecules. From ESR measurements /16, 147-149/ it has been derived that below 16 K the exciton motion is coherent. Exciton motion in single crystals of 1,2,4,5-tetrachlorobenzene (TCB) is one-dimensional, too, and ODMR measurements were first discussed in the coherent picture /150-153/. Recent experiments /154-157/ have been interpreted by allowing a scattering of the excitons between k states. A discussion of the interaction between traps and band states is given in /158-161/.

ESR measurements in naphthalene /162-164/ and anthracene single crystals /162-164,165/ at room temperature are interpreted in the picture of incoherent exciton motion. Analogous measurements have been carried out in tetracene /166,167/ and in pyrene /168,169/. Recent ESR measurements in benzophenone crystals at liquid helium temperatures are discussed in the coherent picture /170/.

The theoretical investigation of the influence of the dynamics of triplet excitons on their ESR line shapes proceeded along the same lines as that one of their motion. The ESR line shape of incoherent excitons may be calculated using the methods introduced by ANDERSON and by KUBO and TOMITA /123,171-176/. The influence of the coherent motion is investigated following STERNLICHT and McCONNEL /152,153,177-179/. For excitations with spin 1/2 moving in a dimer, the ESR line shape has been calculated on the basis of the Haken-Strobl model both numerically /180/ and analytically /122,181,182/. The ESR line shape of the quasi-incoherent motion of triplet excitons in molecular crystals with two differently oriented molecules in the unit cell has been investigated by REINEKER /183-186/, and the influence of the coupled coherent and incoherent motion on the ESR line shape of dimers is represented in /38,187/.

The investigation of naphthalene and anthracene crystals by nuclear magnetic resonance (NMR) under irradiation of light shows that the longitudinal relaxation time of the protons is shortened by the presence of triplet excitons /188,189/. The theoretical problem has been treated by HAKEN, REINEKER, and SCHWARZER /37,42, 181, 190,191/.

A class of experiments, which will not be considered in later sections, deals with sensitized fluorescence in doped crystals /192/. The investigation of the energy transfer in these materials has experienced an increasing interest since the application of picosecond techniques /193-195/. Recent theoretical approaches are given, for example in /196-199/. Furthermore, the interaction between triplet excitons and the influence of a magnetic field /200-207/ will not be considered. Recent theoretical treatments /208,209/ are based on the ideas of MERRIFIELD /210/ and SUNA /211/. A review is given by SWENBERG and GEACINTOV /212/. Another field of current interest, which will not be discussed in this article, is motion of excitations in disordered materials /213-219/. Finally, it should be mentioned that the model of the coupled coherent and incoherent motion /34-36/ has recently been used to describe the motion of chemisorbed atoms /220/ and to determine the mobility of electrons /221-224/ in photoconductivity experiments /225,226/.

In the first of the following sections the model of the coupled coherent and incoherent exciton motion is represented and its stochastic Liouville equation is derived. From the solution of this equation, in Sect. 3 the exact time dependence of the density matrix for the exciton motion in molecular pairs (dimers) and on a linear chain of molecules is obtained. In Sect. 4 the mean-square displacement for the exciton motion on a linear chain is calculated exactly and a diffusion equation describing the quasi-incoherent exciton motion is derived. The line shape for optical absorption experiments in crystals with two differently oriented molecules in the unit cell is determined in Sect. 5 and extensions to include non-Markovian fluctuations are discussed. Section 6 presents the application of the stochastic model for the coupled coherent and incoherent exciton motion to the calculation of its influence on ESR line shapes. A model system for an excitation with spin 1/2 moving in a dimer is considered and analytical expressions for the line shapes are discussed. Additionally the line shapes of triplet excitons moving in a coupled coherent and incoherent manner in the same system are evaluated numerically. Then the ESR of quasi-incoherent triplet excitons moving in an anthracene type crystal is discussed numerically and analytically and investigations of the spin resonance of coherent one-dimensional excitons are reviewed shortly. In Sect. 7 the influence of the exciton motion on the longitudinal relaxation time of protons in organic molecular crystals is derived, and finally the connection between the SLE and GME treatments is discussed in Sect. 8.

2. Model of the Coupled Coherent and Incoherent Exciton Motion and Its Stochastic Liouville Equation

In this section the basic assumptions of the model of the coupled coherent and incoherent exciton motion are summarized and its stochastic Liouville equation is derived in a perturbative manner. An exact derivation of this equation is given in Appendix A. First, in the following subsection the Hamiltonian describing the coherent motion of the excitons is discussed, then in Sect. 2.2 the Hamiltonian responsible for the incoherent part of the motion and its stochastic properties are given, and in the final section the stochastic Liouville equation is derived.

2.1 The Coherent Motion of the Excitons

In the introduction we have remarked that in a first approximation the molecules of an organic molecular crystal keep their identity on account of the weak van der Waals interaction between them. In an excitation process, caused, e.g., by irradiation with light, only those electrons take part which occupy the highest energy levels of the molecular ground states. An excited state of the crystal is thus given as a superposition of these degenerate excited molecular states. The many other electrons of an organic molecule, which do not take part in the excitation process, together with its nuclei may be considered as a core giving rise to a potential for the electrons in the highest energy levels. The Hamiltonian of the electrons, taking part in dynamical processes, is given by

$$H_0 = \sum_i \left[\frac{p_i^2}{2m} + V(\underline{x}_i)\right] + \frac{1}{2}\sum_{i \neq j}\sum \frac{e^2}{|\underline{x}_i - \underline{x}_j|} \quad . \tag{2.1}$$

The sums in this expression run over the dynamical electrons of all molecules. The first term is the kinetic energy of these electrons, the second their potential energy in the potential generated by the cores and by an external field, which in some experimental situations might be applied. The last term finally describes the Coulomb interaction between the electrons.

The wave function of the Schrödinger equation belonging to the Hamiltonian above depends on the coordinates of all electrons (on the order of 10^{23}). This many-particle wave function may be represented by a series using a suitable complete set of state vectors. Such a complete set is given by Slater determinants, which are constructed using molecular wave functions or Wannier functions, one for each molecule. The Slater determinant with the lowest energy is constructed using merely molecular wave functions, describing the ground state of the molecule. Many-particle states with higher energy are constructed by replacing one or several of the molecular ground state wave functions by wave functions describing excited states of a single molecule.

Much more convenient than the description of the problem in terms of these Slater wave functions - and therefore used almost exclusively in many-particle physics - is to utilize second quantization /227/. Using this formulation of many-particle physics, the Hamiltonian (2.1) may be written in the following form:

$$H_0 = \int \psi^+(\underline{x})\, h\, \psi(\underline{x})\, d^3x + \frac{1}{2} \iint \psi^+(\underline{x})\, \psi^+(\underline{x}')\, \frac{e^2}{|\underline{x} - \underline{x}'|}\, \psi(\underline{x}')\, \psi(\underline{x})\, d^3x\, d^3x' \,, \qquad (2.2)$$

where

$$h = -\frac{1}{2m} \Delta + V(\underline{x}) \quad . \qquad (2.3)$$

The field operators $\psi^+(\underline{x})$ and $\psi(\underline{x})$ may be expanded according to localized functions $w_{n\mu}(\underline{x})$, using, e.g., the molecular functions or Wannier functions mentioned above. n describes the site of the localization of the function and μ the quantum state at that site (the vector character of n will not be denoted especially):

$$\psi^+(\underline{x}) = \sum_{n,\mu} a^+_{n\mu}\, w^*_{n\mu}(\underline{x}) \;, \qquad \psi(\underline{x}) = \sum_{n,\mu} a_{n\mu}\, w_{n\mu}(\underline{x}) \;. \qquad (2.4)$$

The operator $a^+_{n\mu}$ creates an electron at site n in the state μ, $a_{n\mu}$ is the corresponding annihilation operator. We have mentioned in the introduction and above that on account of the weak interaction between the molecules in the organic crystals, the molecular properties are transferred to the crystal at least in a first approximation,and that the higher excited states relax very rapidly towards the lowest excited one, either in the singlet or in the triplet system. On account of the first property, in the expansion of the field operators we have used the localized Wannier functions instead of delocalized Bloch functions. The second property will be used when in our model description at each molecule only two electronic states are taken into account: $\mu = 0$ denotes the ground state of the molecule, $\mu = 1$ its lowest excited state either in the singlet or in the triplet system. As we want to treat Frenkel excitons, we allow only such excitations with the hole in the state $\mu = 0$

and the electron in the excited state localized at the same molecule. Therefore, the molecules always remain neutral, and charge transfer excitons /10,25/ are excluded from the considerations. With these assumptions the Hamiltonian H_0 becomes

$$H_0 = \sum_n \varepsilon b_n^+ b_n + \sum_{n \neq n'} \sum H_{n-n'} b_n^+ b_{n'} \; . \tag{2.5}$$

b_n^+ and b_n are now creation and annihilation operators for a localized electron-hole pair at site n and may be expressed by the electron operators in the following way:

$$b_n^+ = a_{n1}^+ a_{n0} \; , \quad b_n = a_{n0}^+ a_{n1} \; . \tag{2.6}$$

The terms of the first sum of (2.5) represent the energy of the exciton when sitting on the molecule at site n. ε is the excitation energy between states with $\mu = 0$ and $\mu = 1$. The terms of the double sum annihilate an exciton at site n' and generate one at site n; therefore, these terms are responsible for the (coherent) energy transport. The mechanism of the exciton transport is described by the matrix element $H_{n-n'}$ which, taking the electron spin into account, in this model is given by

$$\begin{aligned} H_{n-n'} = & \iint \phi^*_{n1s}(\underline{x})\, \phi^*_{n'0s'}(\underline{x}')\, \frac{e^2}{|\underline{x} - \underline{x}'|}\, \phi_{n'1s''}(\underline{x}')\, \phi_{n0s'''}(\underline{x})\, d^3x\, d^3x' \\ & - \iint \phi^*_{n1s}(\underline{x})\, \phi^*_{n'0s'}(\underline{x}')\, \frac{e^2}{|\underline{x} - \underline{x}'|}\, \phi_{n0s'''}(\underline{x}')\, \phi_{n'1s''}(\underline{x})\, d^3x\, d^3x' \; . \end{aligned} \tag{2.7}$$

The first term (Coulomb term) describes the annihilation of an electron at the site n' in the excited state and its creation at the same site in the ground state, while at the site n an electron is annihilated in the ground state and created in the excited state. By this interaction the exciton has moved from the site n' to the site n without an electron having changed its place. The second term (exchange term) describes the transition of an excited electron from site n' to site n and the transition of an electron in the ground state from site n to site n'. By this mechanism, too, the excitation has moved from site n' to site n but now at the same time two electrons have changed their sites. This term will only be important if the wave functions $\phi_{n'}$ and ϕ_n overlap. Its main contribution stems from nearest neighbours and is usually less important than the Coulomb term. It is very easy to show that for triplet excitons the Coulomb term disappears and the only contribution to the matrix element is given by the exchange term, which further on we shall denote by $J_{n-n'}$.

2.2 The Incoherent Motion of the Excitons

The Hamiltonian H_0 of (2.5) may immediately be diagonalized by a simple Fourier transformation. The solutions of its Schrödinger equation are exciton states with definite wave number k and energy E(k). These excitons are collective excited states of the whole crystal and are delocalized throughout it. A localized excited state is obtained as a superposition of these eigenstates with definite phase relations and moves through the crystal with its group velocity. This coherent, wave-like motion is perturbed by the vibrations of the crystal, which scatter the exciton waves and thus destroy the phase relations in the course of time. Therefore, to the Hamiltonian H_0 of (2.5) we have to add the Hamiltonian of the vibrations of the molecular crystal and their interaction with the excitons:

$$H_{ex,ph} = H_0 + \sum_q \omega_q c_q^+ c_q + \sum_{n_1} \sum_{n_2} \sum_q g_{n_1 n_2 q}(c_q^+ + c_{-q}) b_{n_1}^+ b_{n_2} \ . \tag{2.8}$$

In writing (2.8) we have assumed that the sum over q also includes a summation over the various phonon branches. The interaction operator with coupling constant $g_{n_1 n_2 q}$ describes the annihilation of an electron-hole pair at site n_2 and its creation at site n_1 and the simultaneous absorption or emission of a phonon. The transformation to the interaction representation with respect to the phonons results in the following Hamiltonian:

$$\tilde{H}_{ex,ph} = H_0 + \sum_{n_1} \sum_{n_2} \sum_q g_{n_1 n_2 q}(c_q^+ \ e^{i\omega_q t} + c_{-q} \ e^{-i\omega_q t}) b_{n_1}^+ b_{n_2} \ . \tag{2.9}$$

The time dependence of the interaction term is rather complicated, because it is determined by the frequencies of all vibrations of the crystal. In the general case these include internal vibrations of the molecules, and librational and translational motions of the whole molecules. Furthermore, this classification into three kinds of vibrations is valid only in lowest approximation; generally, the coupling between the various kinds of motions of the molecules in the crystal has to be taken into account. For these reasons it seems appropriate to consider the time dependence of the interaction operator of (2.9) as a stochastic process /41,228-230/.

In the Haken-Strobl model for the coupled coherent and incoherent exciton motion /34-36/ the interaction between excitons and phonons is treated in this way. By doing this the molecular vibrations are treated as a heat bath pushing the excitonic system in a stochastic manner. This procedure may be justified if the molecules are not too small and the temperature is not too low. However, it is not possible to take into account the reaction of the exciton on the molecular vibrations. The Hamiltonian of this incoherent part of the motion is given by

$$H_1(t) = \sum_{n_1} \sum_{n_2} h_{n_1 n_2}(t) \ b_{n_1}^+ b_{n_2} \ , \tag{2.10}$$

where $h_{n_1n_2}(t)$ is assumed to be a Gaussian stochastic process with disappearing mean value. The diagonal matrix elements of $H_1(t)$, i.e. $h_{nn}(t)$, describe fluctuations of the excitation energy ε at lattice site n, the non-diagonal ones $h_{n_1n_2}(t)$ represent the stochastic variations of the coherent interaction matrix element $H_{n_1-n_2}$ of (2.5).

Mathematically, a Gaussian stochastic process /231,232/ is characterized by its mean values and its second-order correlation functions. In the treatment of the coupled coherent and incoherent exciton motion according to HAKEN, REINEKER, and STROBL /34-36/, these quantities are assumed to be

$$\langle h_{n_1n_2}(t)\rangle = 0 \ , \tag{2.11}$$

$$\langle h_{n_1n_1'}(t_1)\, h_{n_2n_2'}(t_2)\rangle = 2\Lambda(n_1,n_1',n_2,n_2')\ \delta(t_1-t_2) \ , \tag{2.12}$$

with /35,36/

$$\Lambda(n_1,n_1',n_2,n_2') = \gamma_{|n_1-n_1'|}\ \delta_{n_1n_2'}\ \delta_{n_1'n_2} + \bar{\gamma}_{n_1-n_1'}\ \delta_{n_1n_2}\ \delta_{n_1'n_2'}(1-\delta_{n_1n_1'}) \ . \tag{2.13}$$

This means that the mean value of the fluctuations disappears or that a possibly existing mean value is taken into account in the coherent part H_0 of the Hamiltonian. From (2.12) we see that the correlation function is assumed to be proportional to a δ function. Therefore, the correlation time of the process is infinitely short and the fluctuations have a white spectrum. Physically this requires that the frequency distribution of the phonons should be considerably broader than the bandwidth of the excitons. An approach to the consideration of the influence of finite correlation times has been given recently /125,126,233,234/ and will be discussed later on. Finally, the site dependence of the correlation function as given in (2.13) shows that $\langle h_{n_1n_1}(t_1)\, h_{n_2n_2}(t_2)\rangle = 0$ for $n_1 \neq n_2$, i.e. energy fluctuations at different molecules are not correlated, and the vanishing of $\langle h_{n_1n_2}(t_1)\, h_{n_2n_3}(t_2)\rangle$ ($n_1 \neq n_2 \neq n_3$ or $n_1 \neq n_2 = n_3$, etc.) implies the independence of transitions caused by fluctuations between different pairs of molecules.

2.3 The Stochastic Liouville Equation for the Exciton Density Operator

The total Hamiltonian reads

$$H = H_0 + H_1(t) \ , \tag{2.14}$$

where H_0 and $H_1(t)$ are defined in (2.5) and (2.10), respectively. The equation of motion for the density operator W is

$$\dot{W} = -i[H,W] \equiv -iH^{\times}W = -iLW \ , \tag{2.15}$$

where (Ω is an arbitrary operator) /234/

$$H^{x}\Omega = L\Omega = [H,\Omega] \ . \tag{2.16}$$

In the same way as the Hamiltonian H, these operators are composed of a time-independent and a stochastically time-dependent part

$$H^{x} = H_0^{x} + H_1^{x}(t) \ , \tag{2.17}$$

$$L = L_0 + L_1(t) \ . \tag{2.18}$$

Introducing the density operator in the interaction picture $\widetilde{W}(t)$ by

$$W(t) = e^{-iL_0 t}\,\widetilde{W}(t) \ , \tag{2.19}$$

the equation of motion for $\widetilde{W}(t)$ becomes

$$\dot{\widetilde{W}}(t) = -i\widetilde{L}_1(t)\,\widetilde{W}(t) \ , \tag{2.20}$$

where $\widetilde{L}_1(t)$ is defined by

$$\widetilde{L}_1(t) = e^{iL_0 t}\,L_1(t)\,e^{-iL_0 t} \ . \tag{2.21}$$

We are not interested in the operator W(t) still containing the fluctuations but in the quantity $\rho(t) = \langle W(t)\rangle$ averaged over the fluctuations. Using the definition of the interaction picture we have:

$$\rho(t) = \langle W(t)\rangle = e^{-iL_0 t}\,\langle\widetilde{W}(t)\rangle = e^{-iL_0 t}\,\widetilde{\rho}(t) \ . \tag{2.22}$$

In the following we shall derive an equation of motion for $\widetilde{\rho}(t)$ starting from (2.20). From the mathematical point of view (2.20) is a stochastic differential equation of a multiplicative stochastic process /228,229,235/ with the additional complication that q numbers (or matrices) occur in it. The exact derivation of this equation is given in Appendix A using generalized cumulants. However, in order not to overload the presentation of the derivation of this equation with mathematical details, in this section we content ourselves with a perturbative derivation /45,236,237/. This simplified procedure may be justified by the fact that for infinitely short correlation times both derivations result in the same equation of motion as may be seen by comparison with (A.23).

Iterating (2.20) twice, we arrive at

$$\widetilde{W}(t) = \widetilde{W}(0) - i\int_0^t d\tau_1\,\widetilde{L}(\tau_1)\widetilde{W}(0) + i^2\int_0^t d\tau_1\int_0^{\tau_1} d\tau_2\,\widetilde{L}(\tau_1)\widetilde{L}(\tau_2)\widetilde{W}(0) \ . \tag{2.23}$$

Averaging over the statistical ensemble, with the help of (2.11) we get

$$\langle\widetilde{W}(t)\rangle - \widetilde{W}(0) = -\int_0^t d\tau_1 \int_0^{\tau_1} d\tau_2 \langle\widetilde{L}(\tau_1)\,\widetilde{L}(\tau_2)\rangle\,\widetilde{W}(0)\ . \tag{2.24}$$

We now differentiate this expression once with respect to time and arrive at

$$\frac{d}{dt}\,\widetilde{\rho}(t) = -\int_0^t d\tau\,\langle\widetilde{L}_1(t)\,\widetilde{L}_1(\tau)\rangle\,\widetilde{\rho}(t)\ . \tag{2.25}$$

In writing (2.25) with $\widetilde{\rho}(t)$ instead of $\widetilde{\rho}(0)$ on the right-hand side, it is assumed implicitly that the factorization of the correlation function in (2.24) is valid not only at the initial time but in the same way at an arbitrary time t. Therefore, the time development of the density operator is described in a correct manner for time intervals large as compared to the decay times of the correlation. Inserting L_1 from (2.10,17,18) and using (2.12) the equation of motion becomes

$$\frac{d}{dt}\,\widetilde{\rho}(t) = -\Lambda(n_1,n_1',n_2,n_2')\left(\widetilde{b}^+_{n_1}(t)\,\widetilde{b}_{n_1'}(t)\right)^{\times}\left(\widetilde{b}^+_{n_2}(t)\,\widetilde{b}_{n_2'}(t)\right)^{\times}\widetilde{\rho}(t)\ , \tag{2.26}$$

where the meaning of the crossed operators is defined in (2.16) and, according to a summation convention, repeated indices n_i are to be summed. Transforming back to the Schrödinger picture, we get

$$\dot{\rho}(t) = -iL_0\rho(t) - \Lambda(n_1,n_1',n_2,n_2')\,(b^+_{n_1}b_{n_1'})^{\times}\,(b^+_{n_2}b_{n_2'})^{\times}\,\rho(t)\ . \tag{2.27}$$

If we insert (2.13), we get the equation of motion for the density operator /37,38/:

$$\begin{aligned}\dot{\rho} = -iL_0\rho &- \sum_{n_1}\sum_{n_2}\gamma_{|n_1-n_2|}(b^+_{n_1}b_{n_2})^{\times}\,(b^+_{n_2}b_{n_1})^{\times}\,\rho(t)\\ &- \sum_{n_1}\sum_{n_2}\bar{\gamma}_{n_1-n_2}(1-\delta_{n_1n_2})\,(b^+_{n_1}b_{n_2})^{\times}\,(b^+_{n_1}b_{n_2})^{\times}\,\rho(t)\ .\end{aligned} \tag{2.28}$$

Assuming in addition that the density of excitons is low enough in order to neglect their interaction, we arrive at the following equation of motion for the density matrix $\rho_{nn'} = \langle n|\rho|n'\rangle$, where $|n\rangle$ denotes an excitation at site n /34-36/:

$$\begin{aligned}\dot{\rho}_{nn'} = &-i[H_0,\rho]_{nn'} - \delta_{nn'}\sum_{n_1}2\gamma_{|n-n_1|}(\rho_{nn}-\rho_{n_1n_1})\\ &-(1-\delta_{nn'})\,(2\Gamma\rho_{nn'} - 2\bar{\gamma}_{n-n'}\rho_{n'n})\ .\end{aligned} \tag{2.29}$$

In this equation we have introduced the abbreviation

$$\Gamma = \sum_{n_1}\gamma_{|n-n_1|}\ . \tag{2.30}$$

Instead of representing the equation of motion for the density operator in the local basis $|n\rangle$, we can introduce the basis $|k\rangle$, which diagonalizes the coherent part H_0 of the Hamiltonian:

$$|k\rangle = N^{-1/2} \sum_n e^{-ikn} |n\rangle \, . \tag{2.31}$$

N is the number of unit cells. If we also Fourier transform the operators and the parameters describing the coherent and incoherent interactions,

$$b_k = N^{-1/2} \sum_n e^{-ikn} b_n \, , \tag{2.32}$$

$$\widetilde{H}_k = \sum_m e^{-ikm} H_m \, , \tag{2.33}$$

$$\widetilde{\gamma}_k = N^{-1} \sum_m e^{-ikm} \gamma_m \, , \tag{2.34}$$

and an analogous expression for $\widetilde{\widetilde{\gamma}}_k$, we arrive at the following equation of motion for the density operator:

$$\begin{aligned} \dot{\rho} = &-i \sum_{k_1} \widetilde{H}_{k_1} (b^+_{k_1} b_{k_1})^x \rho \\ &- \sum_{k_1} \sum_{k_2} \sum_{k_3} (\widetilde{\gamma}_{k_2-k_3} + \widetilde{\widetilde{\gamma}}_{k_1+k_3}) (b^+_{k_1} b_{k_2})^x (b^+_{k_3} b_{k_1+k_3-k_2})^x \rho \, . \end{aligned} \tag{2.35}$$

In the delocalized basis $|k\rangle$ this equation reads /238/

$$\dot{\rho}_{k+q,k} = [-i(\widetilde{H}_{k+q} - \widetilde{H}_k) - 2\Gamma] \rho_{k+q,k} + \sum_{k_1} (\widetilde{\gamma}_q + \widetilde{\gamma}_{-q} + 2\widetilde{\widetilde{\gamma}}_{k+q+k_1}) \rho_{k_1+q,k_1} \, , \tag{2.36}$$

and especially for the diagonal elements we get

$$\dot{\rho}_{kk} = \sum_{k_1} 2(\widetilde{\gamma}_0 + \widetilde{\widetilde{\gamma}}_{k+k_1})(\rho_{k_1k_1} - \rho_{kk}) \, . \tag{2.37}$$

In writing (2.36,37) we have used $\Gamma = \sum_m \gamma_m = N\widetilde{\gamma}_0$, and from (2.13) we obtain $\sum_{k_1} \widetilde{\widetilde{\gamma}}_{k+k_1} = 0$, which allows us to write $\widetilde{\widetilde{\gamma}}_{2k} = - \sum_{k_1(\neq k)} \widetilde{\widetilde{\gamma}}_{k+k_1}$.

From (2.28) we see that in the local description the coherent part of the Hamiltonian results in the usual commutator, which generally contains diagonal and non-diagonal elements of the density operator. Considering the equation for the diagonal elements ρ_{nn}, we remark that the contribution of the incoherent part of the Hamiltonian may be interpreted as a rate equation. The hopping rate between sites a distance $n-n_1$ apart is given by $2\gamma_{|n-n_1|}$. In the equation of motion for the non-diagonal elements $\rho_{nn'}$, which describe phase relations between the sites n and n', the stochastic part of the Hamiltonian results in a decay of these phase relations with a decay constant 2Γ, which is independent of the sites. Furthermore, the fluctuating part of the Hamiltonian gives a coupling to $\rho_{n'n}$ determined by $\overline{\gamma}_{n-n'}$.

The equation of motion (2.36) in the delocalized representation shows that the wave number q is the same throughout the equation. This means that the original problem of dimension N^2 factorizes into N problems with fixed q and dimension N. The coherent Hamiltonian gives rise to a periodic motion with frequency $\tilde{H}_{k+q} - \tilde{H}_k$. This periodic motion is damped with a damping constant 2Γ, which is independent of k as well as of q. The coupling to the other matrix elements with the same q consists of two parts: the first one depends on q only and is therefore the same for one set of equations; the second one, $2\bar{\gamma}_{k+q+k_1}$, depends also on k_1 and may thus be different for each k_1.

The equation of motion for the diagonal elements ρ_{kk}, describing the probability of finding an excitation in the excitonic band state with wave number k, is given by (2.37) and may be interpreted as a master equation. The transition rate between states $|k_1\rangle$ and $|k\rangle$ again consists of two parts: the first one is an overall scattering rate, independent of k and k_1, the second one describes an individual scattering between initial and final band states. Another important point in this representation is that the equations of motion for diagonal and non-diagonal matrix elements are completely decoupled.

3. Coupled Coherent and Incoherent Exciton Motion in Molecular Pairs and on a Linear Chain

In this section the equation of motion for the density matrix (2.29) will be solved analytically for a pair of molecules /35,37,238/ and for an infinite linear chain of molecules /239-241/. Energy transfer via the motion of excitons in finite molecular aggregates has been investigated by ASLANGUL and KOTTIS /40/. On the basis of the analytical solutions the transition from the coherent to the incoherent range of motion is discussed. The theoretical results are of relevance in the dimer problem /132,136/ and for excitons moving in linear chains of 1,4-dibromonaphthalene /149/ and 1,2,4,5-tetrachlorobenzene /154-157/.

3.1 Exciton Motion in Molecule Pairs

We specialize the density matrix equation (2.29) for a system of two molecules, i.e. n,n' = {1,2}, and arrive at /37,38,238/ ($H_1 = J$)

$$\dot{\rho}_{11} = -2\gamma_1(\rho_{11} - \rho_{22}) - iJ(\rho_{21} - \rho_{12}) \tag{3.1a}$$

$$\dot{\rho}_{22} = -2\gamma_1(\rho_{22} - \rho_{11}) - iJ(\rho_{12} - \rho_{21}) \tag{3.1b}$$

$$\dot{\rho}_{12} = -2(\gamma_0 + \gamma_1)\rho_{12} + 2\bar{\gamma}_{-1}\rho_{21} - iJ(\rho_{22} - \rho_{11}) \tag{3.1c}$$

$$\dot{\rho}_{21} = -2(\gamma_0 + \gamma_1)\rho_{21} + 2\bar{\gamma}_1\rho_{12} + iJ(\rho_{22} - \rho_{11}) \tag{3.1d}$$

The first two equations describe the change of the occupation numbers of the excitons at the two molecules. From the last equations we get the change of the phase relation between them.

Let us first consider the limiting case of vanishing exchange interaction integral J, i.e. we consider the purely incoherent motion. Then the equations for the diagonal elements of the density matrix ρ_{11} and ρ_{22} are completely decoupled from the equations for the non-diagonal elements ρ_{12} and ρ_{21}.

The equations for the diagonal elements represent a set of rate equations for the occupation numbers of the excitons. The transition probability for an exciton between the two molecules is given by $2\gamma_1$, which means that it is determined by nonlocal fluctuations. On the other hand, from the equations for the non-diagonal elements ρ_{12} and ρ_{21} of the density matrix, we see that the excitonic phase relation between different sites decays exponentially with the exponent given by $2(\gamma_0+\gamma_1) = 2\Gamma$. Thus the phase of the exciton is destroyed both by local and non-local fluctuations. In this discussion we have neglected the influence of $\bar{\gamma}_{-1}$ and $\bar{\gamma}_1$, which is responsible for a coupling of the time evolution of ρ_{12} and ρ_{21}.

Now we consider the situation of non-vanishing exchange interaction integral. Then the equations for the diagonal and non-diagonal matrix elements are coupled and we have to solve a four-dimensional eigenvalue problem.

First, however, it is instructive to represent the equation of motion still in another basis,

$$|\pm\rangle = 2^{-1/2}(|1\rangle \pm |2\rangle) \,, \tag{3.2}$$

which diagonalizes the coherent part of the Hamiltonian

$$H_0 = \varepsilon(b_1^+b_1 + b_2^+b_2) + J(b_2^+b_1 + b_1^+b_2) \,. \tag{3.3}$$

Its eigenvalues are given by

$$E_\pm = \varepsilon \pm J \,. \tag{3.4}$$

The eigenvectors (3.2) correspond to the delocalized basis (2.31) and the equations of motion for the density matrix now read

$$\dot{\rho}_{++} = -(\gamma_0 + \gamma_1 - \bar{\gamma}_1)(\rho_{++} - \rho_{--}) \tag{3.5a}$$

$$\dot{\rho}_{--} = (\gamma_0 + \gamma_1 - \bar{\gamma}_1)(\rho_{++} - \rho_{--}) \tag{3.5b}$$

$$\dot{\rho}_{+-} = -(\gamma_0 + 3\gamma_1 + \bar{\gamma}_1 + i2J)\rho_{+-} + (\gamma_0 - \gamma_1 + \bar{\gamma}_1)\rho_{-+} \tag{3.5c}$$

$$\dot{\rho}_{-+} = -(\gamma_0 + 3\gamma_1 + \bar{\gamma}_1 - i2J)\rho_{-+} + (\gamma_0 - \gamma_1 + \bar{\gamma}_1)\rho_{+-} \,. \tag{3.5d}$$

In writing (3.5) we have assumed that $\bar{\gamma}_1$ is real, i.e. $\bar{\gamma}_1 = \bar{\gamma}_{-1}$. In this basis, the equations for the diagonal and non-diagonal elements of the density matrix are decoupled. Equations (3.5a,b) represent a set of rate equations that describe the change of the occupation probabilities of the eigenstates $|\pm\rangle$ of H_0. The transition rate is given by $(\gamma_0+\gamma_1-\bar{\gamma}_1)$. From the comparison with experimental data, $\gamma_0 \gg \gamma_1, \bar{\gamma}_1$, except at very low temperatures /42,43/. Therefore, the transition rate between the upper and lower energy eigenstates of H_0 is determined by local (energy) fluctuations. The fluctuation parameters as well as the exchange interaction integral enter into the equations for the non-diagonal elements.

The eigenvalues of (3.1) or (3.5) are given by

$$\lambda_1 = 0 \, , \tag{3.6a}$$

$$\lambda_2 = 2(\gamma_0 + \gamma_1 - \bar{\gamma}_1) \, , \tag{3.6b}$$

$$\left.\begin{matrix}\lambda_3 \\ \lambda_4\end{matrix}\right\} = \gamma_0 + 3\gamma_1 + \bar{\gamma}_1 \pm \left[(\gamma_0 - \gamma_1 + \bar{\gamma}_1)^2 - 4J^2\right]^{1/2} \, , \tag{3.6c,d}$$

and the (unnormalized) eigenvectors are represented in Table 3.1 in the delocalized basis and in Table 3.2 in the localized system /35,238/. The square root is given by the square bracket of (3.6c,d).

Table 3.1. Eigenvectors in the delocalized basis system

	$\rho_d^{(1)}$	$\rho_d^{(2)}$	$\rho_d^{(3)}$	$\rho_d^{(4)}$
ρ_{++}	1	1	0	0
ρ_{--}	1	-1	0	0
ρ_{+-}	0	0	$\gamma_0-\gamma_1+\bar{\gamma}_1$	$\gamma_0-\gamma_1+\bar{\gamma}_1$
ρ_{-+}	0	0	$i2J - \sqrt{}$	$i2J + \sqrt{}$

Table 3.2. Eigenvectors in the localized basis system

	$\rho_\ell^{(1)}$	$\rho_\ell^{(2)}$	$\rho_\ell^{(3)}$	$\rho_\ell^{(4)}$
ρ_{11}	1	0	$i2J$	$i2J$
ρ_{22}	1	0	$-i2J$	$-i2J$
ρ_{12}	0	1	$-(\gamma_0-\gamma_1+\bar{\gamma}_1+\sqrt{})$	$-(\gamma_0-\gamma_1+\bar{\gamma}_1-\sqrt{})$
ρ_{21}	0	1	$(\gamma_0-\gamma_1+\bar{\gamma}_1+\sqrt{})$	$(\gamma_0-\gamma_1+\bar{\gamma}_1-\sqrt{})$

The general solution of the density matrix equation is given by a superposition of these eigensolutions

$$\rho(t) = \sum_{i=1}^{4} a_i \rho^{(i)} e^{-\lambda_i t} . \tag{3.7}$$

We shall now consider two different initial situations. First, we assume that at the initial time the two molecule system is in its upper eigenstate with energy E_+. In this case $\rho_{++}(0) = 1$ and all the other matrix elements of the density operator in the delocalized basis vanish. The time evolution of the system is then described by the following density operator:

$$\rho(t) = \frac{1}{2} \rho_d^{(1)} + \frac{1}{2} \rho_d^{(2)} e^{-\lambda_2 t} . \tag{3.8}$$

This equation shows that the initial difference in the occupation probabilities between the states $|+\rangle$ and $|-\rangle$ decays exponentially. For sufficiently long times the upper and lower states have the same occupation probability. From Table 3.2, however, we see that for the same initial condition for all times the excitation is distributed among the two molecules with the same probability and that phase relations between the two molecules, described by the non-diagonal matrix elements ρ_{12} and ρ_{21}, decay exponentially.

As a second example we now assume that at the initial time the excitation is completely localized at molecule 1, i.e. $\rho_{11}(0) = 1$ and all other matrix elements disappear. We then get /35,238/

$$\rho(t) = \frac{1}{2} \rho_\ell^{(1)} + i \frac{\gamma_0 - \gamma_1 + \bar{\gamma}_1 - \sqrt{}}{8J\sqrt{}} \rho_\ell^{(3)} e^{-\lambda_3 t} - i \frac{\gamma_0 - \gamma_1 + \bar{\gamma}_1 + \sqrt{}}{8J\sqrt{}} \rho_\ell^{(4)} e^{-\lambda_4 t} . \tag{3.9}$$

The eigenvalues λ_3 and λ_4 are given by (3.6c,d) and

$$\sqrt{} = [(\gamma_0 - \gamma_1 + \bar{\gamma}_1)^2 - 4J^2]^{1/2} . \tag{3.10}$$

Considering λ_3 and λ_4, we remark that for small values of the fluctuations γ_0, γ_1, $\bar{\gamma}_1$, i.e. for $(\gamma_0 - \gamma_1 + \bar{\gamma}_1) < 2J$, the square root (3.10) becomes imaginary, the eigenvalues λ_3 and λ_4 become complex, and the time development of the density matrix shows damped oscillations. This is especially true for the occupation probabilities ρ_{11} and ρ_{22} at the two molecules, which means that they approach their steady-state value $\rho_{11} = \rho_{22} = 1/2$ by damped oscillations. On the other hand, for $\gamma_0 - \gamma_1 + \bar{\gamma}_1 > 2J$ the square root is real and the occupation numbers and the phase relations between the molecules reach their steady-state values without oscillations. Therefore, it is reasonable to denote the exciton motion in the first case as coherent and in the second as incoherent. This criterion for the classification of the exciton motion is due to HAKEN and STROBL. Originally /34-36/, it was assumed that $\gamma_1 = \bar{\gamma}_1$, and in this case the criterion simplifies to $\gamma_0 \leq 2J$. For small values of the local fluctuations γ_0 - and with respect to (2.9-13) this means for low temperatures - the exciton motion is coherent, in the other case it is incoherent. Considering, for completeness, the delocalized representation also for this initial condition, we see that $\rho_d^{(3)}$ and $\rho_d^{(4)}$ have no diagonal components. Therefore, ρ_{++} and ρ_{--} are time independent and the exciton is found with equal probability in the states $|+\rangle$ and $|-\rangle$.

3.2 Exciton Motion on a Linear Chain of Molecules

3.2.1 The Equation of Motion

We start from the density matrix equation (2.29) for the coupled coherent and incoherent exciton motion, which will now be applied to a linear chain of molecules. We assume periodic boundary conditions, where N is the number of molecules in the range of periodictiy, and let N increase to infinity.

The ansatz .

$$\rho_{nn'}(t) = e^{Rt} \rho_{nn'} \tag{3.11}$$

transforms the set of equations of motion into an eigenvalue problem. This set of N^2 equations is invariant with respect to translations along the chain. Using Bloch's theorem, we can write /34,36/

$$\rho^k_{nn'} = e^{ikn} \rho^k_{n-n'} , \quad -\pi < k \leq \pi , \tag{3.12}$$

where k is a wave vector of the first Brillouin zone. With the help of (3.12) the set of N^2 coupled equations is reduced to N uncoupled sets of N equations, where each set is characterized by a definite wave vector. With respect to the following calculations, the transformation /44,239/

$$\rho^k_{n-n'} = i^{n-n'} e^{-ik(n-n')/2} \bar{\rho}^k_{n-n'} \tag{3.13}$$

will prove convenient. From (2.29) we then get

$$\begin{aligned} R^k \bar{\rho}^k_m = &- \sum_{m'} 2i^{m'-m} H_{m-m'} \sin[\tfrac{k}{2}(m-m')] \bar{\rho}^k_{m'} - 2\Gamma\bar{\rho}^k_m \\ &+ 2\delta_{m0} \sum_{m'} \gamma_{m'} e^{-ikm'} \bar{\rho}^k_0 + 2(1-\delta_{m0}) \bar{\gamma}_m (-1)^m \bar{\rho}^k_{-m} . \end{aligned} \tag{3.14}$$

Taking into account the interaction between nearest neighbours only and using

$$r^k = R^k + 2\Gamma , \tag{3.15}$$

we get the following non-hermitean eigenvalue problem:

$$r^k \bar{\rho}^k = {}^k_1D\, \bar{\rho}^k + {}^k_2D\, \bar{\rho}^k \,. \tag{3.16}$$

${}^k_1D$ and ${}^k_2D$ are N-dimensional matrices describing the coherent and incoherent interactions ($H = H_1$):

$${}^k_1D_{mm'} = \frac{i}{2}\, C_k(\delta_{m',m+1} + \delta_{m',m-1}) \,, \tag{3.17}$$

$${}^k_2D_{mm'} = 2\tilde{\tilde{\gamma}}_k\, \delta_{m0}\, \delta_{m'0} - 2\bar{\gamma}_1(\delta_{m,1} + \delta_{m,-1})\, \delta_{m',-m} \,, \tag{3.18}$$

$$C_k = 4H \sin\frac{k}{2} \,,$$

$$\tilde{\tilde{\gamma}}_k = N\tilde{\gamma}_k = \gamma_0 + 2\gamma_1 \cos k \,.$$

3.2.2 Solution of the Eigenvalue Equation

On account of transformation (3.13) the matrices ${}^k_1D$ and ${}^k_2D$ are invariant with respect to inversion, and therefore from (3.16) we obtain symmetric and antisymmetric solutions. Furthermore, ${}^k_2D$ has matrix elements only in the neighbourhood of $m = m' = 0$, which suggests that (3.16) could be solved by a Green's function method /34/. Inserting

$$\bar{\rho}^k = -\, G\, {}^k_2D\, \bar{\rho}^k \tag{3.19}$$

into (3.16), the Green's function G has to satisfy the equation

$$({}^k_1D - r^k)G = I \,, \tag{3.20}$$

where I is the N-dimensional unit matrix. The formal solution of (3.20) is given by

$$G = ({}^k_1D - r^k)^{-1} \,. \tag{3.21}$$

This expression shows that G, just as ${}^k_1D$, satisfies periodic boundary conditions and is invariant with respect to inversion. With these conditions the Green's function is determined in Appendix B /44/ from (3.20). Expressing the eigenvalue r^k by the complex variable $z = x + iy$,

$$r^k = iC_k \cos z \,, \tag{3.22}$$

and using (3.17), equation (3.20) for the Green's function becomes

$$G_{m+1}(z) + G_{m-1}(z) - 2\cos z\, G_m(z) = \frac{2}{iC_k}\, \delta_{m0} \,. \tag{3.23}$$

The solution of this inhomogeneous difference equation is given in (B.8) of Appendix B:

$$G_m(z) = (iC_k \sin z \sin \tfrac{N}{2}z)^{-1} \cos[(\tfrac{N}{2} - |m|)z] \ . \tag{3.24}$$

Inserting (3.18) and (3.24) into (3.19) and neglecting $\bar{\gamma}_1$ as compared to $\tilde{\bar{\gamma}}_k$, we get

$$\bar{\rho}_m^k = -G_m(z)\, 2\tilde{\bar{\gamma}}_k\, \bar{\rho}_0^k \ . \tag{3.25}$$

For m = 0 we arrive at an equation for the eigenvalues z:

$$2\tilde{\bar{\gamma}}_k\, G_0(z) + 1 = 0 \ , \tag{3.26}$$

which reads explicitly

$$\tan \tfrac{N}{2}z = i\, \frac{2\tilde{\bar{\gamma}}_k}{C_k \sin z} = iA(k,z) \ . \tag{3.27}$$

In the following we shall see that the quantity A(k,z), the complex degree of incoherence /240,241/, plays a decisive role for the discussion of the transition from coherent to incoherent exciton motion. For $z = \pi/2$ and small values of k we may write

$$A(k\,,\tfrac{\pi}{2}) = \frac{2\tilde{\bar{\gamma}}_0}{2Hk} = \frac{1}{\tau v} \ . \tag{3.28}$$

$\tau = (2\tilde{\bar{\gamma}}_0)^{-1} = (2\Gamma)^{-1}$ is the time between two scattering events of an exciton, as may be seen from (2.36), and v is the group velocity. The exciton motion may be denoted as coherent if $\tau v > 1$, i.e. if the distance between two scattering events of the exciton is larger than the lattice constant. In the case of coherent exciton motion $A(k,\pi/2) < 1$; for an incoherently moving exciton $A(k,\pi/2) > 1$. The transition from coherent to incoherent exciton motion occurs at $A(k,\pi/2) = 1$. For given parameters $\tilde{\bar{\gamma}}_0$ and H this transition still depends on k, which means that in a given crystal and at a given temperature some excitons move in a coherent manner, whereas for excitons with other k values coherence has been lost on account of phase-destroying processes.

We now proceed in solving the eigenvalue problem. Splitting (3.27) into real and imaginary parts, we get

$$\begin{aligned} 2\tilde{\bar{\gamma}}_k \cos\tfrac{N}{2}x \cosh\tfrac{N}{2}y - C_k(&\sin x \ \cosh y \ \cos\tfrac{N}{2}x \ \sinh\tfrac{N}{2}y \\ &+ \cos x \ \sinh y \ \sin\tfrac{N}{2}x \ \cosh\tfrac{N}{2}y) = 0 \ , \end{aligned} \tag{3.29}$$

$$\begin{aligned} 2\tilde{\bar{\gamma}}_k \sin\tfrac{N}{2}x \sinh\tfrac{N}{2}y - C_k(&\sin x \ \cosh y \ \sin\tfrac{N}{2}x \ \cosh\tfrac{N}{2}y \\ &- \cos x \ \sinh y \ \cos\tfrac{N}{2}x \ \sinh\tfrac{N}{2}y) = 0 \ . \end{aligned} \tag{3.30}$$

We consider a chain of molecules of infinite length and look for solutions first with y = 0 and then with $y \neq 0$.

(a) In the case $y = 0$ we get from (3.29)

$$x = (2n + 1)\frac{\pi}{N} \tag{3.31}$$

and from (3.30)

$$x = 2n\frac{\pi}{N} \quad , \tag{3.32}$$

where $-N/2 < n \leq N/2$. For finite values of N the two solutions are not compatible which each other and no solutions with $y = 0$ exist. In the limit $N \to \infty$, however, the difference between the two solutions disappears and we get a continuum of solutions with $-\pi < x \leq +\pi$, $y = 0$. From the investigation of the eigenvectors we shall see that eigenvectors with $x = \pm\alpha$ are linearly dependent. Therefore we have to consider only the following range of eigenvalues:

$$0 < x < \pi \ , \quad y = 0 \quad . \tag{3.33}$$

(To complete the set of eigenvectors, we need those solutions which are antisymmetric with respect to inversion. They are not influenced by ${}^{k}_{2}D$ in the approximation considered here ($\bar{\gamma}_1 = 0$).)

In this case from (3.15,22) we get a continuum of eigenvalues

$$r^k = iC_k \cos z \ , \tag{3.34}$$

$$R^k = -2\Gamma + i4H \sin\frac{k}{2} \cos z \quad . \tag{3.35}$$

(b) In the second case $y \neq 0$ we get from (3.29,30)

$$\cos x = 0 \ , \quad \cosh y = |2\tilde{\tilde{\gamma}}_k / C_k| \ , \tag{3.36}$$

and with the help of (3.15,22) we obtain for this additional eigenvalue[1]

$$r^k = \left[(2\tilde{\tilde{\gamma}}_k)^2 - C_k^2\right]^{1/2} \tag{3.37}$$

$$R^k = -2\Gamma + \left[(2\tilde{\tilde{\gamma}}_k)^2 - C_k^2\right]^{1/2} \tag{3.38}$$

On account of $\cosh y > 1$ for $y \neq 0$ and (3.36), this additional solution exists only for

$$|2\tilde{\tilde{\gamma}}_k| > |4H \sin\frac{k}{2}| \quad . \tag{3.39}$$

(b1) If $2\tilde{\tilde{\gamma}}_k > 4H$, the additional solution exists for all values of k, i.e. $-\pi < k \leq \pi$.

(b2) If $2\tilde{\tilde{\gamma}}_k < 4H$, we obtain a value $k = \bar{k}$ from

$$|2\tilde{\tilde{\gamma}}_{\bar{k}}| = |4H \sin\frac{\bar{k}}{2}| \ , \tag{3.40}$$

and the inequality (3.39) is satisfied for all values of k in the range

$$-\bar{k} < k < +\bar{k} \quad . \tag{3.41}$$

---

[1] In the case $\gamma_0 > \gamma_1$, which is considered in the present section, only the positive square root represents an eigenvalue. A detailed investigation (K. KASSNER, private communication) shows that in the physically less important parameter range of $\gamma_1 > \gamma_0$ the negative square root may become a solution.

Now we have to determine the eigenvectors. Using (3.24,27), the equation for the eigenvectors (3.25) may be written in the following way /240,241/:

$$\bar{\rho}_m^k = \frac{1}{2}\{[1+A(k,z)]\, e^{i|m|z} + [1-A(k,z)]\, e^{-i|m|z}\}\, \bar{\rho}_0^k \quad . \tag{3.42}$$

Inserting the eigenvalues corresponding to (3.33,36), we get a continuous set of eigenvectors and the eigenvector corresponding to the discrete eigenvalue, respectively.

Because the operator ${}_1^kD + {}_2^kD$ is not hermitean, the eigenvectors $\bar{\rho}^{k\ell}$ together with the eigenvectors $\bar{\eta}^{k\ell'}$ of the hermitean adjoint operator $({}_1^kD + {}_2^kD)^+$ form a biorthogonal set. On account of the symmetry of the matrices $({}_1^kD + {}_2^kD)_{mm'} = ({}_1^kD + {}_2^kD)_{m'm}$ it may easily be shown that $\bar{\eta}_m^{k\ell} = (\bar{\rho}_m^{k\ell})^*$. Therefore, the normalization condition of the continuous set of eigenvectors may be written in the following way:

$$(\bar{\eta}^{k\ell})^+\, \bar{\rho}^{k\ell'} = \sum_m (\bar{\eta}_m^{k\ell})^*\, \bar{\rho}_m^{k\ell'} = \sum_m \bar{\rho}_m^{k\ell}\, \bar{\rho}_m^{k\ell'} = \delta(\ell - \ell') \; , \tag{3.43}$$

whereas the eigenvector of the additional eigenvalue is normalized to 1. In this way the normalized eigensolutions are determined and summarized in the following.

Normalized symmetric band solutions

$$R^{k\ell} = -2\Gamma + i4H \sin\frac{k}{2} \cos\ell \quad , \tag{3.44}$$

$${}^s\bar{\rho}_m^{k\ell} = N_{k\ell}^{-1}\,[(4H \sin\frac{k}{2} \sin\ell + 2\tilde{\tilde{\gamma}}_k)e^{i\ell|m|} + (4H \sin\frac{k}{2} \sin\ell - 2\tilde{\tilde{\gamma}}_k)e^{-i\ell|m|}] \; , \tag{3.45}$$

$$N_{k\ell} = (4\pi)^{1/2}\,[(4H \sin\frac{k}{2} \sin\ell)^2 - (2\tilde{\tilde{\gamma}}_k)^2]^{1/2} \quad , \tag{3.46}$$

$$-\pi < k \leqslant \pi \, , \qquad 0 \leqslant \ell \leqslant \pi \quad . \tag{3.47}$$

Normalized symmetric additional solution

$$R^{ks} = -2\Gamma + [(2\tilde{\tilde{\gamma}}_k)^2 - (4H \sin\frac{k}{2})^2]^{1/2} \quad , \tag{3.48}$$

$$\bar{\rho}_m^{ks} = N_{ks}'^{-1}(i4H \sin\frac{k}{2})^{-|m|} \left\{[(2\tilde{\tilde{\gamma}}_k)^2 - (4H \sin\frac{k}{2})^2]^{1/2} - 2\tilde{\tilde{\gamma}}_k\right\}^{|m|} \quad , \tag{3.49}$$

$$N_{ks}' = (2\tilde{\tilde{\gamma}}_k)^{-1/2}\,[(2\tilde{\tilde{\gamma}}_k)^2 - (4H \sin\frac{k}{2})^2]^{1/4} \quad . \tag{3.50}$$

Normalized antisymmetric band solution

$$R^k = -2\Gamma + i4H \sin\frac{k}{2} \cos\ell \quad , \tag{3.51}$$

$${}^a\bar{\rho}_m^{k\ell} = \pi^{-1/2} \sin\ell m \quad . \tag{3.52}$$

The general solution is given as a superposition of the eigensolutions:

$$\bar{\rho}^k_m(t) = a_{ks}\, e^{R^{ks}t}\, \bar{\rho}^{ks}_m + \int_0^\pi d\ell\, e^{R^{k\ell}t}\, (a_{k\ell}\, {}^s\bar{\rho}^{k\ell}_m + b_{k\ell}\, {}^a\bar{\rho}^{k\ell}_m) \quad . \tag{3.53}$$

The first term is present only if the inequality (3.39) is satisfied. With the help of (3.12,13) we finally get for the general solution

$$\begin{aligned}\rho_{nn'}(t) = &\int_{-\bar{k}}^{\bar{k}} dk\, e^{ik(n+n')/2}\, i^{(n-n')}\, a_{ks}\, e^{R^{ks}t}\, \bar{\rho}^{ks}_{n-n'} \\ &+ \int_{-\pi}^{\pi} dk \int_0^\pi d\ell\, e^{ik(n+n')/2}\, i^{(n-n')}\, e^{R^{k\ell}t}\, (a_{k\ell}\, {}^s\bar{\rho}^{k\ell}_{n-n'} + b_{k\ell}\, {}^a\bar{\rho}^{k\ell}_{n-n'}) \;,\end{aligned} \tag{3.54}$$

where $\bar{k}$ is determined from (3.39,40). The expansion coefficients have to be determined from the initial condition.

3.2.3 Time Development of an Initially Localized Excitation

Let us consider the time development of an excitation which is initially localized at the origin, i.e. we assume $\rho_{nn'}(0) = \delta_{n0}\delta_{n'0}$. This transforms to

$$\bar{\rho}^k_m(0) = (2\pi)^{-1}\, \delta_{m,0} \quad , \tag{3.55}$$

and using the orthonormality condition (3.43), we obtain

$$a_{ks} = (2\pi)^{-1}\, \rho^{ks}_0 \;, \tag{3.56a}$$

$$a_{k\ell} = (2\pi)^{-1}\, {}^s\rho^{k\ell}_0 \;, \tag{3.56b}$$

$$b_{k\ell} = 0 \quad . \tag{3.56c}$$

We consider in the following only the diagonal matrix elements of the density operator, which represent the time-dependent probability of finding an exciton at lattice site n. With the help of (3.56) we obtain

$$\begin{aligned}\rho_{nn}(t) = &\int_{-\bar{k}}^{\bar{k}} dk\, e^{ikn}\, (2\pi)^{-1}\, 2\tilde{\tilde{\gamma}}_k \left[(2\tilde{\tilde{\gamma}}_k)^2 - (4H \sin\tfrac{k}{2})^2\right]^{-1/2} e^{R^{ks}t} \\ &+ \int_{-\pi}^{\pi} dk \int_0^\pi d\ell\, e^{ikn}\, \tfrac{1}{8}\, \pi^{-2} (8H \sin\tfrac{k}{2}\, \sin\ell)^2 \left[(4H \sin\tfrac{k}{2}\, \sin\ell)^2\right. \\ &\left. - (2\tilde{\tilde{\gamma}}_k)^2\right]^{-1} \cdot e^{R^{k\ell}t} \quad .\end{aligned} \tag{3.57}$$

In the purely coherent case the first term disappears. Evaluating the double integral /242/, we get /31/

$$\rho_{nn}(t) = J_n^2(2Ht) \quad , \tag{3.58}$$

where $J_n(2Ht)$ is the Bessel function of order n. In the purely incoherent case, the second term disappears and the eigenvalue R^{ks} becomes for $H = 0$

$$R^{ks} = -4\gamma_1(1 - \cos k) \quad . \tag{3.59}$$

Evaluating the integral for $\bar{k} = \pi$, we have /242/

$$\rho_{nn}(t) = e^{-4\gamma_1 t} I_n(4\gamma_1 t) \quad . \tag{3.60}$$

In this expression $I_n(4\gamma_1 t)$ is the modified Bessel function.

Expanding the eigenvalue to lowest order in the coherent interaction matrix element H, we get

$$R^{ks} = -2(2\gamma_1 + H^2/\tilde{\tilde{\gamma}}_k)(1 - \cos k) \quad . \tag{3.61}$$

The comparison with (3.59) shows that for $\rho_{nn}(t)$ we arrive again at (3.60) if $\tilde{\tilde{\gamma}}_k$ may be approximated by its value at $k = 0$; only the hopping rate $2\gamma_1$ has to be replaced by $2\gamma_1 + H^2/\tilde{\tilde{\gamma}}_0$. In the continuum approximation /240/ we may replace $2(1-\cos k)$ by k^2 and the evaluation of the integral for $\bar{k} \to \infty$ (which is a good approximation for long enough times) results in

$$\rho_{nn}(t) = \frac{1}{2}\,\frac{1}{\sqrt{\pi D t}}\, e^{-n^2/4Dt} \quad , \tag{3.62}$$

with the diffusion constant D given by

$$D = 2\gamma_1 + H^2/\Gamma \quad . \tag{3.63}$$

The discussion above shows that for not too long times the occupation probability is determined mainly by the second term of (3.57) if the exciton motion is predominantly coherent. In the incoherent limit the main contribution to $\rho_{nn}(t)$ stems from the first integral in (3.57). For long times the exciton motion is determined by the first term in (3.57) because the real part of the band eigenvalues is larger than the real part of additional solution. This, however, means that for long enough times the exciton motion is diffusion-like.

4. Mean-Square Displacement and Diffusion Equation of Excitons

In this section a set of equations of motion for the moments $\langle n^s \rangle = \sum_n n^s \rho_{nn}$ of the exciton motion is derived /92/ starting from the density matrix equation (2.29). From this set the second moment, the mean-square displacement, is calculated explicitly and the general case as well as the limiting cases of the purely coherent and the purely incoherent motion are discussed. In the second part of this section the conditions are derived, under which the exciton motion may be described by a diffusion equation /35,93,93/.

4.1 Moments of the Exciton Motion

In Sect. 2 we have derived the density matrix equation (2.29) which describes the coupled coherent and incoherent exciton motion. Multiplying the equation for the diagonal element ρ_{nn} by n^s and summing over all sites, for a molecular chain of infinite length with inversion symmetry, i.e. $H_m = H_{-m}$, we get /92/

$$\begin{aligned} \frac{d}{dt} \sum_n n^s \rho_{nn} &= 2 \sum_{t=1}^{s} \binom{s}{t} \sum_m m^t \gamma_m \sum_n n^{s-t} \rho_{nn} \\ &+ \sum_{t=1}^{s} \binom{s}{t} \sum_{m>0} m^t H_m \sum_n n^{s-t} i(\rho_{n+m,n} - \rho_{n,n+m}) \ . \end{aligned} \tag{4.1}$$

In this equation the sth moment is coupled to all lower moments, but on account of the coherent part of the motion, also to the quantities $\sum_n n^{s-t} i(\rho_{n+m,n} - \rho_{n,n+m})$. The equation for this quantity reads

$$\begin{aligned} \frac{d}{dt} \sum_n n^s i(\rho_{n+m,n} - \rho_{n,n+m}) &= -2(\Gamma + \overline{\gamma}_m) \sum_n n^s i(\rho_{n+m,n} - \rho_{n,n+m}) \\ &-2 \sum_{t=1}^{s} \binom{s}{t} (-m)^t H_m \sum_n n^{s-t} \rho_{nn} \\ &- \sum_{t=1}^{s} \binom{s}{t} \sum_{r(\neq -m)} r^t H_r \sum_n n^{s-t} (\rho_{n+m+r,n} + \rho_{n,n+m+r}) \ . \end{aligned} \tag{4.2}$$

To arrive at a complete set of equations, another equation is needed for the quantity in the last term of (4.2)

$$\begin{aligned} \frac{d}{dt} \sum_n n^s (\rho_{n+m,n} + \rho_{n,n+m}) &= -2(\Gamma - \overline{\gamma}_m) \sum_n n^s (\rho_{n+m,n} + \rho_{n,n+m}) \\ &+ \sum_{t=1}^{s} \binom{s}{t} \sum_r r^t H_r \sum_n n^{s-t} i(\rho_{n+m+r,n} - \rho_{n,n+m+r}) \ . \end{aligned} \tag{4.3}$$

Equations (4.1-3) represent the complete set of equations which allows calculation of the moments of the exciton motion in a recursive manner.

An important quantity in considering transport processes is the second moment, i.e. the mean-square displacement, $\sum_n n^2 \rho_{nn}$. Using the notation

$$M_2 = \sum_n n^2 \rho_{nn} \ , \tag{4.4}$$

$$V_m = -\sum_n (2n+m)\, i\, (\rho_{n+m,n} - \rho_{n,n+m}) \quad , \tag{4.5}$$

$$W_m = \sum_n (\rho_{n+m,n} + \rho_{n,n+m}) \quad , \tag{4.6}$$

we obtain from (4.1-3) with s = 2

$$\dot{M}_2 = 2\sum_m m^2\gamma_m - \frac{1}{2}\sum_m m\, H_m V_m \quad , \tag{4.7}$$

$$\dot{V}_m = -2(\Gamma+\bar{\gamma}_m)V_m - 4m\, H_m + \sum_{r(\neq -m)} 2r\, H_r W_{m+r} \quad , \tag{4.8}$$

$$\dot{W}_m = -2(\Gamma-\bar{\gamma}_m)W_m \quad . \tag{4.9}$$

The general solution of this system of differential equations is

$$\begin{aligned} M_2(t) = M_2(0) &+ \left[2\sum_m m^2\gamma_m + \sum_m m^2\, H_m^2(\Gamma+\bar{\gamma}_m)^{-1}\right]t \\ &+ \frac{1}{2}\sum_m m^2\, H_m^2(\Gamma+\bar{\gamma}_m)^{-2}\, (e^{-2(\Gamma+\bar{\gamma}_m)t} - 1) \\ &+ \frac{1}{4}\sum_m m\, H_m(\Gamma+\bar{\gamma}_m)^{-1}\, V_m(0)\, (e^{-2(\Gamma+\bar{\gamma}_m)t} - 1) \\ &+ \frac{1}{4}\sum_m \sum_{r(\neq -m)} m\, r\, H_m H_r(\Gamma-\bar{\gamma}_{m+r})^{-1}\, (\bar{\gamma}_m+\bar{\gamma}_{m+r})^{-1}\, W_{m+r}(0)\, (e^{-2(\Gamma-\bar{\gamma}_{m+r})t} - 1) \\ &- \frac{1}{4}\sum_m \sum_{r(\neq -m)} m\, r\, H_m\, H_r(\Gamma+\bar{\gamma}_m)^{-1}\, (\bar{\gamma}_m+\bar{\gamma}_{m+r})^{-1}\, W_{m+r}(0)\, (e^{-2(\Gamma+\bar{\gamma}_m)t} - 1) \; . \end{aligned} \tag{4.10}$$

This general solution for the mean-square displacement is of importance in the calculation of photoconductivity in organic molecular crystals /222-224/. In the special case of the exciton being at site n = 0 at the initial time, i.e. $\rho_{nn'}(0) = \delta_{n0}\,\delta_{n'0}$ and therefore $M_2(0) = V_m(0) = W_m(0) = 0$, we arrive at

$$\begin{aligned} M_2(t) &= \left[2\sum_m m^2\gamma_m + \sum_m m^2\, H_m^2\, (\Gamma+\bar{\gamma}_m)^{-1}\right]t \\ &+ \frac{1}{2}\sum_m m^2\, H_m^2\, (\Gamma+\bar{\gamma}_m)^{-2}\, (e^{-2(\Gamma+\bar{\gamma}_m)t} - 1) \quad . \end{aligned} \tag{4.11}$$

If only the interaction with nearest neighbours is important, we get

$$M_2(t) = [4\gamma_1 + 2H_1^2(\Gamma+\bar{\gamma}_1)^{-1}]t + H_1^2(\Gamma+\bar{\gamma}_1)^{-2}\, (e^{-2(\Gamma+\bar{\gamma}_1)t} - 1) \quad . \tag{4.12}$$

This result has also been derived in /91/ starting from the Fourier transform of the solution of the density matrix.

From (4.11) we obtain the mean-square displacement for the purely incoherent motion as

$$M_2(t) = 2 \sum_m m^2 \gamma_m t \quad , \tag{4.13}$$

and in the completely coherent case we have

$$M_2(t) = \sum_m m^2 H_m^2 t^2 \quad . \tag{4.14}$$

This quadratic time dependence is well known from MERRIFIELD's result /31/. For large times we may neglect the exponential expression in (4.11) and arrive at

$$M_2(t) \approx \left[2 \sum_m m^2 \gamma_m + \sum_m m^2 H_m^2 (\Gamma + \bar{\gamma}_m)^{-1}\right] t - \frac{1}{2} \sum_m m^2 H_m^2 (\Gamma + \bar{\gamma}_m)^{-2} \quad . \tag{4.15}$$

Using the definition of the diffusion constant

$$2D = \lim_{t \to \infty} t^{-1} M_2(t) \quad , \tag{4.16}$$

from (4.15) we obtain

$$2D = \sum_m m^2 \left[2\gamma_m + H_m^2 (\Gamma + \gamma_m)^{-1}\right] \quad . \tag{4.17}$$

The first part of the diffusion constant depends only on the strength of the fluctuations and increases when their strength increases, i.e. with increasing temperature. The second term is also influenced by the coherent interaction matrix element and decreases with increasing strength of the fluctuations.

Expanding the exponential to second order in t, from (4.11) we get

$$M_2(t) \approx \left(\sum_m 2m^2 \gamma_m\right) t + \left(\sum_m m^2 H_m^2\right) t^2 \quad . \tag{4.18}$$

The linear time dependence for very short times is an artifact of the model having its origin in the infinitely short correlation time of the δ-correlated stochastic process /40,73/. For finite correlation times the exciton motion for very short time intervals is coherent and therefore the mean-square displacement has to increase quadratically with time. The influence of finite correlation times on the diffusion constant, i.e. on the long-time behaviour, has been investigated in /233/. Replacing (2.12) by

$$\langle h_{n_1 n_1'}(t_1)\, h_{n_2 n_2'}(t_2) \rangle = \Lambda(n_1, n_1', n_2, n_2') \tau_c^{-1} \exp(-|t_1 - t_2|/\tau_c) \quad , \tag{4.19}$$

to lowest order in τ_c the diffusion constant becomes

$$2D = J^2/\gamma_0 + (\tau_c J)^2\, 3J^2/\gamma_0 \tag{4.20}$$

if only local fluctuations are taken into account, and

$$2D = 4\gamma_1 + (\Gamma + \bar{\gamma}_1)^{-1} J^2 - \tau_c(\Gamma + \bar{\gamma}_1)^{-1} 8J^2\bar{\gamma}_1 \quad , \tag{4.21}$$

if also non-local fluctuations are considered. It is seen that in connection with local fluctuations the finite correlation time increases the diffusion constant, whereas it decreases if non-local fluctuations are taken into account /233/.

4.2 Derivation of a Diffusion Equation

4.2.1 Master Equation for the Exciton Motion

We separate (2.29) for diagonal and non-diagonal elements of the density operator and arrive at

$$\dot{\rho}_{\underline{n}\underline{n}} = -i \sum_{\underline{n}'} H_{\underline{n}-\underline{n}'}\rho_{\underline{n}'\underline{n}} + i \sum_{\underline{n}'} \rho_{\underline{n}\underline{n}'}H_{\underline{n}'-\underline{n}} - 2\Gamma\rho_{\underline{n}\underline{n}} + \sum_{\underline{n}'} 2\gamma_{\underline{n}-\underline{n}'}\rho_{\underline{n}'\underline{n}'} \tag{4.22}$$

$$\dot{\rho}_{\underline{n}\underline{n}'} = -i \sum_{\underline{n}_2} H_{\underline{n}-\underline{n}_2}\rho_{\underline{n}_2\underline{n}'} + i \sum_{\underline{n}_2} \rho_{\underline{n}\underline{n}_2}H_{\underline{n}_2-\underline{n}'} - 2\Gamma\rho_{\underline{n}\underline{n}'} + 2\bar{\gamma}_{\underline{n}-\underline{n}'}\rho_{\underline{n}'\underline{n}} \tag{4.23}$$

and a corresponding equation for $\rho_{\underline{n}'\underline{n}}$. We assume that phase relations of the exciton motion, which are described by the non-diagonal elements of the density matrix, are destroyed very rapidly under the influence of the vibrations within the crystal. In such a situation

$$\dot{\rho}_{\underline{n}\underline{n}'} \ll 2\Gamma\rho_{\underline{n}\underline{n}'} \quad , \tag{4.24}$$

and we may neglect the time derivative on the left-hand side of (4.23). Solving (4.23) and its corresponding equation for $\rho_{\underline{n}\underline{n}'}$, we get /93/

$$\begin{aligned}\rho_{\underline{n}\underline{n}'} = i(|\bar{\gamma}_{\underline{n}-\underline{n}'}|^2 - \Gamma^2)^{-1} (&\bar{\gamma}_{\underline{n}-\underline{n}'} \sum_{\underline{n}_2} H_{\underline{n}'-\underline{n}_2}\rho_{\underline{n}_2\underline{n}} - \bar{\gamma}_{\underline{n}-\underline{n}'} \sum_{\underline{n}_2} \rho_{\underline{n}'\underline{n}_2}H_{\underline{n}_2-\underline{n}} \\ &+ \Gamma \sum_{\underline{n}_2} H_{\underline{n}-\underline{n}_2}\rho_{\underline{n}_2\underline{n}'} - \Gamma \sum_{\underline{n}_2} \rho_{\underline{n}\underline{n}_2}H_{\underline{n}_2-\underline{n}'})/2 \quad .\end{aligned} \tag{4.25}$$

Equation (4.25) represents a set of equations that allows one to express the non-diagonal matrix elements of the density operator by the diagonal ones. This set may, for example, be solved by an iteration procedure. Considering only time intervals larger than the decay time of the phase of the exciton motion, on the right-hand side of (4.25) we may neglect non-diagonal matrix elements as compared to the diagonal ones. In this way we obtain in lowest approximation

$$\begin{aligned}\rho_{\underline{n}\underline{n}'} = i(|\gamma_{\underline{n}-\underline{n}'}|^2 - \Gamma^2)^{-1} (&\bar{\gamma}_{\underline{n}-\underline{n}'}H_{\underline{n}'-\underline{n}}\rho_{\underline{n}\underline{n}} - \bar{\gamma}_{\underline{n}-\underline{n}'}\rho_{\underline{n}'\underline{n}'}H_{\underline{n}'-\underline{n}} \\ &+ \Gamma H_{\underline{n}-\underline{n}'}\rho_{\underline{n}'\underline{n}'} - \Gamma\rho_{\underline{n}\underline{n}}H_{\underline{n}-\underline{n}'})/2 \quad .\end{aligned} \tag{4.26}$$

Inserting this result into (4.22), we arrive at the following master equation:

$$\dot{\rho}_{\underline{n}\underline{n}} = \sum_{\underline{n}'} W_{\underline{n}-\underline{n}'} (\rho_{\underline{n}'\underline{n}'} - \rho_{\underline{n}\underline{n}}) \quad . \tag{4.27}$$

The transition rates $W_{\underline{n}-\underline{n}'}$ are given by /93/

$$W_{\underline{n}-\underline{n}'} = 2\gamma_{\underline{n}-\underline{n}'} + (|H_{\underline{n}-\underline{n}'}|^2 \Gamma - H^2_{\underline{n}-\underline{n}'}\bar{\gamma}_{\underline{n}'-\underline{n}} / 2 - H^2_{\underline{n}'-\underline{n}}\bar{\gamma}_{\underline{n}-\underline{n}'} / 2) \cdot (\Gamma^2 - |\bar{\gamma}_{\underline{n}-\underline{n}'}|^2)^{-1} \quad . \tag{4.28}$$

If the crystal is invariant under inversion, we have $H_{\underline{n}-\underline{n}'} = H_{\underline{n}'-\underline{n}}$ and $\bar{\gamma}_{\underline{n}-\underline{n}'} = \bar{\gamma}_{\underline{n}'-\underline{n}}$, and (4.28) reduces to

$$W_{\underline{n}-\underline{n}'} = 2\gamma_{\underline{n}-\underline{n}'} + H^2_{\underline{n}-\underline{n}'} (\Gamma + \bar{\gamma}_{\underline{n}-\underline{n}'})^{-1} \quad . \tag{4.29}$$

We introduce the Fourier transform $\tilde{\rho}(\underline{k},t)$ of the distribution $\rho_{\underline{n}\underline{n}}(t)$ by

$$\rho_{\underline{n}\underline{n}}(t) = \int_{-\pi}^{\pi} d^m k \; e^{i\underline{k}\underline{n}} \, \tilde{\rho}(\underline{k},t) \tag{4.30}$$

(m = 1 for exciton motion on a linear chain, m = 2 and 3 when the excitons are moving in two- and three-dimensional lattices, respectively). With the help of

$$\tilde{W}(\underline{k}) = \sum_{\underline{n}'} W_{\underline{n}-\underline{n}'} \, e^{-i\underline{k}(\underline{n}-\underline{n}')} \tag{4.31}$$

the equation of motion (4.27) becomes

$$\dot{\tilde{\rho}}(\underline{k},t) = -\,[\tilde{W}(0) - \tilde{W}(\underline{k})]\, \tilde{\rho}(\underline{k},t) \quad , \tag{4.32}$$

and the solution is given by /93,240/

$$\rho_{\underline{n}\underline{n}}(t) = \int_{-\pi}^{\pi} d^m k \; e^{i\underline{k}\underline{n}} \, e^{-[\tilde{W}(0)-\tilde{W}(\underline{k})]t} \, \tilde{\rho}(\underline{k}) \quad . \tag{4.33}$$

With the initial condition $\rho_{\underline{n}\underline{n}}(t=0) = \delta_{\underline{n}0}$ we have $\tilde{\rho}(\underline{k}) = (2\pi)^{-m}$.

In the case of nearest-neighbour interaction all $W_{\underline{n}-\underline{n}'}$ disappear except for $n_i = n_i'$ and $n_i = n_i' \pm 1$. Then (4.31) becomes

$$\tilde{W}(\underline{k}) = W_0 + 2W_{1x} \cos k_x + 2W_{1y} \cos k_y + 2W_{1z} \cos k_z \quad , \tag{4.34}$$

and using an integral representation of modified Bessel functions /242/ from (4.33) we get

$$\rho_{n_x,n_y,n_z}(t) = \exp[-(2W_{1x} + 2W_{1y} + 2W_{1z})t] \cdot I_{n_x}(2W_{1x}t)\, I_{n_y}(2W_{1y}t)\, I_{n_z}(2W_{1z}t) \quad . \tag{4.35}$$

The asymptotic expansion of the modified Bessel functions for large times gives

$$\rho_{n_x,n_y,n_z}(t) \approx (4\pi t)^{-3/2} \, (W_{1x}W_{1y}W_{1z})^{-1/2} \exp\left(-\frac{n_x^2}{4W_{1x}t} - \frac{n_y^2}{4W_{1y}t} - \frac{n_z^2}{4W_{1z}t}\right) \quad . \tag{4.36}$$

Thus, in the limit of large times the exact expression (4.35) is approximated by the solution of a three-dimensional diffusion equation.

In order to obtain the continuum approximation of the solution (4.33), we expand its exponent for small values of k in a power series /93/:

$$\widetilde{W}(0) - \widetilde{W}(\underline{k}) = -\sum_{\mu} \left[\frac{\partial}{\partial k_\mu} \widetilde{W}(\underline{k})\right]_0 k_\mu - \sum_{\mu}\sum_{\nu} \frac{1}{2} \left[\frac{\partial^2}{\partial k_\mu \, \partial k_\nu} \widetilde{W}(\underline{k})\right]_0 k_\mu k_\nu \quad . \tag{4.37}$$

Using the definition (4.31) of $\widetilde{W}(\underline{k})$ and the translational and inversion symmetry of $W_{\underline{n}-\underline{n}'}$, we get

$$\left[\frac{\partial}{\partial k_\mu} \widetilde{W}(\underline{k})\right]_{\underline{k}=0} = -i \sum_{\underline{n}'} W_{\underline{n}-\underline{n}'} (\underline{n}-\underline{n}')_\mu = 0 \tag{4.38}$$

The second derivative becomes

$$\left[\frac{\partial^2}{\partial k_\mu \, \partial k_\nu} \widetilde{W}(\underline{k})\right]_{\underline{k}=0} = -\sum_{\underline{n}'} W_{\underline{n}-\underline{n}'} (\underline{n}-\underline{n}')_\mu \, (\underline{n}-\underline{n}')_\nu = -2D_{\mu\nu} \quad . \tag{4.39}$$

The $D_{\mu\nu}$ are the components of the diffusion tensor and distances are measured in units of lattice vectors. With (4.37-39) we obtain from (4.33)

$$\rho_{\underline{n}\underline{n}}(t) = (2\pi)^{-m} \int_{-\pi}^{\pi} d^m k \; e^{i\underline{k}\underline{n}} \exp[-(\sum_{\mu}\sum_{\nu} D_{\mu\nu} \, k_\mu k_\nu)t] \quad . \tag{4.40}$$

On account of the Gaussian behaviour of the integrand, the limits of the integration may be extended to infinity if the tensor $D_{\mu\nu}$ is positive definite, and we arrive at

$$\rho_{\underline{n}\underline{n}}(t) = (4\pi t)^{-m/2} \, |D|^{-1/2} \exp(-\widetilde{\underline{n}} \, \underline{\underline{D}}^{-1} \, \underline{n} \, / \, 4t) \quad . \tag{4.41}$$

$|D|$ is the determinant of the diffusion tensor and $\widetilde{\underline{n}}$ is the transpose of the vector $\underline{n}$.

4.2.2 Diffusion Equation for the Exciton Motion

In this section we wish to derive a diffusion equation starting from the master equation (4.27). Therefore, it is convenient to denote the probability $\rho_{\underline{n}\underline{n}}(t)$ of finding an exciton at site $\underline{n}$ by $\rho(\underline{X}_{\underline{n}})$. Expanding $\rho(\underline{X}_{\underline{n}'})$ into a Taylor series at $\underline{X}_{\underline{n}}$ up to second order, we have /94/

$$\rho(\underline{X}_{\underline{n}'}) = \rho(\underline{X}_{\underline{n}}) + \sum_{\mu} (X_{\underline{n}'\mu} - X_{\underline{n}\mu}) \frac{\partial}{\partial X_{\underline{n}\mu}} \rho(\underline{X}_{\underline{n}})$$

$$+ \sum_{\mu}\sum_{\nu} \frac{1}{2} (X_{\underline{n}'\mu} - X_{\underline{n}\mu})(X_{\underline{n}'\nu} - X_{\underline{n}\nu}) \frac{\partial^2}{\partial X_{\underline{n}\mu}\, \partial X_{\underline{n}\nu}} \rho(\underline{X}_{\underline{n}}) \quad . \tag{4.42}$$

Inserting this expression into the master equation (4.27), using (4.38), and writing $\underline{X}$ instead of $\underline{X}_{\underline{n}}$, we arrive at

$$\dot{\rho}(\underline{X},t) = \sum_{\mu}\sum_{\nu} D_{\mu\nu} \frac{\partial^2}{\partial X_{\mu}\, \partial X_{\nu}} \rho(\underline{X},t) \quad . \tag{4.43}$$

The components of the diffusion tensor are given by

$$D_{\mu\nu} = \sum_{\underline{n}'} \frac{1}{2} W_{\underline{n}-\underline{n}'}(X_{\underline{n}'\mu} - X_{\underline{n}\mu})(X_{\underline{n}'\nu} - X_{\underline{n}\nu}) \tag{4.44}$$

in close analogy to (4.39). The transition probabilities have to be determined from (4.28) in the general case and from (4.29) if the crystal is invariant under inversion. The solution of (4.43) with the initial condition $\rho(\underline{X},0) = \delta(\underline{X})$ is given by (4.41) with $\underline{n}$ replaced by $\underline{X}$. Master and diffusion equations for the quasi-incoherent exciton motion in molecular crystals with two molecules in the unit cell have been derived in /93,94/.

5. Optical Absorption of Molecular Crystals with Two Differently Oriented Molecules in the Unit Cell

As already mentioned in the introduction a lot of the information on the interaction between electrons and between electrons and phonons is derived from optical experiments. Therefore, in this section the line shape for optical absorption experiments is calculated on the basis of the Haken-Strobl model. In the first and second subsections two-time correlation functions and their equations of motion in Liouville space are represented. The third subsection gives the line shape of optical absorption in crystals with two differently oriented molecules in the unit cell (anthracene type crystals). In the last subsection of this section the influence of non-Markovian extensions of the Haken-Strobl model on the optical line shape is sketched.

5.1 Two-Time Correlation Functions

A variety of experimental situations may be described by two-time correlation functions. For the calculation of such correlation functions it is convenient to introduce a scalar product in an operator space by

$$(\Omega_1^\dagger,\Omega_2) \equiv \mathrm{Tr}\{\Omega_1\Omega_2\} \quad . \tag{5.1}$$

In this expression Ω_1 and Ω_2 are two operators in the Hilbert space of quantum mechanics ($\Omega_1^\dagger$ is the hermitean adjoint of Ω_1 with respect to the scalar product in this space). On the left-hand side of (5.1), these quantities are considered elements in a superspace, frequently called Liouville space.

With this definition of the scalar product in the superspace, two-time correlation functions may be written in the following way (H is the Hamiltonian of the system):

$$\begin{aligned} \langle\Omega_1(t)\,\Omega_2\rangle &= \mathrm{Tr}\{e^{iHt}\Omega_1\, e^{-iHt}\Omega_2\rho(0)\} \\ &= \mathrm{Tr}\{\Omega_1 e^{-iHt}\,\Omega_2\rho(0)e^{iHt}\} \equiv \mathrm{Tr}\{\Omega_1 e^{-iLt}\,\Omega_2\rho(0)\} \quad . \end{aligned} \tag{5.2}$$

In arriving at the first expression of the second line, the cyclic property of the trace has been used and the second equality defines the Liouville operator L. Using the definition of the scalar product (5.1), the correlation function reads

$$\langle\Omega_1(t)\,\Omega_2\rangle = \left(\Omega_1^\dagger, e^{-iLt}\Omega_2\rho(0)\right) = \left(e^{iL^+t}\Omega_1^\dagger,\Omega_2\rho(0)\right) \quad . \tag{5.3}$$

In (5.2) the Liouville operator is defined as the commutator with the Hamiltonian $L\ldots = [H,\ldots]$ and the equation of motion for the density operators reads

$$\dot{\rho} = -i\,L\,\rho \quad . \tag{5.4}$$

In the more general case of a system coupled to a heat bath, the density operator equation may (under certain approximations) also be written in the form (5.4) /236/. The Liouville operator, however, is no longer simply the commutator with the Hamiltonian of the system. Using the quantum regression theorem /243,244/, it may be shown that the correlation function of two system operators keeps the form (5.3).

5.2 Equations of Motion for Correlation Functions

The calculation of a correlation function may frequently be carried out using its equation of motion. Differentiating (5.3) with respect to time, we arrive at

$$\frac{d}{dt}\langle\Omega_1(t)\,\Omega_2\rangle = -i\left(e^{iL^+t}\,L^+\Omega_1^\dagger,\Omega_2\rho(0)\right) \quad . \tag{5.5}$$

In the scalar product on the right-hand side $L^+\Omega_1^\dagger$ usually is a new operator, and we have to write down an equation of motion for the new correlation function $\langle(L^+\Omega_1^\dagger)^\dagger(t)\,\Omega_2\rangle$, and so on, i.e. we arrive at a hierarchy of equations of motion.

For an evaluation of the right-hand side of (5.5) we have to know the adjoint L^+ of the Liouvillian. If L represents a commutator /246/

$$L_1 \ldots = [\Omega_1, \ldots] = \Omega_1^x \ldots \quad , \tag{5.6}$$

we have

$$L_1^+ = [\Omega_1^\dagger, \ldots] = (\Omega_1^\dagger)^x \ldots \quad . \tag{5.7}$$

The adjoint of a double commutator

$$L_2 \ldots = [\Omega_1, [\Omega_2, \ldots]] = \Omega_1^x \Omega_2^x \ldots \tag{5.8}$$

is given by

$$L_2^+ \ldots = [\Omega_2^\dagger, [\Omega_1^\dagger, \ldots]] = (\Omega_2^\dagger)^x (\Omega_1^\dagger)^x \ldots \quad . \tag{5.9}$$

5.3 Optical Line Shape of Anthracene Type Molecular Crystals

Treating the influence of an external time-dependent electric field within linear response theory /245/, the line shape of optical absorption is determined by the imaginary part of the dielectric susceptibility tensor. Assuming that the system at the initial time is in its ground state and taking into account that the wave vector of light is very small, we get

$$\begin{aligned} \chi''_{\alpha\alpha'}(k,\omega) &= \mathrm{Re}\left\{ \sum_{jj'} \int_0^\infty d\tau\, \mu_{\alpha j'} \mu^*_{\alpha' j} \langle b_{kj}\, b^\dagger_{kj'}(\tau) \rangle\, e^{-i\omega t} \right\} \\ &\quad - \mathrm{Re}\left\{ \sum_{jj'} \int_0^\infty d\tau\, \mu^*_{\alpha j} \mu_{\alpha' j'} \langle b_{-kj}(\tau)\, b^\dagger_{-kj'} \rangle\, e^{-i\omega t} \right\} \quad . \end{aligned} \tag{5.10}$$

In this expression k and ω are the wave vector and the frequency of light, respectively. $\mu_{\alpha j} = \langle \phi_{nj1} | x_\alpha | \phi_{nj0} \rangle$ is the component of the dipole matrix element between the ground and the excited states of molecule j in the nth unit cell. $b^\dagger_{kj}$ and b_{kj} are the Fourier transforms of the operators $b^\dagger_{nj}$ and b_{nj} which create and annihilate an excitation at molecule j in the nth unit cell, and Re means real part.

The dynamics of the excitons is described by the stochastic Hamiltonian of the Haken-Strobl model, and from (2.5,28) the Liouville operator is given by $L = L_0 + L_1$:

$$L_0 = \sum_n \varepsilon (b_n^\dagger b_n)^x + \sum_{n_1 \neq n_2} H_{n_1 n_2} (b^\dagger_{n_1} b_{n_2})^x \quad , \tag{5.11a}$$

$$\begin{aligned} L_1 &= -i \sum_{n_1} \sum_{n_2} \gamma_{n_1 n_2} (b^\dagger_{n_1} b_{n_2})^x (b^\dagger_{n_2} b_{n_1})^x \\ &\quad -i \sum_{n_1} \sum_{n_2} \bar{\gamma}_{n_1 n_2} (1 - \delta_{n_1 n_2}) (b^\dagger_{n_1} b_{n_2})^x (b^\dagger_{n_1} b_{n_2})^x \quad . \end{aligned} \tag{5.11b}$$

From now on we denote the unit cells in the crystal by n and the two molecules in the unit cell by j = {1,2}. Additionally, we introduce the Fourier transforms of b_{nj}, $H_{n-n',j-j'}$, $\gamma_{n-n',j-j'}$, and $\bar{\gamma}_{n-n',j-j'}$ in an analogous manner as in (2.32-34). The Liouville operator then transforms into

$$L = \sum_{k_1}\sum_{j_1} \varepsilon(b^\dagger_{k_1j_1}b_{k_1j_1})^\times + \sum_{k_1}\sum_{j_1j_2} \widetilde{H}_{k_1,j_1-j_2}(b^\dagger_{k_1j_1}b_{k_1j_2})^\times$$

$$-i \sum_{k_1k_2k_3}\sum_{j_1j_2} \widetilde{\gamma}_{k_2-k_3,j_1-j_2}(b^\dagger_{k_1j_1}b_{k_2j_2})^\times(b^\dagger_{k_3j_2}b_{k_1+k_3-k_2,j_1})^\times \tag{5.12}$$

$$-i \sum_{k_1k_2k_3}\sum_{j_1j_2} \widetilde{\bar{\gamma}}_{k_1+k_3,j_1-j_2}(b^\dagger_{k_1j_1}b_{k_2j_2})^\times(b^\dagger_{k_3j_1}b_{k_1+k_3-k_2,j_2})^\times \quad .$$

With the help of (5.7,9) the adjoint of L becomes

$$L^+ = \sum_{k_1}\sum_{j_1} \varepsilon(b^\dagger_{k_1j_1}b_{k_1j_1})^\times + \sum_{k_1}\sum_{j_1j_2} \widetilde{H}^*_{k_1,j_1-j_2}(b^\dagger_{k_1j_2}b_{k_1j_1})^\times$$

$$+i \sum_{k_1k_2k_3}\sum_{j_1j_2} \widetilde{\gamma}_{k_2-k_3,j_1-j_2}(b^\dagger_{k_1+k_3-k_2,j_1}b_{k_3j_2})^\times (b^\dagger_{k_2j_2}b_{k_1j_1})^\times \tag{5.13}$$

$$+i \sum_{k_1k_2k_3}\sum_{j_1j_2} \widetilde{\bar{\gamma}}_{k_1+k_3,j_1-j_2}(b^\dagger_{k_1+k_3-k_2,j_2}b_{k_3,j_1})^\times(b^\dagger_{k_2j_2}b_{k_1j_1})^\times \quad .$$

From (5.10) we see that the optical line shape is determined by the following two correlation functions:

$$<b_{kj}\; b^\dagger_{kj'}(\tau)> = \left(e^{iL^+\tau}\; b_{kj'},\rho(0)b_{kj}\right) \quad , \tag{5.14}$$

$$<b_{-kj}(\tau)\; b^\dagger_{-kj'}> = \left(e^{iL^+\tau}\; b^\dagger_{-kj},b^\dagger_{-kj'}\rho(0)\right) \quad . \tag{5.15}$$

We consider states with a single exciton only and get

$$L^+b_{ki} = -\varepsilon b_{ki} - \sum_j \widetilde{H}^*_{k,j-i}b_{kj} + i\Gamma b_{ki} \quad , \tag{5.16}$$

$$L^+b^\dagger_{ki} = \varepsilon b^\dagger_{ki} + \sum_j \widetilde{H}^*_{k,i-j}b^\dagger_{kj} + i\Gamma b^\dagger_{ki} \quad . \tag{5.17}$$

Using these results, from (5.5) we arrive at the following equations of motion for the correlation functions:

$$\frac{d}{dt} <b_{kj}\; b^\dagger_{kj'}(t)> = \{i(\varepsilon + \widetilde{H}_{k,0}) - \Gamma\} <b_{kj}\; b^\dagger_{kj'}(t)>$$
$$+ \sum_{j''} i(1-\delta_{j'j''})\widetilde{H}_{k,j''-j'} <b_{kj}\; b^\dagger_{kj''}(t)> \quad , \tag{5.18}$$

$$\frac{d}{dt} \langle b_{-kj}(t)\, b^{\dagger}_{-kj'} \rangle = \{-i(\varepsilon + \widetilde{H}_{-k,0}) - \Gamma\} \langle b_{-kj}(t)\, b^{\dagger}_{-kj'} \rangle$$
$$- \sum_{j''} i(1-\delta_{j'j''})\widetilde{H}_{-k,j-j''} \langle b_{-kj''}(t)\, b^{\dagger}_{-kj'} \rangle \quad . \tag{5.19}$$

In arriving at these equations, we have used $\sum_k \widetilde{\widetilde{\gamma}}_k = 0$ and $\Gamma = \sum_{kj} \widetilde{\gamma}_{0,j-i}$. Equations (5.18,19) have been derived in /122/ by applying the method of Appendix A.

With the initial condition $\rho(0) = |0\rangle\langle 0|$ for the density operator, which means that at $t = 0$ all molecules are in the ground state, we obtain as initial conditions for the correlation functions

$$\langle b_{kj}\, b^{\dagger}_{kj'} \rangle = \langle b_{-kj}\, b^{\dagger}_{-kj'} \rangle = \delta_{jj'} \quad . \tag{5.20}$$

The solution of (5.18,19) is then given by ($i = 1,2$) /122/

$$\langle b_{ki}\, b^{\dagger}_{ki}(t) \rangle = \frac{1}{2} K_1^* \;, \quad \langle b_{-ki}(t)\, b^{\dagger}_{-ki} \rangle = \frac{1}{2} K_1 \;, \tag{5.21}$$

$$\langle b_{ki}\, b^{\dagger}_{kj}(t) \rangle = \frac{1}{2} \frac{\widetilde{H}_{k,i-j}}{|\widetilde{H}_{k,1}|} K_2^* \quad , \tag{5.22}$$

$$\langle b_{-k,i}(t)\, b^{\dagger}_{-k,j} \rangle = \frac{1}{2} \frac{\widetilde{H}_{k,i-j}}{|\widetilde{H}_{k,1}|} K_2 \quad . \tag{5.23}$$

K_1 and K_2 are given by

$$K_1 = \exp\{[-i(\varepsilon + \widetilde{H}_{k,0} - |\widetilde{H}_{k,1}|) - \Gamma]t\} + \exp\{[-i(\varepsilon + \widetilde{H}_{k,0} + |\widetilde{H}_{k,1}|) - \Gamma]t\} \quad , \tag{5.24}$$

$$K_2 = -\exp\{[-i(\varepsilon + \widetilde{H}_{k,0} - |\widetilde{H}_{k,1}|) - \Gamma]t\} + \exp\{[-i(\varepsilon + \widetilde{H}_{k,0} + |\widetilde{H}_{k,1}|) - \Gamma]t\} \quad , \tag{5.25}$$

Using these results, from (5.10) the line shape of optical absorption is given by ($\mathcal{N}$ is a normalization factor)

$$\begin{aligned} \chi''_{\alpha\alpha'} = {} & \frac{\mathcal{N}}{2|\widetilde{H}_{k,1}|} (\sqrt{\widetilde{H}_{k,-1}}\, \mu^*_{\alpha 1} + \sqrt{\widetilde{H}_{k,1}}\, \mu^*_{\alpha 2})(\sqrt{\widetilde{H}_{k,1}}\, \mu_{\alpha' 1} + \sqrt{\widetilde{H}_{k,-1}}\, \mu_{\alpha' 2}) \\ & \cdot \Gamma\{[\big(\omega - (\varepsilon + \widetilde{H}_{k,0} + |\widetilde{H}_{k,1}|)\big)^2 + \Gamma^2]^{-1} - [\big(\omega + (\varepsilon + \widetilde{H}_{k,0} + |\widetilde{H}_{k,1}|)\big)^2 + \Gamma^2]^{-1}\} \\ & + \frac{\mathcal{N}}{2|\widetilde{H}_{k,1}|} (\sqrt{\widetilde{H}_{k,-1}}\, \mu^*_{\alpha 1} - \sqrt{\widetilde{H}_{k,1}}\, \mu^*_{\alpha 2})(\sqrt{\widetilde{H}_{k,1}}\, \mu_{\alpha' 1} - \sqrt{\widetilde{H}_{k,-1}}\, \mu_{\alpha' 2}) \\ & \cdot \Gamma\{[\big(\omega - (\varepsilon + \widetilde{H}_{k,0} - |\widetilde{H}_{k,1}|)\big)^2 + \Gamma^2]^{-1} - [\big(\omega + (\varepsilon + \widetilde{H}_{k,0} - |\widetilde{H}_{k,1}|)\big)^2 + \Gamma^2]^{-1}\} \; . \end{aligned} \tag{5.26}$$

For the discussion of this expression we take into account that the two molecules in the unit cell of anthracene type crystals differ only in orientation. Therefore, the transition dipole moments at the two sites in the unit cell have the same amount and are arranged symmetrically with respect to the b axis of the crystal (Fig. 5.1).

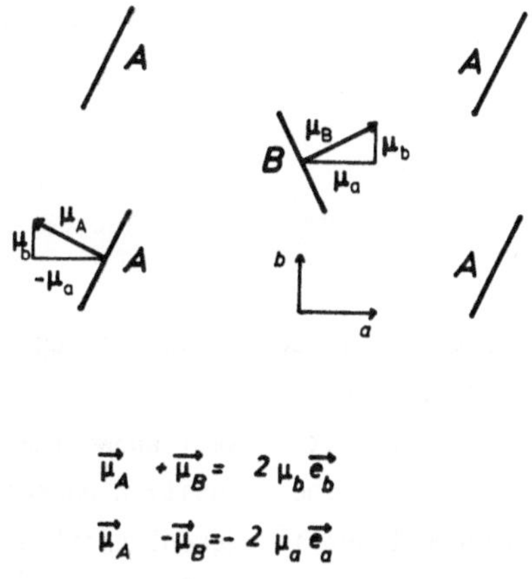

Fig. 5.1. Orientation of molecules and dipole moments in anthracene

Furthermore, we assume the dipole moments to be real. With these assumptions and $\{A,B\} = \{1,2\}$ we have

$$\underline{\mu}_2 + \underline{\mu}_1 = \underline{\mu}_2^* + \underline{\mu}_1^* = 2\mu_b\underline{e}_b \quad ; \quad \underline{\mu}_2 - \underline{\mu}_1 = \underline{\mu}_2^* - \underline{\mu}_1^* = 2\mu_a\underline{e}_a \quad . \tag{5.27}$$

The only non-vanishing components of the tensor of the dielectric susceptibility are given by

$$\begin{aligned} \chi''_{aa} = 2\mu_a^2\Gamma\{&[(\omega - (\varepsilon + \widetilde{H}_{k,0} - |\widetilde{H}_{k,1}|))^2 + \Gamma^2]^{-1} \\ &- [(\omega + (\varepsilon + \widetilde{H}_{k,0} - |\widetilde{H}_{k,1}|))^2 + \Gamma^2]^{-1}\} \quad , \end{aligned} \tag{5.28}$$

$$\begin{aligned} \chi''_{bb} = 2\mu_b^2\Gamma\{&[(\omega - (\varepsilon + \widetilde{H}_{k,0} + |\widetilde{H}_{k,1}|))^2 + \Gamma^2]^{-1} \\ &- [(\omega + (\varepsilon + \widetilde{H}_{k,0} + |\widetilde{H}_{k,1}|))^2 + \Gamma^2]^{-1}\} \quad . \end{aligned} \tag{5.29}$$

These expressions for the optical line shape have been derived in /121,122/. The stochastic model thus gives an absorption line polarized parallel to the a axis with its maximum at $\varepsilon + \widetilde{H}_{k,0} - |\widetilde{H}_{k,1}|$ and a second absorption line polarized parallel to the b axis with its maximum at $\varepsilon + \widetilde{H}_{k,0} + |\widetilde{H}_{k,1}|$. As compared to the distance of the energy levels of the isolated molecules, the lines are displaced by $\widetilde{H}_{k,0} - |\widetilde{H}_{k,1}|$ and by $\widetilde{H}_{k,0} + |\widetilde{H}_{k,1}|$. Thus the distance between the lines, the Davydov-splitting, is given by $2|\widetilde{H}_{k,1}|$. The line width of the Lorentzian lines is determined by Γ, which is the sum of all fluctuation parameters. In our model, we have assumed that the fluctuations are caused by the phonons. Therefore, their strengths should increase with increasing temperature leading to a broadening of the absorption line.

Considering a two-molecule system /35/ instead of a crystal, the parameters in (5.28,29) assume the following values:

$$\Gamma = \gamma_0 + \gamma_1 \; ; \quad \tilde{H}_{k,0} = 0 \; ; \quad \tilde{H}_{k,1} = J \; . \qquad (5.30)$$

Here J is the interaction between the molecules.

5.4 Extensions to Non-Markovian Fluctuations

The optical absorption line shape of the previous paragraph is Lorentzian, independent of the amplitude of the fluctuations. From the purely incoherent treatment of absorption line shapes by ANDERSON /171/ and KUBO /123,173/, it is well known that such a line shape is the result of very rapid modulations of the transition energies. In the Haken-Strobl model /34-36/ for the coupled coherent and incoherent exciton motion such rapid fluctuations are implied by the δ function form (2.12) of the correlation functions.

To investigate the influence of non-δ correlated fluctuations, SUMI /124/, following TOYOZAWA /247,248/, considered the optical absorption of a crystal with a single molecule in the unit cell. The influence of the lattice vibrations is taken into account by Gaussian local fluctuations with an exponentially decaying correlation function. The model is described by three parameters: the energy band width B, determined by the coherent interaction between the molecules, the amplitude D of the fluctuations, and the decay constant γ of the correlation function. The optical line shape is calculated in the dynamical coherent potential approximation (CPA). For rapid fluctuations the Lorentzian line shape is obtained again; for slow fluctuations the line shape turns out to be a Gaussian. A discussion of intermediate cases is carried out numerically.

Starting from the same model, BLUMEN and SILBEY /217/ calculated the optical line shape using a generalized cumulant method /234,228,249-251/. The expression for the line shape, which has been evaluated up to the third cumulant /217/, results also in Gaussian and Lorentzian line shapes in the limiting cases of slow and fast modulations, respectively. Furthermore, the numerical evaluation of intermediate cases gives line widths which are very close to SUMI's results from the dynamical CPA. Recently, RIPS and CAPEK /126/ have determined the fourth generalized cumulant, which takes into account that the Hamiltonian for different times does not commute. With this cumulant they discuss the range of validity of the BLUMEN-SILBEY result. Another approach to the treatment of non-Markovian effects on exciton dynamics is under investigation /252/.

6. Electron-Spin Resonance in Molecular Crystals with Two Differently Oriented Molecules in the Unit Cell

For triplet excitons information is obtained not only from diffusion and optical measurements, but also from ESR. From the investigations of ANDERSON /171/, KUBO /123,172/, and HUDSON and McLACHLAN /174/ it is well known that the line shape and width of ESR depends on the hopping rate of the excitation between states with different magnetic properties. This section discusses how the line shape is influenced by the way of the exciton motion, i.e. whether the exciton moves coherently or incoherently.

From a theoretical point of view the simplest kind of exciton motion occurs in a dimer. This kind of motion has been investigated by WOLF and SCHWOERER /136/, and HINKEL et al. /144,145/ for pairs of differently oriented naphthalene-h_8 molecules in a deuterated matrix and by BOTTER et al. /134/ for pairs of equivalent molecules. For pairs of equivalently oriented molecules of 1,2,4,5-tetrachlorobenzene, theoretical and experimental investigations have been carried out by ZEWAIL and HARRIS /132,133/. ESR investigations of exciton transport in linear chains of molecules have been reported by WOLF and SCHMIDBERGER /147-149/ in dibromonaphthalene and by FAYER and HARRIS /151-159/ in tetrachlorobenzene. The theoretical problem of the exciton motion in linear chains of molecules has been discussed by BREILAND et al. /156,253/ and its motion in finite molecular aggregates has been calculated by KOTTIS and LEMAISTRE /179/. The motion of triplet excitons in molecular crystals of anthracene and naphthalene was detected first by HAARER and WOLF /163,164/. Recently, the investigation of these and related materials and a discussion of the earlier experiments of HAARER and WOLF have been reported by BERK et al. /165,167-169,254, 255/.

The ESR line shape is determined from linear response theory by the Fourier transform of two-time correlation functions of the magnetization operators. The time dependence of these correlation functions is calculated from (5.5). The Liouville operator L is composed of two parts

$$L = L_0 + L_1 \quad , \tag{6.1}$$

where $L_0 = H_0^x$ is the commutator with the Hamiltonian H_0 of the coherent part of the motion given explicitly below. L_1 describes the incoherent motion and has been derived in (5.11b).

The Hamiltonian of the coherent motion again consists of two parts:

$$H_0 = H_{ex,0} + H_{s,0} \quad , \tag{6.2}$$

where $H_{ex,0}$ results in the Liouvillian (5.11a) and $H_{s,0}$ is given by

$$H_{s,0} = \sum_n \sum_{i=1}^{2} b_{ni}^+ b_{ni} \, g\mu_B \underline{H} \cdot \underline{S} + \sum_n b_{n1}^+ b_{n1} \underline{S} \cdot \underline{\underline{F}}_1 \cdot \underline{S} + \sum_n b_{n2}^+ b_{n2} \underline{S} \cdot \underline{\underline{F}}_2 \cdot \underline{S} \; . \quad (6.3)$$

Here the first sum is the Zeeman energy of a spin $\underline{S}$ in an external magnetic field $\underline{H}$; the two following sums comprise the fine-structure interaction of the two electronic spins forming the triplet spin. $\underline{\underline{F}}_1$ and $\underline{\underline{F}}_2$ are the fine-structure tensors of the two differently oriented molecules. Introducing the sum and the difference of the fine-structure tensors,

$$\underline{\underline{M}} = \frac{1}{2}(\underline{\underline{F}}_1 + \underline{\underline{F}}_2) \; ; \quad \underline{\underline{D}} = \frac{1}{2}(\underline{\underline{F}}_1 - \underline{\underline{F}}_2) \quad (6.4)$$

and the difference of the occupation numbers at the two differently oriented molecules

$$\Delta b = \sum_n (b_{n1}^+ b_{n1} - b_{n2}^+ b_{n2}) \quad , \quad (6.5)$$

$H_{s,0}$ may be rewritten in the following form:

$$H_{s,0} = g\mu_B \underline{H} \cdot \underline{S} + \underline{S} \cdot \underline{\underline{M}} \cdot \underline{S} + \Delta b \, \underline{S} \cdot \underline{\underline{D}} \cdot \underline{S} \quad . \quad (6.6)$$

Starting from these expressions, the ESR line shape will be determined for different situations. In the first subsection of this section we discuss the influence of the coupled coherent and incoherent motion on the ESR line shape of excitations with spin 1/2 in a dimer. The second subsection comprises the line shape of triplet states in pairs of differently oriented molecules. The final subsection is devoted to the quasi-incoherent (diffusional) motion of excitons in molecular crystals with two differently oriented molecules in the unit cell, e.g. in anthracene or naphthalene.

6.1 ESR Line Shape of Excitons with Spin 1/2 in Dimers of Inequivalent Molecules

6.1.1 Effective Hamiltonian and Equations of Motion for Correlation Functions

For dimers the Hamiltonian (6.3) may be written in the following way:

$$H_{s,0} = (b_1^+ b_1 + b_2^+ b_2) g\mu_B \underline{H} \cdot \underline{S} + b_1^+ b_1 \underline{S} \cdot \underline{\underline{F}}_1 \cdot \underline{S} + b_2^+ b_2 \underline{S} \cdot \underline{\underline{F}}_2 \cdot \underline{S} \quad (6.7)$$

$$= b_1^+ b_1 (g\mu_B \underline{H} + \underline{S} \cdot \underline{\underline{F}}_1) \cdot \underline{S} + b_2^+ b_2 (g\mu_B \underline{H} + \underline{S} \cdot \underline{\underline{F}}_2) \cdot \underline{S} \quad . \quad (6.8)$$

In this subsection we interpret the two brackets of (6.8) as the effective magnetic fields at molecules 1 and 2, respectively, thereby neglecting the operator character of $\underline{S}$ in the brackets. Assuming, furthermore, that these fields have only a z component and that $\underline{S}$ is the operator of a spin 1/2 that may be represented by

Pauli operators $\underline{\sigma}$, from (6.8) we arrive at the following Hamiltonian:

$$H_{s,0} = A\, b_1^+ b_1 \sigma_z + B\, b_2^+ b_2 \sigma_z \quad . \tag{6.9}$$

Using linear response theory /245/, the ESR line shape is given by

$$\chi''(\omega) = i \int_0^\infty d\tau \, \sin\omega\tau \, [\langle\sigma_x(\tau)\, \sigma_x\rangle - \langle\sigma_x\, \sigma_x(\tau)\rangle] \quad . \tag{6.10}$$

Assuming that initially the spin at the two molecules is aligned in -z direction, the density operator of the dimer is

$$\rho(0) = \frac{1}{2} \left(|1,-1\rangle \langle 1,-1| + |2,-1\rangle \langle 2,-1|\right) \quad . \tag{6.11}$$

The state $|j,s\rangle$ describes an excitation at molecule j with eigenvalue s of σ_z. If this initial condition is taken into account in (6.10), we get

$$\chi''(\omega) = i \int_0^\infty d\tau \, \sin\omega\tau \, [\langle\sigma^-(\tau)\, \sigma^+\rangle - \langle\sigma^-\, \sigma^+(\tau)\rangle] \quad . \tag{6.12}$$

The ESR line shape is therefore determined by the correlation functions $\langle\sigma^-(\tau)\, \sigma^+\rangle$ and $\langle\sigma^-\, \sigma^+(\tau)\rangle$. For the Liouville operator (6.1) with L_1 given by (5.11b), from (5.5) we arrive at the following equation of motion for the correlation function of two arbitrary operators Ω_1 and Ω_2:

$$\begin{aligned} \frac{d}{dt} \langle\Omega_1(t)\, \Omega_2\rangle = {} & i \langle(L_0\Omega_1)_t\, \Omega_2(0)\rangle - \Big\{ \sum_{\lambda,\lambda'} \gamma_{|\lambda-\lambda'|} \langle(b_\lambda^+ b_\lambda \Omega_1)_t\, \Omega_2\rangle \\ & + \sum_{\lambda,\lambda'} \gamma_{|\lambda-\lambda'|} \langle(\Omega_1 b_\lambda^+ b_\lambda)_t\, \Omega_2\rangle - 2 \sum_{\lambda,\lambda'} \gamma_{|\lambda-\lambda'|} \langle(b_\lambda^+ b_{\lambda'} \Omega_1 b_\lambda^+ b_{\lambda'})_t\, \Omega_2\rangle \\ & -2 \sum_{\lambda,\lambda'} \gamma_{|\lambda-\lambda'|} \langle(b_\lambda^+ b_{\lambda'} \Omega_1 b_{\lambda'}^+ b_\lambda)_t\, \Omega_2\rangle + 2 \sum_\lambda \gamma_0 \langle(b_\lambda^+ b_\lambda \Omega_1 b_\lambda^+ b_\lambda)_t\, \Omega_2\rangle \Big\} \quad , \end{aligned} \tag{6.13}$$

$$\begin{aligned} \frac{d}{dt} \langle\Omega_1\, \Omega_2(t)\rangle = {} & i \langle\Omega_1\, (L_0\Omega_2)_t\rangle - \Big\{ \sum_{\lambda,\lambda'} \gamma_{|\lambda-\lambda'|} \langle\Omega_1\, (b_\lambda^+ b_\lambda \Omega_2)_t\rangle \\ & + \sum_{\lambda,\lambda'} \gamma_{|\lambda-\lambda'|} \langle\Omega_1\, (\Omega_2 b_\lambda^+ b_\lambda)_t\rangle - 2 \sum_{\lambda,\lambda'} \gamma_{|\lambda-\lambda'|} \langle\Omega_1\, (b_\lambda^+ b_{\lambda'} \Omega_2 b_\lambda^+ b_{\lambda'})_t\rangle \\ & -2 \sum_{\lambda,\lambda'} \gamma_{|\lambda-\lambda'|} \langle\Omega_1\, (b_\lambda^+ b_{\lambda'} \Omega_2 b_{\lambda'}^+ b_\lambda)_t\rangle + 2 \sum_\lambda \gamma_0 \langle\Omega_1\, (b_\lambda^+ b_\lambda \Omega_2 b_\lambda^+ b_\lambda)_t\rangle \Big\} \quad . \end{aligned} \tag{6.14}$$

In writing (6.13,14) we have assumed $\gamma_{\lambda\lambda'} = \bar{\gamma}_{\lambda\lambda'}$.

6.1.2 Calculation of the ESR Line Shape

Writing the equations of motion (6.13,14) for one of the correlation functions in (6.12), we arrive at further correlation functions for which we have to set up additional equations. The complete set of equations comprises nine correlation functions. Introducing the following notation

$$
\begin{aligned}
&G_1(t\,,p{=}1) = \langle\sigma^-(t)\sigma^+\rangle \;, && G_1(t\,,p{=}-1) = \langle\sigma^-\sigma^+(t)\rangle \;,\\
&G_2(t\,,p{=}1) = \langle(b_1^+b_1\sigma^-)_t\sigma^+\rangle \;, && G_2(t\,,p{=}-1) = \langle\sigma^-(b_1^+b_1\sigma^+)_t\rangle \;,\\
&G_3(t\,,p{=}1) = \langle(b_2^+b_2\sigma^-)_t\sigma^+\rangle \;, && G_3(t\,,p{=}-1) = \langle\sigma^-(b_2^+b_2\sigma^+)_t\rangle \;,\\
&G_4(t\,,p{=}1) = \langle(b_1^+b_2\sigma^-)_t\sigma^+\rangle \;, && G_4(t\,,p{=}-1) = \langle\sigma^-(b_1^+b_2\sigma^+)_t\rangle \;,\\
&G_5(t\,,p{=}1) = \langle(b_2^+b_1\sigma^-)_t\sigma^+\rangle \;, && G_5(t\,,p{=}-1) = \langle\sigma^-(b_2^+b_1\sigma^+)_t\rangle \;,\\
&G_6(t\,,p{=}1) = \langle(b_1^+b_2\sigma^-\sigma_z)_t\sigma^+\rangle \;, && G_6(t\,,p{=}-1) = \langle\sigma^-(b_1^+b_2\sigma^+\sigma_z)_t\rangle \;,\\
&G_7(t\,,p{=}1) = \langle(b_2^+b_1\sigma^-\sigma_z)_t\sigma^+\rangle \;, && G_7(t\,,p{=}-1) = \langle\sigma^-(b_2^+b_1\sigma^+\sigma_z)_t\rangle \;,\\
&G_8(t\,,p{=}1) = \langle(b_1^+b_1\sigma^-\sigma_z)_t\sigma^+\rangle \;, && G_8(t\,,p{=}-1) = \langle\sigma^-(b_1^+b_1\sigma^+\sigma_z)_t\rangle \;,\\
&G_9(t\,,p{=}1) = \langle(b_2^+b_2\sigma^-\sigma_z)_t\sigma^+\rangle \;, && G_9(t\,,p{=}-1) = \langle\sigma^-(b_2^+b_2\sigma^+\sigma_z)_t\rangle \;,
\end{aligned}
\tag{6.15}
$$

we obtain from (6.13,14) for the dimer ($\lambda = 1,2$; $\varepsilon = 0$; $H_1 = J$; $\Gamma = \gamma_0 + \gamma_1$)

$$
\begin{aligned}
\dot G_1 &= -p\, i[2AG_2 + 2BG_3] \;,\\
\dot G_2 &= i[J(G_5 - G_4) - 2pAG_2] - 2\gamma_1[G_2 - G_3] \;,\\
\dot G_3 &= i[J(G_4 - G_5) - 2pBG_3] - 2\gamma_1[G_3 - G_2] \;,\\
\dot G_4 &= i[J(G_3 - G_2) + (A - B)G_6 - 2pAG_4] - 2[\Gamma G_4 - \gamma_1 G_5] \;,\\
\dot G_5 &= i[J(G_2 - G_3) - (A - B)G_7 - 2pBG_5] - 2[\Gamma G_5 - \gamma_1 G_4] \;,\\
\dot G_6 &= i[J(G_9 - G_8) + (A - B)G_4 - 2pAG_6] - 2[\Gamma G_6 - \gamma_1 G_7] \;,\\
\dot G_7 &= i[J(G_8 - G_9) - (A - B)G_5 - 2pBG_7] - 2[\Gamma G_7 - \gamma_1 G_6] \;,\\
\dot G_8 &= i[J(G_7 - G_6) - 2pAG_8] - 2\gamma_1[G_8 - G_9] \;,\\
\dot G_9 &= i[J(G_6 - G_7) - 2pBG_9] - 2\gamma_1[G_9 - G_8] \;,
\end{aligned}
\tag{6.16}
$$

with $G_j = G_j(t,p)$.

This set of equations may be solved by a Laplace transformation. With the definition

$$\tilde{G}_j(\omega,p) = \int_0^\infty d\tau\, e^{-i\omega\tau}\, G_j(t,p) \tag{6.17}$$

and the initial values

$$\begin{aligned} &G_1(0,p) = 1 \quad , \\ &G_2(0,p) = G_3(0,p) = \frac{1}{2} \quad , \\ &G_4(0,p) = G_5(0,p) = G_6(0,p) = G_7(0,p) = 0 \quad , \\ &G_8(0,p) = G_9(0,p) = \frac{p}{2} \quad , \end{aligned} \tag{6.18}$$

which are determined with the help of (6.11), we are led to the following set of algebraic equations $\left(\tilde{G}_j = \tilde{G}_j(\omega,p)\right)$:

$$\begin{aligned} &i\omega\tilde{G}_1 + i2pA\tilde{G}_2 + i2pB\tilde{G}_3 &&= 1 \quad , \\ &(i\omega + i2pA + 2\gamma_1)\tilde{G}_2 - 2\gamma_1\tilde{G}_3 + iJ\tilde{G}_4 - iJ\tilde{G}_5 &&= 1/2 \quad , \\ &-2\gamma_1\tilde{G}_2 + (i\omega + i2pB + 2\gamma_1)\tilde{G}_3 - iJ\tilde{G}_4 + iJ\tilde{G}_5 &&= 1/2 \quad , \\ &iJ\tilde{G}_2 - iJ\tilde{G}_3 + (i\omega + i2pA + 2\Gamma)\tilde{G}_4 - 2\gamma_1\tilde{G}_5 - i(A-B)\tilde{G}_6 &&= 0 \quad , \\ &-iJ\tilde{G}_2 + iJ\tilde{G}_3 - 2\gamma_1\tilde{G}_4 + (i\omega + i2pB + 2\Gamma)\tilde{G}_5 + i(A-B)\tilde{G}_7 &&= 0 \quad , \\ &-i(A-B)\tilde{G}_4 + (i\omega + i2pA + 2\Gamma)\tilde{G}_6 - 2\gamma_1\tilde{G}_7 + iJ\tilde{G}_8 - iJ\tilde{G}_9 &&= 0 \quad , \\ &i(A-B)\tilde{G}_5 - 2\gamma_1\tilde{G}_6 + (i\omega + i2pB + 2\Gamma)\tilde{G}_7 - iJ\tilde{G}_8 + iJ\tilde{G}_9 &&= 0 \quad , \\ &iJ\tilde{G}_6 - iJ\tilde{G}_7 + (i\omega + i2pA + 2\gamma_1)\tilde{G}_8 - 2\gamma_1\tilde{G}_9 &&= p/2 \quad , \\ &-iJ\tilde{G}_6 + iJ\tilde{G}_7 - 2\gamma_1\tilde{G}_8 + (i\omega + i2pB + 2\gamma_1)\tilde{G}_9 &&= p/2 \quad . \end{aligned} \tag{6.19}$$

The ESR line shape (6.12) may be expressed directly by the Laplace transforms (6.17):

$$\chi''(\omega) = \frac{1}{2}\,\{\tilde{G}_1(-\omega,1) - \tilde{G}_1(-\omega,-1) - \tilde{G}_1(\omega,1) + \tilde{G}_1(\omega,-1)\} \quad , \tag{6.20}$$

where $\tilde{G}_1(\omega,p)$ is calculated from (6.19) as

$$\tilde{G}_1(\omega,p) = \frac{D_1(\omega,p)}{D(\omega,p)} \quad . \tag{6.21}$$

$D_1(\omega,p)$ and $D(\omega,p)$ are given by

$$D_1 = (P_1P_2 + P_3)(P_4P_2 + P_5) + 2(A-B)^4(P_5 - 2P_1P_4) \quad , \tag{6.22a}$$

$$D = (P_1P_2 + P_3)^2 - 4(A-B)^4 P_1^2 \quad , \tag{6.22b}$$

with

$$P_1 = (i\omega + i2pA)(i\omega + i2pB) + 4\gamma_1[i\omega + ip(A+B)] \quad (6.23a)$$

$$P_2 = 4\gamma_1[i\omega + ip(A+B) + 2\gamma_0] + (i\omega + i2pA + 2\gamma_0)(i\omega + i2pB + 2\gamma_0) + (A-B)^2 \quad , (6.23b)$$

$$P_3 = 4J^2[i\omega + ip(A+B) + 2\gamma_0][i\omega + ip(A+B)] \quad , \quad (6.23c)$$

$$P_4 = i\omega + ip(A+B) + 4\gamma_1 \quad , \quad (6.23d)$$

$$P_5 = 4J^2[i\omega + ip(A+B) + 2\gamma_0] \quad . \quad (6.23e)$$

In the following, from (6.20) $\chi''(\omega)$ is evaluated and explicit expressions for the line shape are obtained in several limiting cases.

6.1.3 ESR Line Shape in the Completely Incoherent Case

In the limiting case of a vanishing exchange interaction integral, we get from (6.23) $P_3 = P_5 = 0$. A short calculation then results in

$$\tilde{G}_1(\omega,p) = P_4/P_1 \quad , \quad (6.24)$$

and from (6.20) we obtain /173,181/

$$\begin{aligned}\chi''(\omega) = {} & 4\gamma_1(A-B)^2\cdot\{(\omega-2A)^2(\omega-2B)^2 + 16\gamma_1^2[\omega-(A+B)]^2\}^{-1} \\ & -4\gamma_1(A-B)^2\cdot\{(\omega+2A)^2(\omega+2B)^2 + 16\gamma_1^2[\omega+(A+B)]^2\}^{-1}\end{aligned} \quad (6.25)$$

This expression is represented in Fig. 6.1 for $A = 1$, $B = 2$ and several values of the parameter γ_1, which - according to (3.1) - is half the hopping rate of the excitation between the two inequivalent molecules. For small values of the hopping rate $2\gamma_1$ - more precisely, if twice the hopping rate is small as compared to the difference in Larmor frequencies, i.e. if $2\gamma_1 \ll A-B$ (slow modulation condition) - Fig. 6.1 shows two Lorentzian lines at $\omega = E = 2A$ und $\omega = E = 2B$. The line width is given by the hopping rate $2\gamma_1$, as may be shown by an expansion of the exact result. On the other hand, for $2\gamma_1 \gg A-B$ (fast modulation condition), the ESR spectrum consists of a single line at $\omega = E = A+B$, whose width is given by $(A-B)^2/4\gamma_1$. The width of this line decreases with increasing values of the hopping rate (motional narrowing).

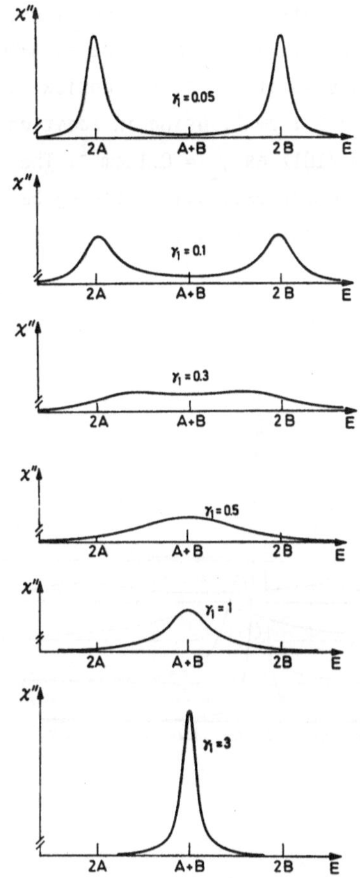

Fig. 6.1. Line shapes of the ESR in the pure incoherent case for A = 1, B = 2, and different values of the fluctuation parameter γ_1 /181/

6.1.4 ESR Line Shape for the Coupled Coherent and Incoherent Exiton Motion

The general expression for the line shape (6.20) contains the parameters of our model, namely A and B, which describe the strengths of the effective magnetic fields at the sites of the two inequivalent molecules, the exchange interaction integral J stemming from the coherent interaction between the molecules , and the quantities γ_0 and γ_1 representing the strengths of the local and non-local fluctuations, respectively. In the model we have assumed that the fluctuations are due to the phonons, and therefore γ_0 and γ_1 should depend on temperature, whereas the other quantities should be constant for a given system. In the following,we wish to discuss especially the dependence of the ESR line shape on the fluctuation parameters γ_0 and γ_1. In order to get a reasonable set of parameters, we consider ESR investigations in anthracene crystals /162-164/. In these crystals the spectral positions of the ESR lines of the two inequivalent molecules in the unit cell depend on the orientation of the external magnetic field and, in our model, are given by 2A and 2B, respectively. For that orientation of the external field with the largest distance of the two lines, i.e. where $2A - 2B$ is largest, we have $A = 0.595\ cm^{-1}$ and $B = 0.63\ cm^{-1}$. The exchange interaction integral J may be obtained from the Davydov splitting of the optical absorption /89,104/. For anthracene crystals we have $J = 2.1\ cm^{-1}$. Using in addition measurements of the diffusion constant, γ_1 is derived /181/ as $\gamma_1 = 0.1\ cm^{-1}$. The same order of magnitude of the parameters is valid also for excitations moving between pairs of inequivalent molecules in naphthalene crystals /136/.

Fig. 6.2. (a) ESR line shapes according to (6.20) for $A = 0.595$, $B = 0.63$, $J = 2.1$, $\gamma_1 = 0$ and several values of γ_0 (units of $A, B, J, \gamma_0, \gamma_1$, and ω in cm^{-1}) /122/ (b) for larger values of γ_0

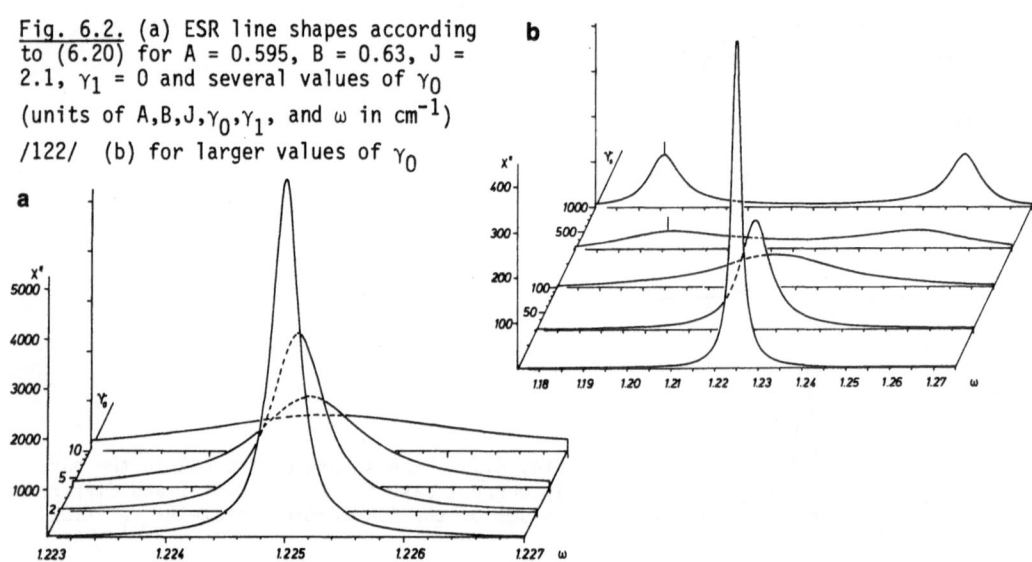

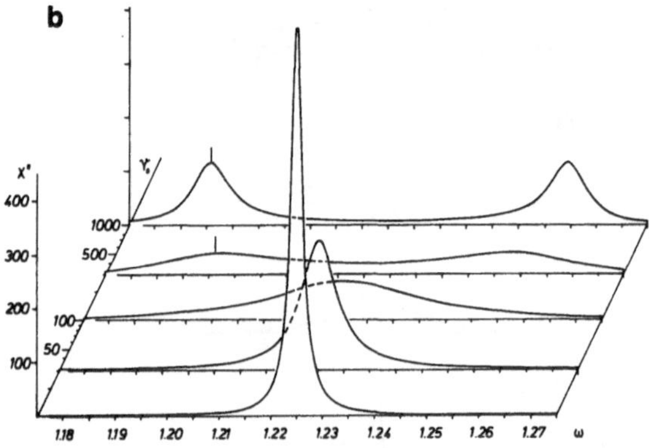

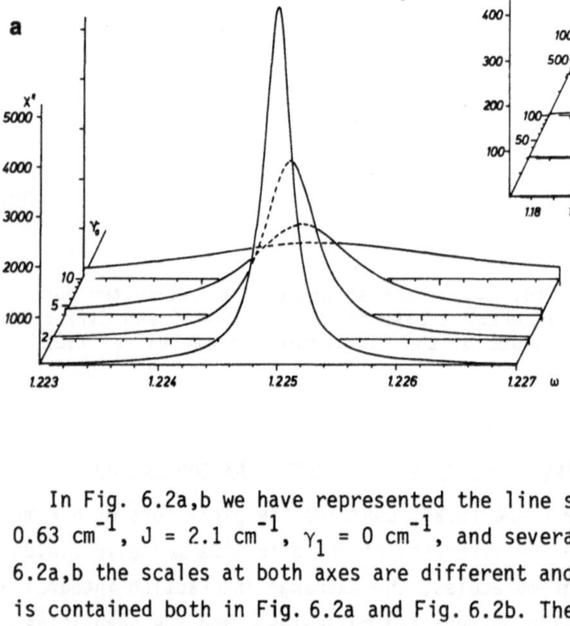

In Fig. 6.2a,b we have represented the line shapes for $A = 0.595\ cm^{-1}$, $B = 0.63\ cm^{-1}$, $J = 2.1\ cm^{-1}$, $\gamma_1 = 0\ cm^{-1}$, and several values of γ_0. Note that in Fig. 6.2a,b the scales at both axes are different and that the line shape for $\gamma_0 = 10\ cm^{-1}$ is contained both in Fig. 6.2a and Fig. 6.2b. The figure shows that for small values of γ_0 we have a single line at $\omega = A + B = 1.225\ cm^{-1}$, which broadens with increasing γ_0. For large values of γ_0 we have two lines which narrow at $2A = 1.19\ cm^{-1}$ and at $2B = 1.26\ cm^{-1}$ when γ_0 increases. The transition from the single line at small values of γ_0 to the two lines at large γ_0 has nothing to do with the transition from coherent to incoherent exciton motion. This transition occurs when $\gamma_0 > 2J$ /34-36,181/, whereas the transition from a single line to two lines happens apparently when

$$J^2/\gamma_0 < |A - B| \quad .$$

In Fig. 6.3 the parameters A, B and J are the same as in Fig. 6.2, but now $\gamma_0 = 100\ cm^{-1}$ and the line shapes are given for several values of γ_1. The line for $\gamma_1 = 0$

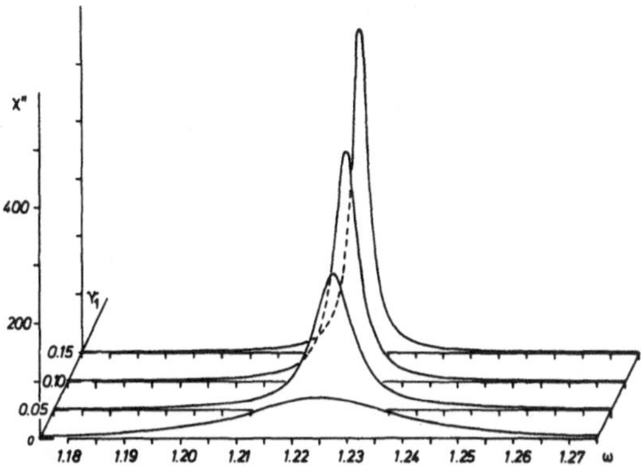

Fig. 6.3. ESR line shapes according to (6.20) for A = 0.595, B = 0.63, J = 2.1, γ_0 = 100, and several values of γ_1 (units of A,B,J,γ_0,γ_1, and ω in cm^{-1}) /122/

is also represented in Fig. 6.2b. We see that the ESR line narrows when γ_1 increases. This is even more obvious when we consider Fig. 6.4, which represents the line shapes for the same values of A, B, and J, but now $\gamma_0 = 1000\ cm^{-1}$. The curve for $\gamma_1 = 0$ is also shown in Fig. 6.2b, having two lines at approximately 2A and 2B. With increasing γ_1, these lines broaden until for $2\gamma_1 > |A - B|$ we find a single line at $\omega = A + B$, which narrows with increasing γ_1. Line shapes for other values of J are given in /122/.

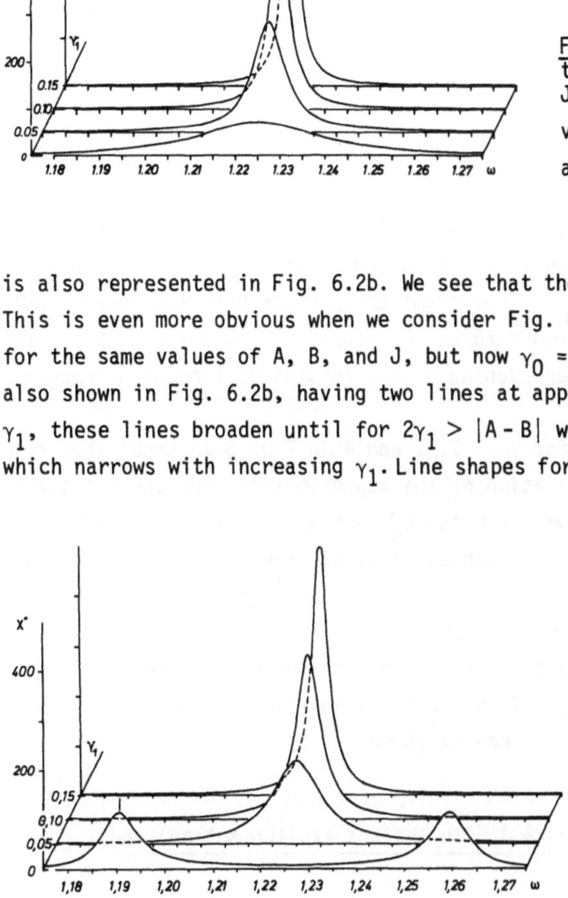

Fig. 6.4. ESR line shapes according to (6.20) for A = 0.595, B = 0.63, J = 2.1, γ_0 = 1000 and several values of γ_1 (units of A,B,J,γ_0,γ_1 and ω in cm^{-1}) /122/

6.1.5 Approximate Expressions for the Line Shapes

In Figs. 6.2-4 we have seen that in limiting cases we have either a single narrow ESR line at $\omega = A + B$ or two lines at $\omega = 2A$ and at $\omega = 2B$ with small line widths. Therefore it seems useful, also with respect to the comparison with experimental data, to obtain approximate analytical expressions for the line shapes in these limiting situations, starting from the exact expression (6.20).

6.1.5a Expansion at $\omega = A + B$ for $\gamma_0 < 2J$ (Coherent Exciton Motion)

The line shape is given by

$$\chi''(\varepsilon) \sim \tilde{\Gamma}/(\varepsilon^2 + \tilde{\Gamma}^2) \tag{6.26}$$

with

$$\varepsilon = \omega - (A + B) \tag{6.27}$$

and

$$\tilde{\Gamma} = (A - B)^2/(2J^2/\gamma_0) \quad . \tag{6.28}$$

Thus the maximum of the ESR line is at $\omega = A + B$, i.e. in the middle between the two lines that one would expect from the isolated molecules at sites 1 and 2, respectively, and the line width is given by $\tilde{\Gamma}$. The parameter γ_0 describing the local fluctuations, increases with increasing temperature; therefore, at low temperature, where $\gamma_0 < 2J$ is fulfilled, we expect a broadening of the absorption line when temperature is raised.

The comparison with the curves for $\gamma_0 = 1,2$, and 4 in Fig. 6.2a shows that the approximation is very good. The deviation of the approximation from the exact curve is less than 0.02% in the range from $\varepsilon = 0$ to $\pm\varepsilon_h$, where $\varepsilon = 0$ and $\pm\varepsilon_h$ are the ε values at maximum and at half maximum, respectively. But the approximation is also very good for the curves with $\gamma_0 = 10\ \mathrm{cm}^{-1}$ and $\gamma_0 = 30\ \mathrm{cm}^{-1}$ of Fig. 6.2b; the deviations between $\varepsilon = 0$ and $\pm\varepsilon_h$ are less than 0.2% and 2%, respectively. This indicates that for the ESR line shape it is not important whether the exciton motion is coherent or incoherent, i.e. whether $\gamma_0 < 2J$ or $\gamma_0 > 2J$. What is of importance in this context is the magnitude of J^2/γ_0 compared to $|A - B|$.

6.1.5b Expansion at $\omega = A + B$ for $\gamma_0 \gg J$ (Incoherent Exciton Motion) and $2\gamma_1,\ J^2/\gamma_0 \gg |A - B|$

For the line shape we have (6.26), where ε is given by (6.27) and

$$\tilde{\Gamma} = (A - B)^2/2(2\gamma_1 + J^2/\gamma_0) \quad . \tag{6.29}$$

Again the maximum of the ESR line is at $\omega = A + B$. The temperature dependence of the line width $\tilde{\Gamma}$ is now determined by γ_0 and by γ_1, where γ_1 describes the strength of the non-local fluctuations. If $2\gamma_1$ increases more rapidly than J^2/γ_0 decreases when temperature rises, the ESR line narrows. In the opposite case we obtain a broadening of the absorption line.

Comparing (6.28) and (6.29), we remark that the first expression is a special case of the second one, obtained for $2\gamma_1 \ll J^2/\gamma_0$, which may be realized for sufficiently low temperatures. But we see once more that for the ESR line shape the magnitude of $2\gamma_1 + J^2/\gamma_0$ as compared to $|A - B|$ is the decisive quantity. The comparison with (4.17) shows that this expression is approximately the hopping rate in the diffusion constant of exciton transport in the nearest-neighbour approximation.

6.1.5c Expansion at $\omega = A + B$ for very large γ_0 (Limiting Case of the Hopping Model for the Exciton Motion)

Expansion to second order in ε gives for the line shape of ESR

$$\chi''(\varepsilon) \sim \frac{4\gamma_1}{(A-B)^2} \cdot \left\{1 - \varepsilon^2\left[16\,\frac{\gamma_1^2}{(A-B)^4} - \frac{2}{(A-B)^2}\right]\right\} . \tag{6.30}$$

Now we consider two cases:

$\gamma_1 \gg |A - B|$:

$$\chi''(\varepsilon) \sim \frac{4\gamma_1}{(A-B)^2}\left\{1 - \varepsilon^2\, 16\,\frac{\gamma_1^2}{(A-B)^4}\right\} \tag{6.31}$$

or

$$\chi''(\varepsilon) \sim \frac{\tilde{\Gamma}}{\varepsilon^2 + \tilde{\Gamma}^2} \,, \qquad \tilde{\Gamma} = \frac{(A-B)^2}{4\gamma_1} \,, \tag{6.32}$$

$\gamma_1 \ll |A - B|$:

$$\chi''(\varepsilon) \sim \frac{4\gamma_1}{(A-B)^4}\left\{(A-B)^2 + 2\varepsilon^2\right\} . \tag{6.33}$$

The transition from (6.31) to (6.32) can be interpreted as the first step of a Padê approximation /256/. In the first case of rapid incoherent motion (remember that $2\gamma_1$ is the hopping rate in the limiting case of the completely incoherent motion) we have a line at $\omega = A + B$ with line width $\tilde{\Gamma}$ from (6.32), whereas in the second case we find a minimum at $\omega = A + B$.

6.1.5d Expansion at $\omega = 2A$ for $\gamma_0 \gg 2J$ (Incoherent Exciton Motion) and $2\gamma_1,\ J^2/\gamma_0 \ll |A - B|$.

Expanding for $\varepsilon = \omega - 2A$ to second order, we arrive at

$$\chi''(\omega) \sim \frac{1}{2(2\gamma_1 + J^2/\gamma_0)^3}\left\{(2\gamma_1 + J^2/\gamma_0)^2 - \left[\varepsilon + \frac{(2\gamma_1 + J^2/\gamma_0)^2}{(A-B)}\right]^2\right\} , \tag{6.34}$$

which may be written as

$$\chi''(\varepsilon) \sim \frac{1}{2}\,\frac{\tilde{\Gamma}}{(\varepsilon - \bar{\varepsilon})^2 + \tilde{\Gamma}^2} \tag{6.35}$$

with

$$\bar{\varepsilon} = -\frac{(2\gamma_1 + J^2/\gamma_0)^2}{(A - B)} \quad , \tag{6.36}$$

$$\tilde{\Gamma} = 2\gamma_1 + J^2/\gamma_0 \quad . \tag{6.37}$$

Equations (6.35,36) show that the maximum of the absorption line is displaced from 2A towards A + B. The line width is given by (6.37). Analogous expressions are obtained by expanding at ω = 2B. In these cases we have a broadening of the line and an increase in the shift, if $2\gamma_1$ is growing more rapidly than J^2/γ_0 is decreasing with increasing temperature.

The positions of the maxima of the ESR lines calculated from (6.36) are shown in the figures 6.2b and 6.4 by small vertical strokes and demonstrate that the relative deviation is small and for none of the represented lines larger than 0.3%.

Before concluding this subsection it should be mentioned that the ESR line shape has also been calculated when the effective magnetic field at one of the two molecules has a component orthogonal to the z direction; this calculation has been carried out for the completely incoherent /174/ and for the coupled coherent and incoherent motion /180/.

6.2 ESR Line Shape of Triplet Excitons in Dimers of Inequivalent Molecules (AB Pairs)

In contrast to Sect. 6.1 we now consider the coupled coherent and incoherent motion of excitons with spin S = 1 in a dimer. Furthermore, the fine-structure interaction is not replaced by an effective field but explicitly taken into account. The Hamiltonian of the coherent motion is given by $H_0 = H_{ex,0} + H_{s,0}$ (6.2), where $H_{ex,0}$ is obtained from the Liouvillian (5.11a) and $H_{s,0}$ is represented in (6.4-6). The incoherent part of the motion is described by the Liouville operator (5.11b). The total Liouville operator then reads

$$L = L_0 + L_1 \quad , \tag{6.38}$$

and the equation of motion for the density operator becomes

$$\dot{\rho} = -iL\rho \quad . \tag{6.39}$$

In our case the Liouville operator may be represented by a non-hermitean 36 × 36 matrix, which is given explicitly in /38/.

6.2.1 The Parameters of the Model

Realizations of such pair systems are obtained by doping a naphthalene-d_8 matrix with naphthalene-h_8 molecules /135,136,144,145/. In this material the guests are embedded substitutionally. The parameters of our equations, which are chosen to model this system, are the exchange interaction integral J, the fine-structure tensors $\underline{\underline{F}}_1$ and $\underline{\underline{F}}_2$ of the two differently oriented molecules, the components of the external magnetic field $\underline{H}$, and γ_0 and γ_1, describing the strengths of the local and non-local fluctuations, respectively. For the naphthalene-h_8 pair embedded in the deuterated host, the exchange interaction integral J is known from optical measurements /98,139/ as

$$J = 1.20\ \mathrm{cm}^{-1} \mathrel{\hat{=}} 1.28\cdot 10^4\ \mathrm{G}\ . \tag{6.40}$$

The fine-structure tensors $\underline{\underline{F}}_1$ and $\underline{\underline{F}}_2$ are diagonal in the principal axes system $\{\xi_i, \eta_i, \zeta_i\}$ of the respective molecule and are given by (i = 1,2)

$$\begin{aligned} F^{(i)}_{\xi_i\xi_i} &= -X = E - \frac{1}{3} D\ , \\ F^{(i)}_{\eta_i\eta_i} &= -Y = -E - \frac{1}{3} D\ , \\ F^{(i)}_{\zeta_i\zeta_i} &= -Z = \frac{2}{3} D\ . \end{aligned} \tag{6.41}$$

The numerical values for the fine-structure parameters are /138/

$$\begin{aligned} D &= 1063.3\ \mathrm{G} \mathrel{\hat{=}} 9.9387\cdot 10^{-2}\ \mathrm{cm}^{-1}\ , \\ E &= -\ 164.7\ \mathrm{G} \mathrel{\hat{=}} -1.534\cdot 10^{-2}\ \mathrm{cm}^{-1}\ . \end{aligned} \tag{6.42}$$

[The D used here should not be mixed up with that of (6.4), which is the difference of the fine-structure tensors.] The orientations of the principal axes systems $\{\xi_i, \eta_i, \zeta_i\}$ of both molecules with respect to the crystal axes system {a,b,c'} are determined by the direction cosines of Table 6.1 and the angles of Table 6.2 /257/. The

Table 6.1. Direction cosines of the two inequivalent molecules with respect to the crystal axes /163/

	ξ_1 ξ_2	η_1 η_2	ζ_1 ζ_2
a	$\cos\chi$	$\cos\chi'$	$\cos\chi''$
b	$\pm\cos\psi$	$\pm\cos\psi'$	$\pm\cos\psi''$
c'	$\cos\omega$	$\cos\omega'$	$\cos\omega''$

Table 6.2. Angles between molecular axes of molecule 1 and crystal axes /257/

	ξ_i	η_i	ζ_i
a	115.97^0	71.29^0	32.87^0
b	102.14^0	29.33^0	116.26^0
c'	29.06^0	68.26^0	71.68^0

strength of the magnetic field is assumed to be

$$|\underline{H}| = 4000 \text{ G} \mathrel{\hat{=}} 0.37388 \text{ cm}^{-1} \quad . \tag{6.43}$$

In all subsequent line shape calculations, $\underline{H}$ is oriented in the y_p-z_p plane of an axes system $\{x_p, y_p, z_p\}$ and forms an angle of $+60^0$ with the y_p axis. The system $\{x_p, y_p, z_p\}$ is defined by the tensor $\underline{\underline{M}}$ of (6.4) being diagonal. In naphthalene crystals the y_p axis coincides with the b axis of the crystal; the angle between the $+z_p$ axis and the +a axis is $+22.4^0$ /145,258/.

6.2.2 ESR Line Positions in the Completely Coherent Case

Before determining the line shapes for the coupled coherent and incoherent motion, in this subsection the positions of the ESR lines are determined in the completely coherent case for the AB pair and for isolated A and B molecules at the two inequivalent sites in the unit cell of the deuterated host. The line positions for the AB pair are obtained as the differences of the eigenvalues of the Hamiltonian H_0 (6.2). For isolated A and B molecules the eigenvalues and line positions are obtained by neglecting the interaction between the molecules, i.e. for $J = 0$.

The solution of the eigenvalue problem of H_0 for the parameter values given above results in the line positions for the AB pair and is pictured together with these for isolated A and B molecules in Fig. 6.5 when the magnetic field is rotated about the x_p, y_p, and z_p axes. The scale of this figure cannot show that the ESR lines of the AB pair consist of two pairs of very closely neighboured lines; therefore, in the coherent limit the ESR spectrum of the AB pairs shows four lines as well. The distance of the lines in each of the two pairs is represented in Fig. 6.6. The figure shows that the distance depends on the orientation of the magnetic high field and is of the order of 10 G.

In the following subsection the ESR spectrum is calculated for the coupled coherent and incoherent motion when the magnetic field is rotated about the x_p axis (i.e. within the y_p z_p plane) and forms an angle of 60^0 with the y_p axis. For this orientation of the magnetic field the eigenvalues and their differences, which are relevant for the ESR spectrum, are given in Tables 6.3,4.

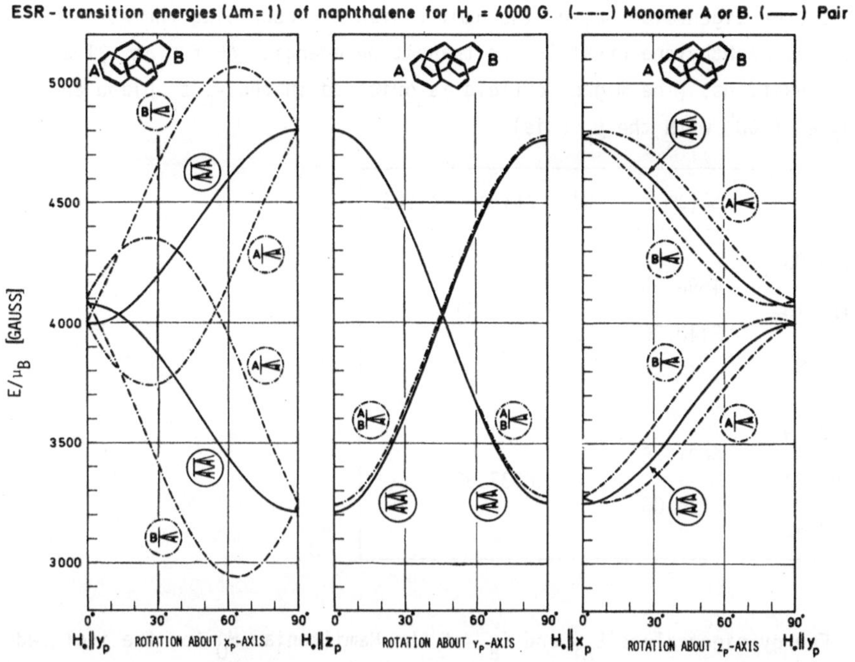

Fig. 6.5. ESR transition energies of monomers and molecular pairs in naphthalene for rotation of the magnetic field about the x_p, y_p, and z_p axes /258/

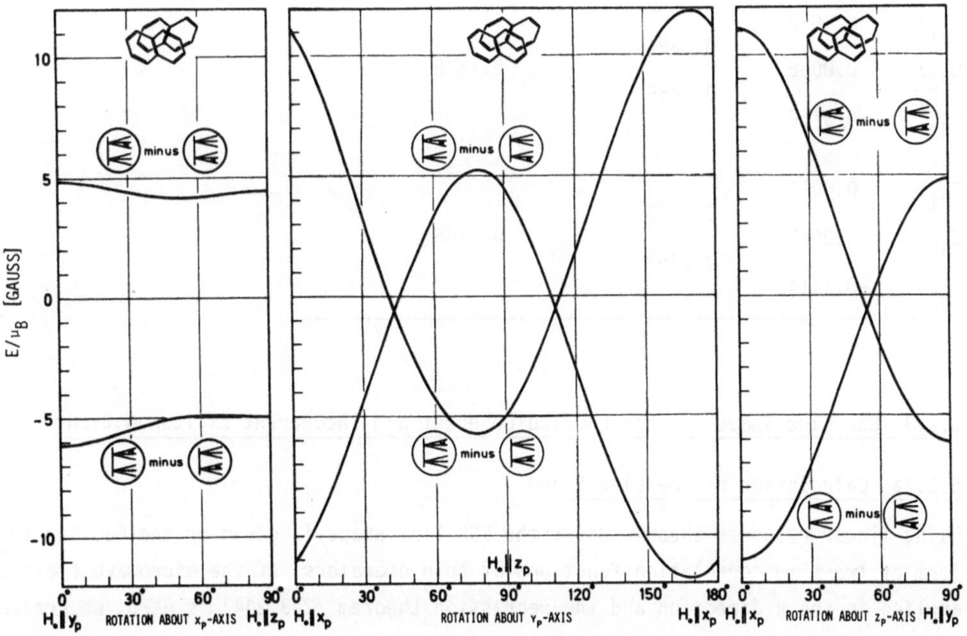

Fig. 6.6. Difference of the fine-structure splitting of the pair lines in naphthalene /258/

Table 6.3. Energy eigenvalues E_i of the Hamiltonian H_0 describing the AB pair in the unit cell of naphthalene crystals and transition energies ΔE for $\Delta m = 1$ and $\Delta m = 2$ ESR transitions. (The magnetic field is oriented in the y_p z_p plane and forms an angle of 60^0 with the y_p axis)

E_i[G]		$\Delta E(\Delta m = 1)$[G]	$\Delta E(\Delta m = 2)$[G]
E_1:	17015.00		
		4596.85	
E_3:	12418.15		8037.11
		3440.26	
E_5:	8977.89		
E_2:	- 8590.49		
		4601.75	
E_4:	-13192.24		8037.83
		3436.08	
E_6:	-16628.32		

Table 6.4. Energy eigenvalues E_{Ai} and E_{Bi} of the Hamiltonian H_0 for the isolated A and B molecules (J = 0) in the unit cell of the naphthalene crystal. (Orientation of the magnetic field as in Table 6.3)

E[cm^{-1}]		$\Delta E(\Delta m = 1)$[cm^{-1}]	$\Delta E(\Delta m = 2)$[cm^{-1}]
E_{A1}:	0.3818		
		0.3884	
E_{A2}:	-0.0066		0.7575
		0.3686	
E_{A3}:	-0.3752		
E_{B1}:	0.4071		
		0.4728	
E_{B2}:	-0.0657		0.7485
		0.2757	
E_{B3}:	-0.3414		

6.2.3 ESR Line Shape for the Coupled Coherent and Incoherent Exciton Motion

6.2.3a Calculation of the Line Shape

Using linear response theory /245/, the ESR line shape is given by the Fourier transform of two-time correlation functions of spin operators. If the microwave field is applied in the x direction and the regression theorem /243,244/ is used, we arrive at the following expression for the line shape /38/:

$$\chi''(\omega) = \beta\omega \int_0^\infty dt\, (S_x, e^{-iL\tau} S_x) \cos\omega\tau \quad , \tag{6.44}$$

where $\beta = (kT)^{-1}$ and the scalar product

$$(A,B) = \mathrm{Tr}\{A^{\dagger} B\} \tag{6.45}$$

is defined in the same way as in (5.1). In arriving at the expression (6.44) for the line shape, a high-temperature approximation /259/ has been used, which is consistent with the stochastic description of the phonons in the Haken-Strobl model. Using the complex eigenvalues $R_i = \omega_i - i\kappa_i$ and the eigenvectors ρ^i of the Liouville operator,

$$L\,\rho^i = R_i\,\rho^i \quad , \quad i = 1,\ldots,36 \; , \tag{6.46}$$

as well as the eigenvectors η^i of the adjoint operator L^+, the line shape may be written as

$$\chi''(\omega) = \frac{i}{N} \sum_{j=1}^{36} (\eta^s\,, S_x\,\rho^j)(\eta^j\,, S_x\,\rho^s) \left(\frac{\omega}{\omega + R_j} - \frac{\omega}{\omega - R_j}\right) \; . \tag{6.47}$$

The normalization constant N is determined from

$$\int_0^{\infty} \chi''(\omega)\; d\omega = 1 \tag{6.48}$$

as

$$N = \sum_j (\eta^s\,, S_x\,\rho^j)(\eta^j\,, S_x\,\rho^s)(-iR_j) \cdot [-\ln(\omega_j^2 + \kappa_j^2) + 2i\,\arctan\,(\kappa_j/\omega_j)] \quad , \tag{6.49}$$

and ρ^s , η^s are the stationary solutions of (6.46).

The eigenvalue problem has been solved numerically using a modified version of a program by GRAD and BREBNER /260/. The numerical calculation /261/ proceeds in three steps. First, the real non-symmetric matrix is transformed to upper Hessenberg form by a similarity transformation. In the second step, the eigenvalues are calculated by an iteration procedure called "Q-R double step method". Then the eigenvectors are calculated by the procedure of the "inverse iteration". With the help of these eigensolutions the line shape is determined from (6.47).

6.2.3b Results and Discussion

ESR line shapes calculated in the way just described are pictured in Figs. 6.7-13 /38/. In all figures the exchange interaction integral, the fine-structure parameters, and the strength and orientation of the static magnetic field have the values of Sect. 6.2.1. In each figure either the strength of the local fluctuations γ_0 or the strength of the non-local fluctuations γ_1 is varied, while the other parameter is fixed. Figure 6.7 gives the ESR line shape for fixed $\gamma_1 = 10^{-5}$ G and for values of γ_0 between 0.1 and 1000 G ($\approx$ 0.1 cm^{-1}).

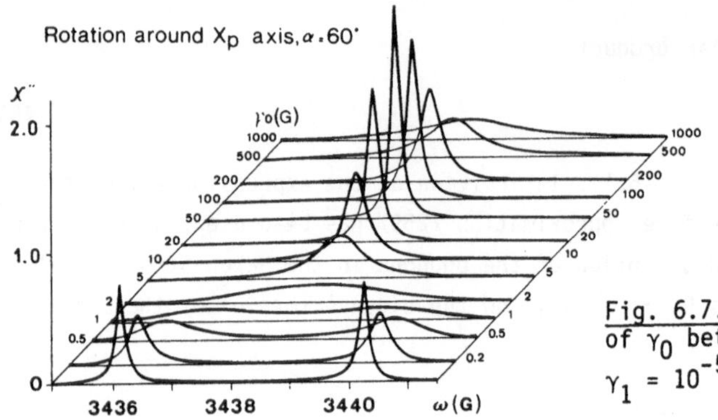

Fig. 6.7. ESR line shapes for values of γ_0 between 0.1 and 1000 G and for $\gamma_1 = 10^{-5}$ G /38/

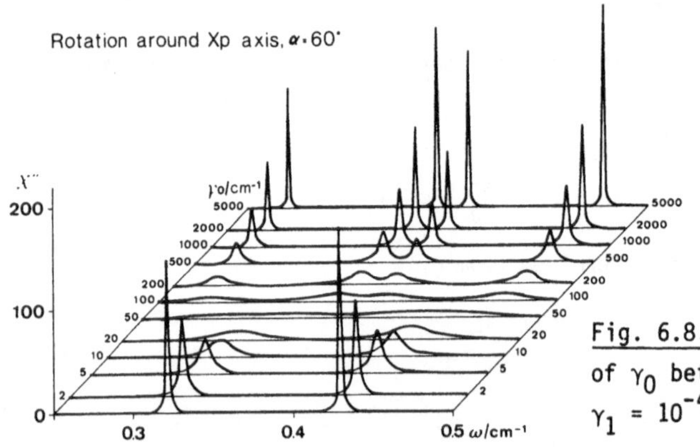

Fig. 6.8. ESR line shapes for values of γ_0 between 1 and 5000 cm^{-1} and for $\gamma_1 = 10^{-4}$ cm^{-1} /38/

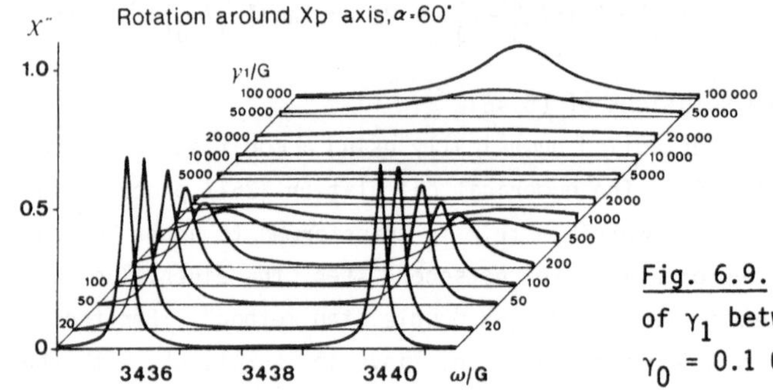

Fig. 6.9. ESR line shapes for values of γ_1 between 10 and 10^5 G and for γ_0 = 0.1 G /38/

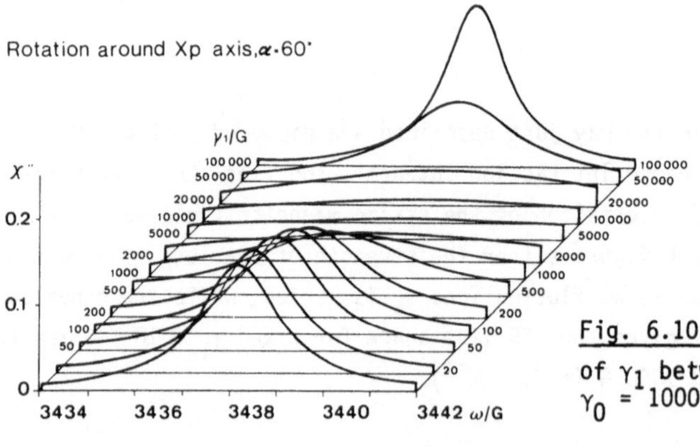

Fig. 6.10. ESR line shapes for values of γ_1 between 10 and 10^5 G and for γ_0 = 1000 G /38/

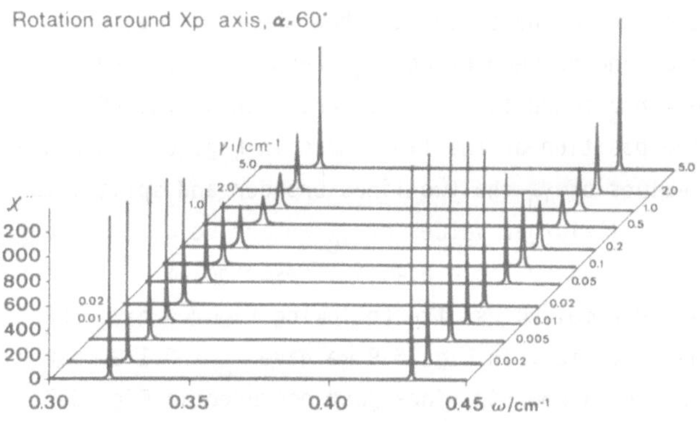

Fig. 6.11.
ESR line shapes for values of γ_1 between 10^{-3} and 5 cm^{-1} and for γ_0 = 0.1 cm^{-1} /38/

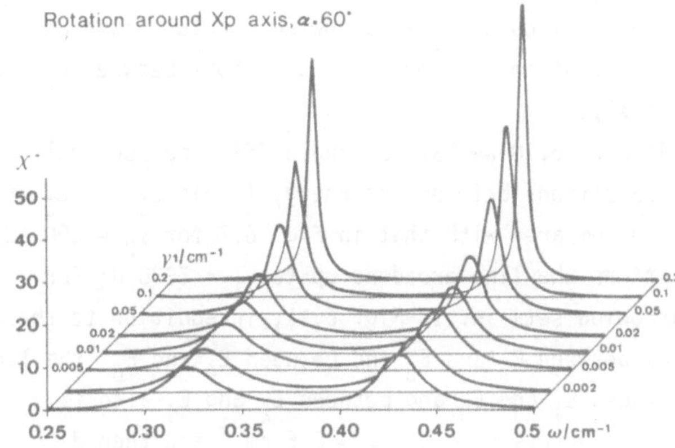

Fig. 6.12.
ESR line shapes for values of γ_1 between 10^{-3} and 0.2 cm^{-1} and for γ_0 = 20 cm^{-1} /38/

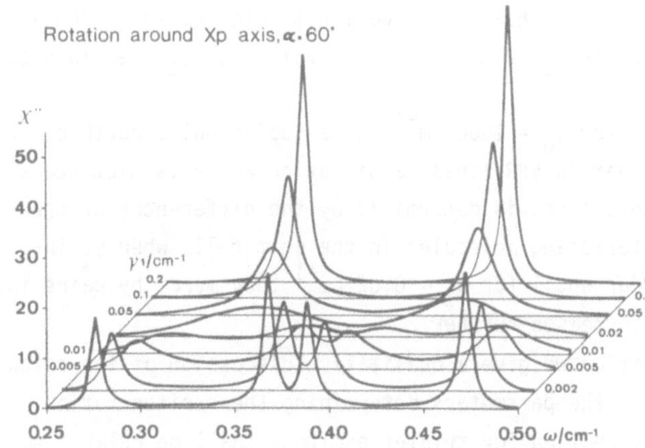

Fig. 6.13.
ESR line shapes for values of γ_1 between 10^{-3} and 0.2 cm^{-1} and for γ_0 = 1000 cm^{-1} /38/

On account of the large ω scale in this figure, only the lines corresponding to transitions between levels E_3 and E_5 and between E_4 and E_6 (compare Table 6.3) in the upper and lower Davydov component are pictured. With increasing values of γ_0 the lines of Fig. 6.7 broaden and move together. For values of γ_0 > 2 G, we have only a single line which narrows with increasing γ_0. For γ_0 > 50 G, however, the line broadens again[2]. A completely analogous behaviour show the ESR lines stemming from transitions between E_2 and E_4 and between E_1 and E_3, respectively. These ESR transitions have also been taken into account in Fig 6.8 which shows the ESR spectrum for larger values of γ_0. The position of the line shown in Fig. 6.7 is now at $\omega \approx 0.32$ cm^{-1}. With increasing values of γ_0 the two lines broaden and split into four lines for $\gamma_0 > 30$ cm^{-1}. For still larger values of γ_0 we arrive at four narrow ESR lines whose positions are determined by Table 6.4, and describe $\Delta m = 1$ ESR transitions in the noninteracting A and B molecules. The following figures give ESR line shapes when γ_0 is held fixed and γ_1 varies. In Fig. 6.9 we have $\gamma_0 = 0.1$ G. For small values of γ_1 we again have the narrow ESR lines just pictured in Fig. 6.7. These lines broaden with increasing γ_1 and coalesce into a single line for $\gamma_1 \approx$ 5000 G (≈ 0.5 cm^{-1}). For still larger values of γ_1 a narrowing starts. Again an analogous behaviour is shown by the lines stemming from the transitions between E_2 and E_4 and between E_1 and E_3, respectively.

For $\gamma_0 = 1000$ G (≈ 0.1 cm^{-1}) in Fig. 6.10 we have a single ESR line for small values of γ_1, stemming from the transitions between the energy levels E_3 and E_5 and between E_4 and E_6. This line may be compared with that in Fig. 6.7 for $\gamma_0 = 1000$ G. With increasing non-local fluctuations the line broadens up to $\gamma_1 = 5000$ G; for still larger values of γ_1 the narrowing sets in. In Fig. 6.11, in addition to the line stemming from the transitions between E_3 and E_5 and between E_4 and E_6, the line originating in the transitions between E_1 and E_3 and between E_2 and E_4 also is shown. The width of both lines first increases up to $\gamma_1 = 0.5$ cm^{-1} and then decreases. The ESR spectrum for $\gamma_0 = 20$ cm^{-1} also shows only two lines (Fig. 6.12), which are relatively broad for small values of γ_1. With increasing values of γ_1 the width of the two lines becomes smaller.

The ESR spectrum in Fig. 6.13 for $\gamma_0 = 1000$ cm^{-1} shows additional structure. For small values of γ_1 we have four narrow ESR lines, a situation which is pictured also in Fig. 6.8. The position of these lines is determined by the differences of the energy eigenvalues of the noninteracting molecules in the unit cell. When γ_1 increases these lines become broader until for $\gamma_1 \approx 0.02$ cm^{-1} they merge by pairs into two lines which narrow when γ_1 increases further.

In the final part of this section we give a qualitative discussion of the calculated ESR line shapes in terms of the parameters determining the exciton motion. Furthermore, the results of this section for triplet excitons shall be related to those for excitations with spin 1/2 of Sect. 6.2.

Remember that according to the discussion in connection with (3.1) and (3.5), $2\gamma_1$ represents the hopping rate between the two sites of a dimer in the completely incoherent case and that γ_0 is the scattering rate between the upper and lower Davydov components in the coherent case. With this in mind it is easy to understand the behaviour of the line shape in Fig. 6.7. For small values of γ_0 we have two

[2] Such a behaviour of the line width with increasing temperature has been observed for linear excitons in TCB /157a/. The author wishes to thank Prof. J. Schmidt for this comment.

$\Delta m = 1$ ESR lines (besides the other two at $\omega \approx 4600$ G) at $\omega_1 \approx 3440$ G and at $\omega_2 \approx 3436$ G corresponding to transitions between E_3 and E_5 and between E_4 and E_6 in the upper and lower Davydov components. With increasing values of γ_0 the two lines broden and coalesce for $\gamma_0 = 2$ G, i.e. $2\gamma_0 = \omega_1 - \omega_2$. For still larger values of γ_0 the single line becomes smaller. This situation may be described as a motional narrowing, where the spin does not move between different sites but is scattered between the two Davydov states with a scattering rate γ_0 /38/. This limiting case of our model has been treated by VAN'T HOF and SCHMIDT /143/. However, in this motional narrowing picture the successive broadening of the ESR lines and the splitting into four lines for still larger values of γ_0 (Fig. 6.8) cannot be explained; in order to discuss also these effects from a unified viewpoint, the whole 36×36 density matrix has to be considered.

However, if we consider in Fig. 6.7 only γ_0 values larger than 50 G, the behaviour of the line shape together with that one of Fig. 6.8 may be related to the line shapes for the spin 1/2 represented in Fig. 6.2a,b. In both situations we have with increasing γ_0 first a broadening of the ESR lines, then a splitting into several lines, ending up with narrow lines at the positions of those of the isolated molecules in the pair. This may be interpreted as a localization of the excitation at the two differently oriented molecules of the pair in the sense that on account of the strong local fluctuations described by the large value of γ_0, phase relations between the molecules have been destroyed completely. From (6.28,37) it is obvious that the line shape and width is determined by the difference in line positions of the isolated molecules and by J^2/γ_0. Starting at very large values of γ_0, with decreasing γ_0 we have a broadening of the lines at the isolated molecules. If $2J^2/\gamma_0$ is equal to the frequency difference of corresponding ESR transitions, the lines merge to a single line, which narrows with γ_0 further decreasing. This narrowing, however, turns into a broadening, when γ_0 is small enough that the range of the inter-Davydov scattering described above is reached.

Another case, which can also be understood in terms of a motional narrowing, is represented in Fig. 6.13. On account of $\gamma_0 = 1000\ \mathrm{cm}^{-1} \gg J$, the exciton motion is incoherent. For very small values of γ_1 we have the four ESR lines corresponding to $\Delta m = 1$ transitions in the A and B molecules (Table 6.4). If $4\gamma_1 \approx 0.08\ \mathrm{cm}^{-1}$ and thus equals the difference in the line positions of corresponding ESR transitions in the two molecules, the lines merge into a single line that narrows when γ_1 increases further. This is exactly the usual case of motional narrowing, where the excitation is hopping between the two molecules as described in (3.1) if the non-diagonal terms ρ_{12} and ρ_{21} may be neglected. The situation is similar in Fig. 6.12, where $\gamma_0 \gg 2J$ and the exciton motion,therefore, is incoherent. In Figs. 6.9-11, however, $\gamma_0 \ll 2J$; and if γ_1 is small too, we have the coherent exciton motion. With increasing values of γ_1 the damping parts of the eigenvalues λ_3 and λ_4 in (3.6) increase. For $\gamma_1 = 0.5\ \mathrm{cm}^{-1}$ (≈ 5000 G), i.e. $4\gamma_1 = 2\mathrm{cm}^{-1}$, the damping part of the eigenvalues has the same magnitude as the oscillatory part, and now we have a situation that can be denoted as overdamped. When γ_1 increases further, the narrowing of the ESR lines starts.

In our calculation we have described the exciton motion within the Haken-Strobl model, which takes into account the influence of the phonons in a stochastic manner. The consequence is that in the stationary state of the model all energy levels in the upper and lower Davydov components are populated with the same occupation probability, whereas in the real crystal we have a Boltzmann distribution. Taking this into account, we expect for $kT < 2J$ the contribution to the ESR spectrum of the transitions between E_3 and E_5 at 3440 G in the upper Davydov component to be weaker than that of the transition between E_4 and E_6 at 3436 G in the lower Davydov component (Fig. 6.7). The details of the transition in Fig. 6.7 from the two lines to the single averaged line will therefore depend on how γ_0 and the Boltzmann factor increase with temperature. For $kT > 2J$, however, the calculations should directly apply to experimental situations. In two recent publications WERTHEIMER and SILBEY /262/ discuss this problem. A recent investigation of the naphthalene pair /144,263/ has shown that the influence of an inhomogenous distribution of electronic excitation energies /145/ has to be taken into account. In the coherent case from the solution of the Schrödinger equation /140/ a relation between the inhomogeneously broadened optical and ESR lines has been derived /145,146/. A saturation dip in the ESR line /144,263/ may be explained within the model of the coupled coherent and incoherent exciton motion /145,258/.

6.3 ESR of Quasi-Incoherent Excitons in Molecular Crystals with Two Differently Oriented Molecules in the Unit Cell

In this section a theory of the ESR line shape will be presented that can be or has been applied to several recently investigated systems such as naphthalene /163,164/, anthracene /164,165/, tetracene /166,167/, and pyrene /168,169/. Several equivalent formulations of this theory exist in the literature /174-176,251,264/, the essential ingredient of which is a stochastic transport of the spin between different magnetic environments, and which are all more or less based on the fundamental papers of ANDERSON /171/ and KUBO /172/. In Sect. 6.3.1 we show - starting from the Haken-Strobl model - that in the case of the quasi-incoherent exciton motion the line shape is determined by an effective hopping rate of the excitation between inequivalent crystal sites. This hopping rate is determined not only by the stochastic part of the Hamiltonian of the model but also by the coherent interaction. Then the ESR line shape is determined from equations of motion for correlation functions and the line position and width are determined perturbationally. A discussion of experimental results concludes this section.

6.3.1 Equation of Motion for Correlation Functions in the Quasi-Incoherent Case of Exciton Motion

In the high-temperature limit the ESR line shape is given by the expression /259/

$$\chi''(\omega) = \frac{\omega}{N'} \int_0^\infty d\tau \cos\omega\tau \; \mathrm{Tr}\{S_x(\tau)\, S_x\} \quad . \tag{6.50}$$

Here N' is a normalization constant and $S_x(\tau)$ is the operator for a triplet spin in the Heisenberg picture. The line shape is thus determined by a two-time correlation function. With the notation

$$\langle\Omega_1(t)\;\Omega_2\rangle = \langle\Omega_1\,|\,\Omega_2\rangle \tag{6.51}$$

from (5.5) together with (2.13,27), we obtain the following equation of motion for the correlation function:

$$\begin{aligned}
\frac{d}{dt}\langle\Omega_1\,|\,\Omega_2\rangle = {} & i\,\langle[H_0,\Omega_1]\,|\,\Omega_2\rangle \\
& -\Big\{\frac{1}{2}\sum_{nj}\sum_{n'j'}(\gamma_{nj;n'j'}+\gamma_{n'j';nj})\,\langle b^+_{n'j'}\,b_{n'j'}\Omega_1\,|\,\Omega_2\rangle \\
& +\frac{1}{2}\sum_{nj}\sum_{n'j'}(\gamma_{nj;n'j'}+\gamma_{n'j';nj})\,\langle\Omega_1 b^+_{n'j'}\,b_{n'j'}|\Omega_2\rangle \\
& -\sum_{nj}\sum_{n'j'}(\gamma_{nj;n'j'}+\gamma_{n'j';nj})\,\langle b^+_{nj}\,b_{n'j'}\Omega_1 b^+_{n'j'}b_{nj}|\Omega_2\rangle \\
& -2\sum_{nj}\sum_{n'j'}\gamma_{nj,n'j'}\,\langle b^+_{nj}b_{n'j'}\Omega_1 b^+_{nj}b_{n'j'}|\Omega_2\rangle \\
& +2\sum_{nj}\gamma_{nj,nj}\,\langle b^+_{nj}b_{nj}\Omega_1 b^+_{nj}b_{nj}|\,\Omega_2\rangle\Big\} \quad .
\end{aligned} \tag{6.52}$$

For the calculation of the line shape we need the correlation function with $\Omega_1 = \Omega_2 = S_x$.

Starting the calculation of this correlation function by setting $\Omega_1 = \Omega_s$, where Ω_s contains only spin operators, a little algebra shows that the terms in the curly bracket of (6.52) cancel. Furthermore, Ω_s commutes with $H_{ex,0}$ and thus from (6.52) we arrive at /265/

$$\frac{d}{dt}\langle\Omega_s\,|\,\Omega_2\rangle = i\,\langle[H_{s,0},\Omega_s]|\,\Omega_2\rangle \quad . \tag{6.53}$$

In (6.6) it was shown that $H_{s,0}$ is composed of two different types of terms. The first type contains only spin operators $\Omega_{s'}$, while the second type has the structure $(\Delta b)\Omega_{s'}$. Therefore, the commutator $[H_{s,0},\Omega_s]$ leads to the following two types of correlation functions:

$$\langle\Omega_{s''}\,|\,\Omega_2\rangle \quad \text{and} \quad \langle\Delta b\Omega_{s''}\,|\,\Omega_2\rangle \quad . \tag{6.54}$$

The equation of motion for the first type is just (6.53), whereas for the second type we obtain from (6.52) with $\Omega_1 = \Delta b\Omega_s$:

$$\frac{d}{dt}\langle\Delta b\Omega_s \mid \Omega_2\rangle = i\,\langle[H_0,\Delta b\Omega_s] \mid \Omega_2\rangle - \Big\{\sum_n 2(\gamma_{n2;\nu1}+\gamma_{\nu1;n2})\,\langle\sum_\nu b^+_{\nu1}b_{\nu1}\Omega_s \mid \Omega_2\rangle - \sum_n 2(\gamma_{n1;\nu2}+\gamma_{\nu2;n1})\,\langle\sum_\nu b^+_{\nu2}b_{\nu2}\Omega_s \mid \Omega_2\rangle\Big\}\ . \tag{6.55}$$

Using the translational symmetry of the crystals, the second term in the braces of (6.55) may be transformed to

$$\sum_n(\gamma_{n1;\nu2}+\gamma_{\nu2;n1}) = \sum_a(\gamma_{\nu-a,1;\nu2}+\gamma_{\nu2;\nu-a,1}) = \sum_a(\gamma_{\nu1;\nu+a,2}+\gamma_{\nu+a,2;\nu,1}) = \sum_n(\gamma_{\nu1;n2}+\gamma_{n2;\nu1})\ . \tag{6.56}$$

With this result (6.55) may be rewritten:

$$\frac{d}{dt}\langle\Delta b\Omega_s \mid \Omega_2\rangle = i\,\langle[H_0,\Delta b\Omega_s] \mid \Omega_2\rangle - \{\sum_n 2(\gamma_{n2;\nu1}+\gamma_{\nu1;n2})\,\langle\Delta b\Omega_s \mid \Omega_2\rangle\}\ . \tag{6.57}$$

The commutator of $H_{s,0}$ with $\Delta b\Omega_s$ gives the two types of terms

$$[\Omega_{s'},\Delta b\Omega_s] = \Delta b\Omega_{s''}\ , \tag{6.58}$$

$$[\Delta b\Omega_{s'},\Delta b\Omega_s] = (\Delta b)^2\Omega_{s''}\ . \tag{6.59}$$

Now from the definition of (Δb) we have

$$(\Delta b)^2 = \sum_\nu\sum_{\nu'}(b^+_{\nu1}b_{\nu1}-b^+_{\nu2}b_{\nu2})(b^+_{\nu'1}b_{\nu'1}-b^+_{\nu'2}b_{\nu'2}) = \sum_\nu(b^+_{\nu1}b_{\nu1}+b^+_{\nu2}b_{\nu2}) = 1\ , \tag{6.60}$$

where we have used the fact that exactly one exciton is in the crystal. With this result from (6.58,59) it is clear that from the Hamiltonian of the spin motion no new correlation functions occur. Furthermore, we have to consider the terms arising from the commutator of the coherent part of the exciton motion:

$$[H_{ex,0},\Delta b\Omega_s] = [\sum_{nj}\sum_{n'j'}H_{nj;n'j'}b^+_{nj}\,b_{n'j'}\,,\ \sum_\nu(b^+_{\nu1}b_{\nu1}-b^+_{\nu2}b_{\nu2})\Omega_s]$$

$$[H_{ex,0},\Delta b\Omega_s] = \sum_{nj}\sum_\nu\{H_{nj;\nu1}b^+_{nj}b_{\nu1}+H_{\nu2;nj}b^+_{\nu2}b_{nj} - H_{nj;\nu2}b^+_{nj}b_{\nu2}-H_{\nu1;nj}b^+_{\nu1}b_{nj}\}\,\Omega_s\ . \tag{6.61}$$

We can normalize the energy of the system in such a way that the diagonal elements of $H_{ex,0}$ disappear. From the commutator with $H_{ex,0}$ the following new type of correlation function occurs:

$$<b^+_{m\ell} b_{m'\ell'} \Omega_s \mid \Omega_2> \quad , \tag{6.62}$$

where the creation and annihilation operators always refer to different molecules. The equation for this type of correlation function is derived as

$$\frac{d}{dt} <b^+_{m\ell} b_{m'\ell'} \Omega_s \mid \Omega_2> = i <[H_0, b^+_{m\ell} b_{m'\ell'} \Omega_s] | \Omega_2> - \{2\Gamma <b^+_{m\ell} b_{m'\ell'} \Omega_s | \Omega_2> - 2\gamma_{m'\ell';m\ell} <b^+_{m'\ell'} b_{m\ell} \Omega_s | \Omega_2>\} . \tag{6.63a}$$

For the following it is useful to consider also

$$\frac{d}{dt} <b^+_{m'\ell'} b_{m\ell} \Omega_s \mid \Omega_2> = i <[H_0, b^+_{m'\ell'} b_{m\ell} \Omega_s] | \Omega_2> - \{2\Gamma <b^+_{m'\ell'} b_{m\ell} \Omega_s \mid \Omega_2> - 2\gamma_{m\ell;m'\ell'} <b^+_{m\ell} b_{m'\ell'} \Omega_s \mid \Omega_2>\}, \tag{6.63b}$$

with

$$2\Gamma = \sum_{n,j} (\gamma_{nj;m\ell} + \gamma_{m\ell;nj}) \quad . \tag{6.64}$$

Up to this point the calculations have been exact within the one exciton picture. But now we introduce the decisive approximation. The non-diagonal ($m,\ell \neq m'\ell'$) quantities $<b^+_{m\ell} b_{m'\ell'} \Omega_s | \Omega_2>$ describe phase relations of the motion of triplet excitons between molecule ℓ in unit-cell number m and molecule ℓ' in unit-cell number m'. We shall assume that these phase relations decay rapidly in a time interval small compared to the time intervals relevant for spin resonance. This finally allows the elimination of the non-diagonal terms from the equations of motion in an adiabatic approximation /266,267/. In mathematical terms we shall assume that

$$\frac{d}{dt} <b^+_{m\ell} b_{m'\ell'} \Omega_s \mid \Omega_2> \ll 2\Gamma <b^+_{m\ell} b_{m'\ell'} \Omega_s \mid \Omega_2> \quad . \tag{6.65}$$

Whether this approximation is justified or not depends not only on the magnitude of Γ, which is a measure for the total fluctuations and, therefore, on the temperature, but also on the magnitude of the magnetic field, which determines the time behaviour of the correlation functions directly describing the ESR line shape. Using optical absorption data /89/, phase relations of the exciton motion have been shown /35,42/ to be destroyed rather rapidly, except at low temperatures. This means that (6.65) is justified in a wide range of temperatures of experimental interest.

On account of (6.65) we may neglect the time derivatives in (6.63a,b) and solve for $<b^+_{m'\ell'} b_{m\ell} \Omega_s | \Omega_2>$:

$$\langle b^+_{m'\ell'} b_{m\ell} \Omega_s \mid \Omega_2 \rangle = \frac{i}{2} \frac{\gamma_{m\ell;m'\ell'}}{\Gamma^2 - \gamma_{m'\ell';m\ell}\,\gamma_{m\ell;m'\ell'}} \langle [H_0, b^+_{m\ell}\, b_{m'\ell'} \Omega_s] | \Omega_2 \rangle$$
$$+ \frac{i}{2} \frac{\Gamma}{\Gamma^2 - \gamma_{m'\ell';m\ell}\,\gamma_{m\ell;m'\ell'}} \langle [H_0, b^+_{m'\ell'} b_{m\ell} \Omega_s] | \Omega_2 \rangle \quad . \tag{6.66}$$

We are considering the situation that phase relations between different molecules are lost rather rapidly. Therefore, we shall assume that correlation functions which are non-diagonal in the exciton operators are much smaller than correlation functions whose exciton operators refer to the same molecule. In this sense (6.66) might be the starting point for an iterative procedure for the calculation of the non-diagonal correlation functions. However, in this paper we shall execute the first step only and express the non-diagonal correlation functions by the diagonal ones. To that end we have to evaluate the commutators.

$H_{s,0}$ is of the form

$$H_{s,0} = \Omega_{s'} + \Delta b \Omega_{s''} \quad , \tag{6.67}$$

where $\Omega_{s'}$ and $\Omega_{s''}$ contain solely spin operators. Thus we have

$$[\Omega_{s'}, b^+_{m\ell} b_{m'\ell'} \Omega_s] = b^+_{m\ell} b_{m'\ell'} \Omega_{s'''} \quad . \tag{6.68}$$

According to (6.62) the creation and annihilation operators refer to different molecules, and, therefore, in the approximation considered we have no contribution from (6.66) and (6.68).

From the second term in $H_{s,0}$ we obtain

$$[\sum_\nu (b^+_{\nu 1} b_{\nu 1} - b^+_{\nu 2} b_{\nu 2}) \Omega_{s''}, b^+_{m\ell} b_{m'\ell'} \Omega_s]$$
$$= \delta_{\ell 1} b^+_{m1} b_{m'\ell'} \Omega_{s''} \Omega_s - \delta_{\ell' 1} b^+_{m\ell} b_{m'1} \Omega_s \Omega_{s''}$$
$$- \delta_{\ell 2} b^+_{m2} b_{m'\ell'} \Omega_{s''} \Omega_s + \delta_{\ell' 2} b^+_{m\ell} b_{m'2} \Omega_s \Omega_{s''} \quad . \tag{6.69}$$

Let us discuss the first term on the right-hand side of (6.69). If $\ell = 1$, then $\ell' = 2$. Thus this term is non-diagonal, and the same is true for the remaining three terms. Therefore, in the lowest approximation there is no contribution to (6.66) from the commutator with $H_{s,0}$.

From $H_{ex,0}$ we have

$$[H_{ex,0}, b^+_{m'\ell'} b_{m\ell} \Omega_s] = \left(\sum_{nj} H_{nj;m'\ell'} b^+_{nj} b_{m\ell} - \sum_{n'j'} H_{m\ell;n'j'} b^+_{m'\ell'} b_{n'j'}\right) \Omega_s \quad , \tag{6.70}$$

$$[H_{ex,0}, b^+_{m\ell} b_{m'\ell'} \Omega_s] = \left(\sum_{nj} H_{nj;m\ell} b^+_{nj} b_{m'\ell'} - \sum_{n'j'} H_{m'\ell';n'j'} b^+_{m\ell} b_{n'j'}\right) \Omega_s \quad . \tag{6.71}$$

Retaining on the right-hand side of (6.70,71) only the terms diagonal in the exciton operators, we arrive at

$$[H_{ex,0}, b^+_{m'\ell'} b_{m\ell} \Omega_s] = H_{m\ell;m'\ell'} (b^+_{m\ell} b_{m\ell} - b^+_{m'\ell'} b_{m'\ell'}) \Omega_s \quad , \tag{6.72}$$

$$[H_{ex,0}, b^+_{m\ell} b_{m'\ell'} \Omega_s] = H_{m'\ell';m\ell} (b^+_{m'\ell'} b_{m'\ell'} - b^+_{m\ell} b_{m\ell}) \Omega_s \quad . \tag{6.73}$$

Inserting (6.72,73) into (6.66) we obtain for the correlation functions, which are non-diagonal in the exciton operators,

$$\begin{aligned} \langle b^+_{m'\ell'} b_{m\ell} \Omega_s | \Omega_2 \rangle = {} & \frac{i}{2} \frac{\gamma_{m\ell;m'\ell'}}{\Gamma^2 - \gamma_{m'\ell';m\ell}\, \gamma_{m\ell;m'\ell'}} \\ & \cdot H_{m'\ell';m\ell} \langle (b^+_{m'\ell'} b_{m'\ell'} - b^+_{m\ell} b_{m\ell}) \Omega_s | \Omega_2 \rangle \\ & - \frac{i}{2} \frac{\Gamma}{\Gamma^2 - \gamma_{m'\ell';m\ell}\, \gamma_{m\ell;m'\ell'}} \\ & \cdot H_{m\ell;m'\ell'} \langle (b^+_{m'\ell'} b_{m'\ell'} - b^+_{m\ell} b_{m\ell}) \Omega_s | \Omega_2 \rangle \quad . \end{aligned} \tag{6.74}$$

From (6.61) we immediately have

$$\begin{aligned} i \langle [H_{ex,0}, \Delta b \Omega_s] | \Omega_2 \rangle = {} & \sum_{nj} \sum_{\nu} \{ i H_{nj;\nu 1} \langle b^+_{nj} b_{\nu 1} \Omega_s | \Omega_2 \rangle \\ & + i H_{\nu 2;nj} \langle b^+_{\nu 2} b_{nj} \Omega_s | \Omega_2 \rangle - i H_{nj;\nu 2} \langle b^+_{nj} b_{\nu 2} \Omega_s | \Omega_2 \rangle - i H_{\nu 1;nj} \langle b^+_{\nu 1} b_{nj} \Omega_s | \Omega_2 \rangle \} \quad . \end{aligned}$$

Using (6.74) and some algebraic operations, we arrive at

$$\begin{aligned} i \langle [H_{ex,0}, \Delta b \Omega_s] | \Omega_2 \rangle = {} & \\ & \sum_{n\nu} \frac{\langle b^+_{\nu 1} b_{\nu 1} \Omega_s | \Omega_2 \rangle}{\Gamma^2 - \gamma_{\nu 1;n2}\, \gamma_{n2;\nu 1}} \cdot \{ \gamma_{n2;\nu 1} H^2_{\nu 1;n2} + \gamma_{\nu 1;n2} H^2_{n2;\nu 1} - 2\Gamma H_{\nu 1;n2} H_{n2;\nu 1} \} \\ & + \sum_{n\nu} \frac{\langle b^+_{\nu 2} b_{\nu 2} \Omega_s | \Omega_2 \rangle}{\Gamma^2 - \gamma_{\nu 2;n1}\, \gamma_{n1;\nu 2}} \cdot \{ - \gamma_{n1;\nu 2} H^2_{\nu 2;n1} - \gamma_{\nu 2;n1} H^2_{n1;\nu 2} + 2\Gamma H_{\nu 2;n1} H_{n1;\nu 2} \} \quad . \end{aligned} \tag{6.75}$$

On account of the translational symmetry of the crystal, the second term in (6.75) may be transformed in the same way as in (6.56) and we finally arrive at

$$\begin{aligned} i \langle [H_{ex,0}, \Delta b \Omega_s] | \Omega_2 \rangle = - \sum_{n} \{ & - \gamma_{n2;\nu 1} H^2_{\nu 1;n2} - \gamma_{\nu 1;n2} H^2_{n2;\nu 1} \\ & + 2\Gamma H_{\nu 1;n2} H_{n2;\nu 1} \} \frac{\langle \Delta b \Omega_s | \Omega_2 \rangle}{\Gamma^2 - \gamma_{\nu 1;n2}\, \gamma_{n2;\nu 1}} \quad . \end{aligned} \tag{6.76}$$

Using this result, (6.57) becomes

$$\frac{d}{dt} \langle \Delta b\Omega_s | \Omega_2 \rangle = i \langle [H_{s,0}, \Delta b\Omega_s] | \Omega_2 \rangle - 4\Gamma_1 \langle \Delta b\Omega_s | \Omega_2 \rangle \quad , \tag{6.77}$$

where

$$4\Gamma_1 = \sum_n 2(\gamma_{n2;\nu 1} + \gamma_{\nu 1;n2}) + \sum_n \left\{ \frac{2\Gamma}{\Gamma^2 - \gamma_{\nu 1;n2}\,\gamma_{n2;\nu 1}} H_{\nu 1;n2} H_{n2;\nu 1} - \frac{\gamma_{n2;\nu 1} H^2_{\nu 1;n2} + \gamma_{\nu 1;n2} H^2_{n2;\nu 1}}{\Gamma^2 - \gamma_{\nu 1;n2}\,\gamma_{n2;\nu 1}} \right\} \tag{6.78}$$

represents an effective hopping rate from an arbitrary molecule to all inequivalent molecules in the crystal. The second sum in (6.78) describes the influence of the coherent interaction.

In (6.54) and the discussion following (6.58-60), it was shown that the commutators with $H_{s,0}$ lead only to correlation functions of the type

$$\langle \Omega_s | \Omega_2 \rangle \quad \text{and} \quad \langle \Delta b\Omega_s | \Omega_2 \rangle \quad .$$

Therefore, from (6.53,77) we may derive a complete set of coupled differential equations determining the ESR line shape. The number of its equations is determined by the dimension of the spin space under consideration and by the number of molecules per unit cell. For example, in the case of triplet excitons in molecular crystals with two molecules per unit cell, we have a set of 16 coupled equations. The essential point of this system of equations is that the dynamics of the electronic degrees of freedom is approximately expressed by the hopping rate Γ_1 between the two kinds of molecules. In this way contact is made to the other procedures for the calculation of ESR line shapes, which have been mentioned at the beginning of this section.

For triplet excitons in anthracene crystals it may be a satisfying approximation to consider solely the interaction with the nearest (inequivalent) neighbours in the ab plane. Then from (6.78) we arrive at

$$4\Gamma_1 = 16 \left(\gamma_1 + \frac{1}{2} \frac{H_1^2}{\Gamma + \gamma_1} \right) \quad . \tag{6.79}$$

From optical absorption experiments, H_1 and Γ have been derived /35,37/ (Sect. 5), and thus from the experimental value of Γ_1 we may determine γ_1.

It has been shown /93,94/ that in this nearest-neighbour approximation, the information derived from ESR and from diffusion measurements is the same. However, if we have to take into account the interactions with other lattice sites, the expressions for Γ_1 and for the diffusion constant become different, because in (6.78) we have to sum over inequivalent molecules only, whereas in the diffusion constant the interactions with all molecules enter.

6.3.2 Exact Calculation of the ESR Line Shape of Quasi-Incoherent Triplet Excitons in Molecular Crystals with Two Molecules in the Unit Cell

In this section we shall exactly evaluate the ESR line shape by a computer calculation. Approximate expressions for the line positions and widths are derived analytically and comparison with experimental results in anthracene is made /183,184,186/.

The ESR line shape is given by $\chi''(\omega)$ in (6.50). This equation may be written in the following way:

$$\chi''(\omega) = \frac{\omega}{2N'}\left(\tilde{K}_1^-(\omega) + \tilde{K}_1^+(\omega)\right) \quad , \tag{6.80}$$

where the Fourier-Laplace transforms of the correlation functions are determined from

$$\tilde{K}_1^\pm(\omega) = \int_0^\infty d\tau\, e^{\pm i\omega\tau}\, \mathrm{Tr}\{S_x(\tau)\, S_x\} \quad . \tag{6.81}$$

The equations of motion for the correlation functions are determined from (6.53,77) where $H_{s,0}$ is given in (6.4-6). The explicit calculation /183/ of the equation of motion needs the evaluation of the commutator with $H_{s,0}$. It may be shortened considerably by using the ring property of triplet spin operators. Using this property, all products of spin operators may be represented by eight operators, e.g. by

$$S_x\ ,\ S_y\ ,\ S_z\ ,\ S_x^2\ ,\ S_y^2\ ,\ S_xS_y\ ,\ S_yS_z\ ,\ S_zS_x \quad . \tag{6.82}$$

For brevity we introduce the following notation:

$$\begin{aligned}
K_1(t) &= \langle S_x|S_x\rangle \ , & K_9(t) &= \langle \Delta b S_x|S_x\rangle \ , \\
K_2(t) &= \langle S_y|S_x\rangle \ , & K_{10}(t) &= \langle \Delta b S_y|S_x\rangle \ , \\
K_3(t) &= \langle S_zS_x + S_xS_z|S_x\rangle \ , & K_{11}(t) &= \langle \Delta b(S_zS_x + S_xS_z)|S_x\rangle \ , \\
K_4(t) &= \langle S_yS_z + S_zS_y|S_x\rangle \ , & K_{12}(t) &= \langle \Delta b(S_yS_z + S_zS_y)|S_x\rangle \ , \\
K_5(t) &= \langle S_z|S_x\rangle \ , & K_{13}(t) &= \langle \Delta b S_z|S_x\rangle \ , \\
K_6(t) &= \langle S_xS_y + S_yS_x|S_x\rangle \ , & K_{14}(t) &= \langle \Delta b(S_xS_y + S_yS_x)|S_x\rangle \ , \\
K_7(t) &= \langle S_y^2 - S_z^2|S_x\rangle \ , & K_{15}(t) &= \langle \Delta b(S_y^2 - S_z^2)|S_x\rangle \ , \\
K_8(t) &= \langle S_z^2 - S_x^2|S_x\rangle \ , & K_{16}(t) &= \langle \Delta b(S_z^2 - S_x^2)|S_x\rangle \ .
\end{aligned} \tag{6.83}$$

As in (6.51) the operators in the first part of the angular bracket have to be understood as Heisenberg operators at time t. With (6.83) the equations of motion for the correlation functions may be written as

Table 6.5. Matrix $\underline{\underline{L}}$. Only non-vanishing elements are represented

	1	2	3	4	5	6	7	8	9	10	11	12	13	14	15	16
1		$-H_z$	M_{xy}	ΔM_y		$-M_{xz}$	$-2M_{yz}$				D_{xy}	ΔD_y		$-D_{xz}$	$-2D_{yz}$	
2	H_z		ΔM_z	$-M_{xy}$		M_{yz}		$-2M_{xz}$			ΔD_z	$-D_{xy}$		D_{yz}		$-2D_{xz}$
3	$-M_{xy}$	$-\Delta M_z$		$-H_z$	M_{yz}				$-D_{xy}$	$-\Delta D_z$			D_{yz}			
4	$-\Delta M_y$	M_{xy}	H_z		$-M_{xz}$				$-\Delta D_y$	D_{xy}			$-D_{xz}$			
5			$-M_{yz}$	M_{xz}		ΔM_x	$2M_{xy}$	$2M_{xy}$			$-D_{yz}$	D_{xz}		ΔD_x	$2D_{xy}$	$2D_{xy}$
6	M_{xz}	$-M_{yz}$			$-\Delta M_x$		$-2H_z$	$-2H_z$	D_{xz}	$-D_{yz}$			$-\Delta D_x$			
7	$2M_{yz}$	$-M_{xz}$			$-M_{xy}$	H_z			$2D_{yz}$	$-D_{xz}$			$-D_{xy}$			
8	$-M_{yz}$	$2M_{xz}$			$-M_{xy}$	H_z			$-D_{yz}$	$2D_{xz}$			$-D_{xy}$			
9			D_{xy}	ΔD_y		$-D_{xz}$	$-2D_{yz}$		$-4\Gamma_1$	$-H_z$	M_{xy}	ΔM_y		$-M_{xz}$	$-2M_{yz}$	
10			ΔD_z	$-D_{xy}$		D_{yz}		$-2D_{xz}$	H_z	$-4\Gamma_1$	ΔM_z	$-M_{xy}$		M_{yz}		$-2M_{xz}$
11	$-D_{xy}$	$-\Delta D_z$			D_{yz}				$-M_{xy}$	$-\Delta M_z$	$-4\Gamma_1$	$-H_z$	M_{yz}			
12	$-\Delta D_y$	D_{xy}			$-D_{xz}$				$-\Delta M_y$	M_{xy}	H_z	$-4\Gamma_1$	$-M_{xz}$			
13			$-D_{yz}$	D_{xz}		ΔD_x	$2D_{xy}$	$2D_{xy}$			$-M_{yz}$	M_{xz}	$-4\Gamma_1$	ΔM_x	$2M_{xy}$	$2M_{xy}$
14	D_{xz}	$-D_{yz}$			$-\Delta D_x$				M_{xz}	$-M_{yz}$			$-\Delta M_x$	$-4\Gamma_1$	$-2H_z$	$-2H_z$
15	$2D_{yz}$	$-D_{xz}$			$-D_{xy}$				$2M_{yz}$	$-M_{xz}$			$-M_{xy}$	H_z	$-4\Gamma_1$	
16	$-D_{yz}$	$2D_{xz}$			$-D_{xy}$				$-M_{yz}$	$2M_{xz}$			$-M_{xy}$	H_z		$-4\Gamma_1$

$$\dot{\underline{K}} = \underline{\underline{L}}\,\underline{K} \quad , \tag{6.84}$$

where $\underline{K}$ is a 16-dimensional column vector and $\underline{\underline{L}}$ the 16×16 matrix of Table 6.5 with

$$\Delta M_x = M_{xx} - M_{yy} \quad , \qquad \Delta M_y = M_{yy} - M_{zz} \quad , \qquad \Delta M_z = M_{zz} - M_{xx} \quad , \tag{6.85}$$

$$\Delta D_x = D_{xx} - D_{yy} \quad , \qquad \Delta D_y = D_{yy} - D_{zz} \quad , \qquad \Delta D_z = D_{zz} - D_{xx} \quad . \tag{6.86}$$

According to (6.80), the line shape may be calculated if we know the Laplace transform of the correlation function $K_1(t)$. Transforming (6.84) we arrive at (summation over repeated indices)

$$\int_0^\infty e^{\mp i\omega\tau}\,\dot{K}_k\,d\tau = L_{k\ell}\int_0^\infty e^{\mp i\omega\tau}\,K_\ell(\tau)\,d\tau = L_{k\ell}\tilde{K}_\ell^{\pm}(\omega) \quad , \qquad k = 1,\ldots,16 \tag{6.87}$$

or after calculating the left-hand side of this equation

$$(L_{k\ell} \mp \delta_{k\ell}i\omega)\,\tilde{K}_\ell^{\pm}(\omega) = -K_k(0) \quad . \tag{6.88}$$

The problem of calculating ESR line shapes is,therefore,reduced to that of solving a system of 16 algebraic equations, which is carried out numerically. The right-hand side of (6.88) is easily calculated:

$$K_1(0) = 2 \quad , \qquad K_k(0) = 0 \quad (k \neq 1) \quad . \tag{6.89}$$

The parameters in the matrix $\underline{\underline{L}}$ are the external magnetic field $\underline{H}$ and the hopping rate Γ_1 between inequivalent molecules. The parameters M_{ij} and D_{ij} are determined from the fine-structure tensors $\underline{\underline{F}}_1$ and $\underline{\underline{F}}_2$ of the two differently oriented molecules at sites 1 (A) and 2 (B). In the principal axes systems of their molecules the fine-struture tensors are diagonal and given by (6.41).

The coordinate system of the calculation (x,y,z) has been chosen in such a way that its z axis is parallel to the direction of the external magnetic field $\underline{H} = (0,0,H_z)$ and that its x axis is identical with the x' axis of the laboratory frame (x',y',z'). The x' axis is also assumed to be parallel to the high-frequency magnetic field. The angle between the z axis, i.e. between the magnetic high field, and the y' axis is called α.

The orientation of the crystal axes (a,b,c') as referred to the laboratory frame (x',y',z') is described by Eulerian angles ψ, ϑ, and φ, as shown in Fig. 6.14. Rotations of the magnetic field around the a, b, and c' axes are then described by the angles ($\psi = 0$, $\vartheta = 0$, $\varphi = 0$), ($\psi = 90^\circ$, $\vartheta = 0$, $\varphi = 0$) and ($\psi = 90^\circ$, $\vartheta = 90^\circ$, $\varphi = 0$), respectively. Further on, any other orientation of the crystal as referred to the axis of rotation may be chosen very simply. Finally, the orientation of the principal axes systems (ξ_1,η_1,ζ_1) and (ξ_2,η_2,ζ_2) of the two anthracene molecules in the unit cell as referred to the crystal axes are given by Tables 6.1 and 6.6 /257/. In the numerical calculation for D, E /164/, and H_z we use the values

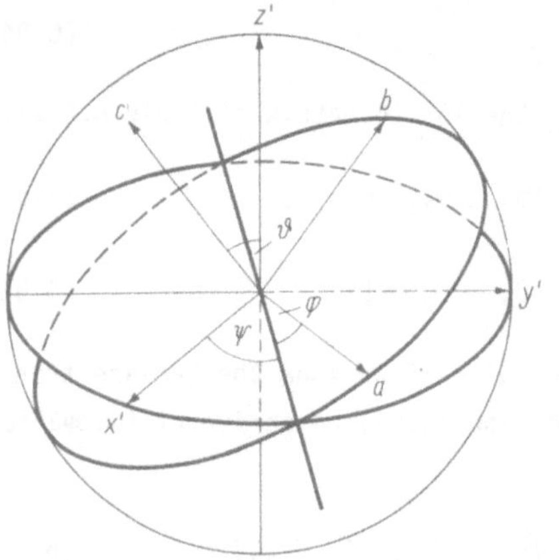

Fig. 6.14. Eulerian angles ψ, ϑ, φ between crystal axes and laboratory frame /183/

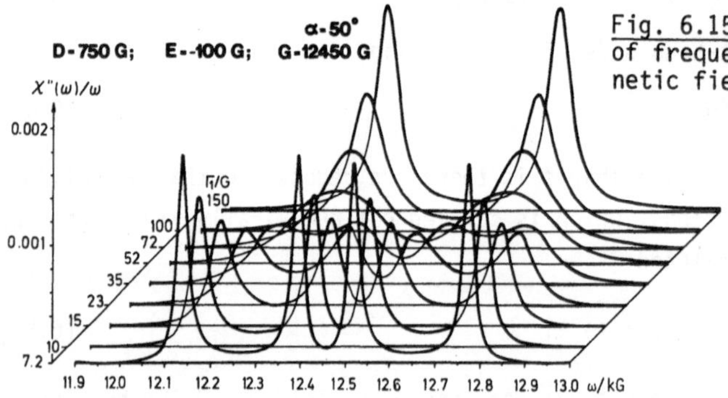

Fig. 6.15. Line shape as a function of frequency for rotation of the magnetic field around the a axis /183/

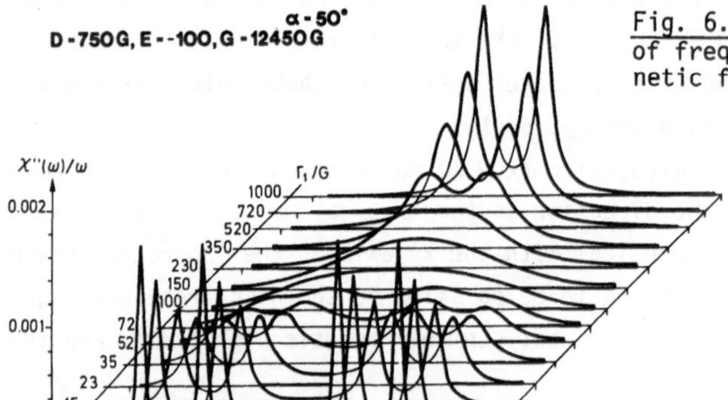

Fig. 6.16. Line shape as a function of frequency for rotation of the magnetic field around the c' axis /183/

$$D = 750\ \text{G} \quad , \qquad E = -100\ \text{G} \quad , \qquad H_z = 12450\ \text{G} \quad .$$

(The equations have been divided by $g\mu_B$ and, therefore, all quantities are measured in G.)

In Fig. 6.15 ($G = |\underline{H}|$) the unnormalized line shape $\chi''(\omega)/\omega$ has been pictured

Table 6.6. Angles between molecular and crystal axes for anthracene /257/

	ξ	η	ζ
a	$\chi = 119.6^{\circ}$	$\chi' = 71.5^{\circ}$	$\chi'' = 36.0^{\circ}$
b	$\psi = 97.3^{\circ}$	$\psi' = 26.6^{\circ}$	$\psi'' = 115.3^{\circ}$
c'	$\omega = 30.6^{\circ}$	$\omega' = 71.3^{\circ}$	$\omega'' = 66.3^{\circ}$

/183/ as a function of the frequency for several values of Γ_1 and for a fixed value of the magnetic field, which is rotated around the a axis with $\alpha = 50^{\circ}$. We see that we have four ESR lines for small values of Γ_1, e.g. for $\Gamma_1 = 10$ G, corresponding to the two lines of molecules 1 and to the two lines of molecule 2. For high values of Γ_1, we have merely two lines lying in the middle between the two lines at lower and at higher frequency. The transition from four lines at low values of Γ_1 to two lines at high values of Γ_1 occurs when $4\Gamma_1$ is approximately equal to the initial distance of those lines that coincide at very large values of Γ_1 (Sect. 6.1.3).

A situation similar to that just described is represented in Fig. 6.16, where the magnetic field has a fixed value and is rotated around the c' axis and again $\alpha = 50^{\circ}$. The only difference to Fig. 6.15 is that now, for large values of Γ_1, the first and the third and the second and the fourth lines coincide, respectively. In both cases, with increasing values of Γ_1, the line widths become smaller and smaller on account of motional narrowing.

Frequently the ESR line shape is not measured as a function of the frequency for a fixed value of the magnetic field but as a function of the strength of the magnetic field for fixed frequency. Obviously there is no problem of solving (6.88) also for this case /183/.

6.3.3 Eigenvalues Describing the ESR Line Shape

With the ansatz

$$\underline{K}(t) = \underline{K}(0)\, e^{Rt} \quad , \tag{6.90}$$

the differential equation (6.84) for the correlation function transforms to an eigenvalue problem

$$\underline{\underline{L}}\, \underline{K}^{i} = R_i\, \underline{K}^{i} \tag{6.91}$$

with eigenvalues R_i and eigenvectors $\underline{K}^i$. According to the dimension of the problem, we have 16 eigensolutions, and the general solution is given by their superposition

$$\underline{K}(t) = \sum_{i=1}^{16} c_i \underline{K}^i e^{R_i t} \quad , \tag{6.92}$$

showing that in general, the exact line shape is the sum of 16 Lorentzian lines. From Figs. 6.15,16 it seems reasonable to assume that for small values of Γ_1 the ESR line shape may be represented approximately by four, and for large values of Γ_1 by two Lorentzian lines only, using four and two suitably chosen eigenvalues, respectively. The positions of the maxima of the line shape are given by the imaginary parts, and the line widths by the real parts of these eigenvalues. The real and imaginary parts of these eigenvalues are pictured in Figs. 6.17-22 as a function of the angle α, which describes the orientation of the magnetic field in the crystal, and for several values of Γ_1. The eigenvalues are numbered from one to four, so that real and imaginary parts of one eigenvalue have the same number.

Let us consider Figs. 6.17,18 in more detail. Here the magnetic field is rotated around the a axis. $\alpha = 50^0$ corresponds to the situation pictured in Fig. 6.15. For $\Gamma_1 = 52$ G and smaller values, we have four different imaginary parts, corresponding

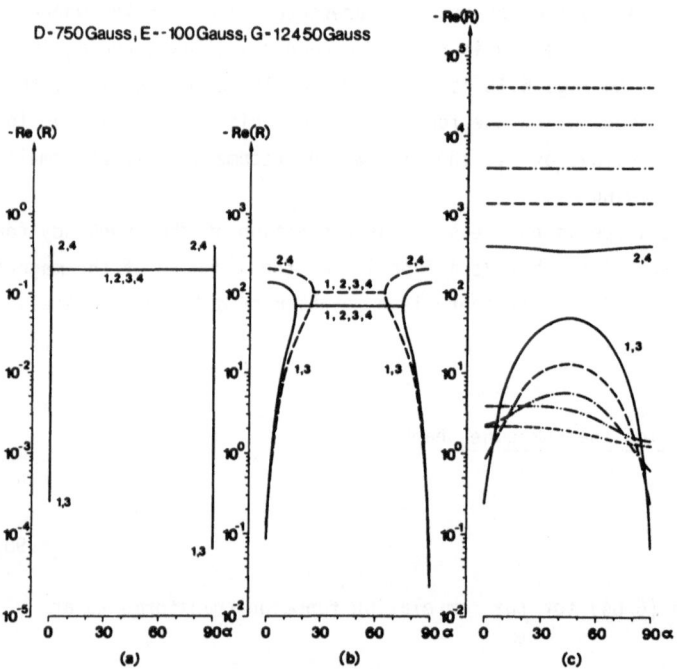

Fig. 6.17. Real parts of the eigenvalue R as a function of α for several values of Γ_1. The magnetic field is rotated around the a axis. a) $\Gamma_1 = 0.1$ G; b) $\Gamma_1 = 35$ G (——), 52 G (-----); c) $\Gamma_1 = 100$ G (——), 350 G (-----), 1000 G (·-·-·), 3500 G (-·····), 10 000 G (--·---) /183/

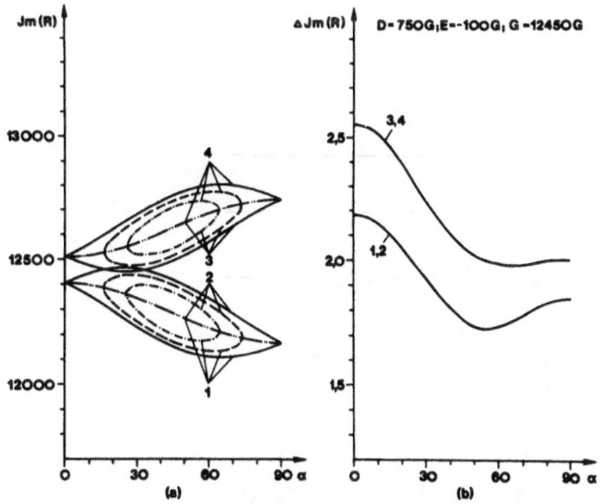

Fig. 6.18. (a) Imaginary parts of the eigenvalues R as a function of α for several values of Γ_1. The magnetic field is rotated around the a axis, Γ_1 = 0.1 G (———), 35 G (-----), 52 G (·-·-·), 100 G (-·····). (b) Difference Δ Im{R} of the imaginary parts of the eigenvalues for Γ_1 = 100 G and Γ_1 = 10 000 G /183/

to the four ESR lines at small Γ_1 values. The real parts of these four eigenvalues are pictured in Figs. 6.17a,b and have, for $\alpha = 50^0$, the same value of $2\Gamma_1$. For larger values of Γ_1, we find only two imaginary parts, corresponding to two ESR lines. The absolute values of the real parts now are different, and by comparison with Fig. 6.15 it may be shown that the smaller one describes the ESR line width. For completeness it should be mentioned that the real parts of the eigenvalues 1 and 3 and of 2 and 4 are not the same, but the difference is so small that it cannot be pictured.

Now we consider the situation for a fixed value of Γ_1, e.g. Γ_1 = 52 G, as a function of α. For $0^0 < \alpha < 25^0$ we have two imaginary parts. Again the smaller one of the real parts describes the linewidth, which increases until $\alpha \approx 25^0$. For $25^0 < \alpha < 65^0$ there is only one real part of magnitude $2\Gamma_1$, but in this range of α there are four different imaginary parts, which describe four ESR lines. For $65^0 < \alpha < 90^0$ the situation is analogous to that in the range $0^0 < \alpha < 25^0$.

Figure 6.18b shows the difference of the imaginary parts for Γ_1 = 100 G and for Γ_1 = 10 000 G, demonstrating that after the transition from four to two different imaginary parts of the eigenvalues considered above, there is only a small shift over a large range of Γ_1.

The discussion of the rotation of the magnetic field around the c' axis (Figs. 6.19,20) is quite analogous and shall not be repeated. We find, however, a completely different behaviour for the rotation of the magnetic field around the b axis (Figs. 6.21,22), which is a two-fold screw axis of the crystal. On account of the symmetry, for every value of Γ_1 and of α, we have only two imaginary parts

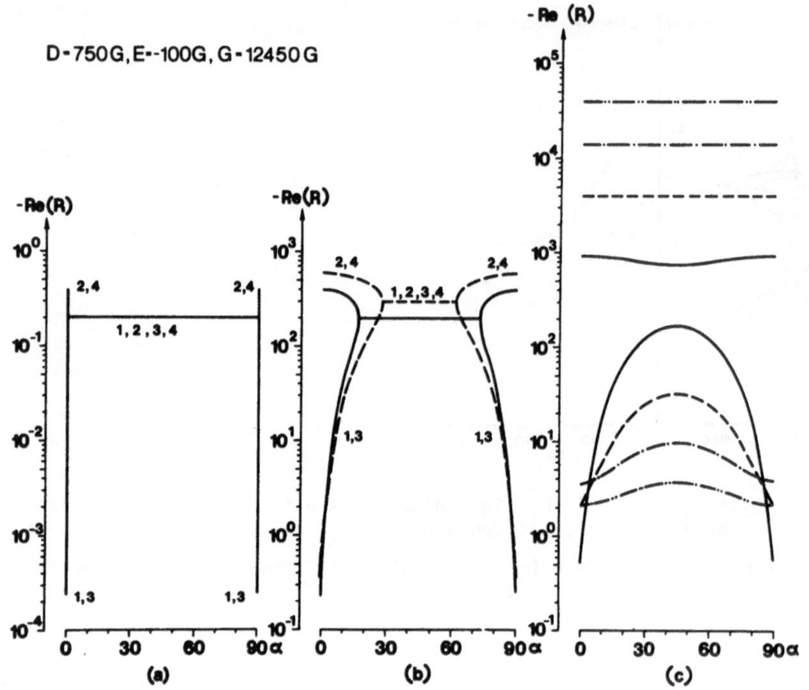

Fig. 6.19a-c. Real parts of the eigenvalues R as a function of α for several values of Γ_1. Rotation of the magnetic field around the c' axis. (a) Γ_1 = 0.1 G; (b) Γ_1 = 100 G (———), 150 G (-----); (c) Γ_1 = 230 G (———), 1000 G (-----), 3500 G (-·-·-), 10 000 G (-··-··) /183/

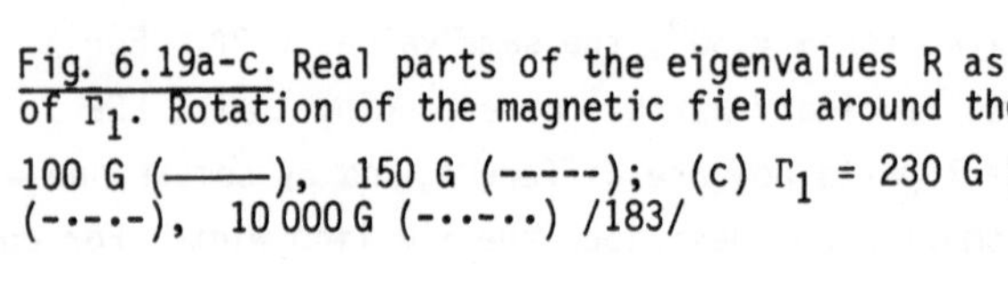

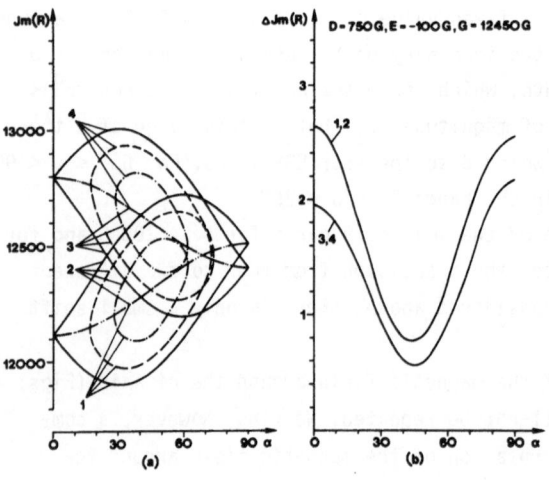

Fig. 6.20. (a) Imaginary parts of the eigenvalues R as a function of α for several values of Γ_1. Rotation of the magnetic field around the c' axis. Γ_1 = 0.1 G (———), 100 G (-----), 150 G (·-·-·), 230 G (-··-··). (b) Difference Δ Im{R} of the imaginary parts of the eigenvalues for Γ_1 = 230 G and Γ_1 = 10 000 G /183/

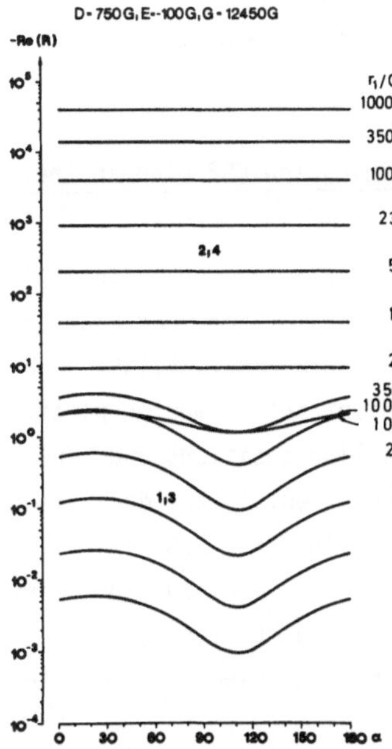

Fig. 6.21. Real parts of the eigenvalues R as a function of α for several values of Γ_1. Rotation of the magnetic field around b axis /183/

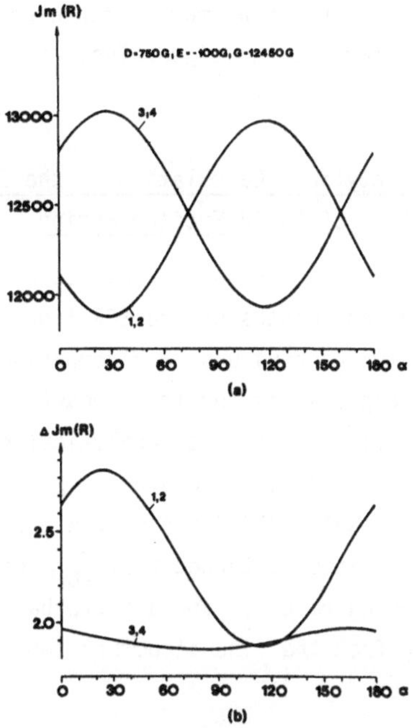

Fig. 6.22. (a) Imaginary parts of the eigenvalues R as a function of α for the same values of Γ_1 as in Fig. 6.21. Rotation of the magnetic field around the b axis. (b) Difference Δ Im{R} of the imaginary parts of the eigenvalues for Γ_1 = 0.1 G and Γ_1 = 10 000 G /183/

but four real parts, consisting of two nearly degenerate pairs. Comparison with exact calculations of the line shape reveals that the smaller one is responsible for the line width. In this case we may expect only two absorption lines.

After this discussion we can try to represent the line shape by a sum of several Lorentzian lines for small and large values of Γ_1:

$$\omega^{-1}\chi''(\omega) \approx \sum_i \frac{c_i}{[\mathrm{Re}\{R_i\}]^2 + [\omega - \mathrm{Im}\{R_i\}]^2} \; . \tag{6.93}$$

The sum over i runs from one to four if there are four different imaginary parts of the eigenvalues, i.e. for values of Γ_1 small enough. If we have only two different imaginary parts of the eigenvalues, i.e. for large values of Γ_1, in the notation of Figs. 6.17-22 i assumes the values 1 and 3. The c_i are determined by fitting the maxima of the line shape. A detailed investigation /183/ shows that the difference between the exact and the approximate line shape for large and small values of Γ_1 is very small. Therefore it is clear that for small enough and large enough values of Γ_1, the line shape may be represented by a sum of four and two Lorentzian lines, i.e. by four and two eigenvalues of the correlation functions, respectively.

6.3.4 Analytical Calculation of the Eigenvalues Determining the ESR Line in the Case of Rapid Exciton Motion

In the previous section we have seen that in the case of rapid exciton motion the positions and widths of the ESR lines are determined by the imaginary and real parts of only two eigenvalues. For comparison with experimental data /164,165,169/, analytical expressions for these quantities are more convenient than computer plots. Therefore, in this section analytical expressions for the line positions and line widths, obtained by second-order perturbation theory, are given. In the derivation of these results /184/ the matrix $\underline{\underline{L}}$ of (6.84), represented in Table 6.5, is split into an unperturbed part $\underline{\underline{L}}_0$, containing the magnetic field H_z and the effective hopping rate Γ_1, and a perturbation $\underline{\underline{L}}_1$, containing the quantities M_{ij} and D_{ij}, stemming from the fine-structure tensors. The calculation is carried out under the assumption

$$H_z\,,\Gamma_1 \gg M_{ij}\,,D_{ij}\quad, \tag{6.94}$$

and results in the following expressions for the line positions and line widths $[\Delta M = (M_{xx}+M_{yy}-2M_{zz})/2]$:

$$\begin{aligned}\mathrm{Im}\begin{Bmatrix}R^3\\R^1\end{Bmatrix} &= H_z \mp \Delta M + \frac{4M_{xy}^2+4M_{yz}^2+4M_{zx}^2+\Delta M_x^2}{8H_z}\\ &+\frac{1}{2}\,\frac{H_z}{16\Gamma_1^2+4H_z^2}\,(4D_{xy}^2+\Delta D_x^2)+\frac{1}{2}\,\frac{H_z}{16\Gamma_1^2+H_z^2}\,(D_{yz}^2+D_{xz}^2)\quad,\end{aligned} \tag{6.95}$$

$$\begin{aligned}-\mathrm{Re}\{R^1\} = -\mathrm{Re}\{R^3\} &= \frac{(\Delta D_y-\Delta D_z)^2}{16\Gamma_1}+\frac{\Gamma_1}{16\Gamma_1^2+4H_z^2}\,(4D_{xy}^2+\Delta D_x^2)\\ &+6\,\frac{\Gamma_1}{16\Gamma_1^2+H_z^2}\,(D_{yz}^2+D_{xz}^2)\quad.\end{aligned} \tag{6.96}$$

Measuring the ESR line shape for a fixed value of the external magnetic field H_z as a function of the frequency of the alternating magnetic field, $\mathrm{Im}\{R^1\}$ and $\mathrm{Im}\{R^3\}$ determine the positions of the maxima of the two lines and $-\mathrm{Re}\{R^1\} = -\mathrm{Re}\{R^3\}$ their line widths.

However, frequently ESR measurements are not done in the way just described, but instead the line shape is measured for fixed frequency of the alternating magnetic field as a function of the amount of the static magnetic field. This situation is also contained in (6.95,96). All we have to do is to solve (6.95) for H_z, i.e. for the value of the magnetic field, at which resonance occurs. Denoting the frequency of the alternating magnetic field by $\tilde{R}$, we arrive by an iteration procedure at equations (6.97)

$$\left.\begin{matrix} H_z^3 \\ H_z^1 \end{matrix}\right\} = \tilde{R} \pm \Delta M - \frac{4M_{xy}^2 + 4M_{yz}^2 + 4M_{zx}^2 + \Delta M_x^2}{8\tilde{R}} - \frac{1}{2}\frac{\tilde{R}}{16\Gamma_1^2 + 4\tilde{R}^2}(4D_{xy}^2 + \Delta D_x^2) - \frac{1}{2}\frac{\tilde{R}}{16\Gamma_1^2 + \tilde{R}^2}(D_{yz}^2 + D_{xz}^2) \quad , \tag{6.97}$$

which are also correct to second order.

Finally, we have to insert this value of the magnetic field in (6.96) and get for the line width (correct to second order)

$$- \mathrm{Re}\{R^1\} = - \mathrm{Re}\{R^3\} = \frac{(\Delta D_y - \Delta D_z)^2}{16\Gamma_1} + \frac{\Gamma_1}{16\Gamma_1^2 + 4\tilde{R}^2}(4D_{xy}^2 + \Delta D_x^2) + 6\frac{\Gamma_1}{16\Gamma_1^2 + \tilde{R}^2}(D_{yz}^2 + D_{xz}^2) \quad . \tag{6.98}$$

A generalization of the above results, taking into account the anisotropy of the g tensor, has been given by REINEKER /185/. The expressions given above for the line positions and the line widths follow also from a recent calculation of BERK et al. /169/, which has been carried out following the procedure of BLUME /175/. In addition to the magnetic field H_z and the frequency $\tilde{R}$, respectively, on the right-hand sides of (6.95-98), we find the hopping rate Γ_1 and the tensor components, M_{ij} and D_{ij}, which describe the fine-structure interaction. On account of this fine-structure interaction, the positions and the widths of the ESR lines depend on the orientation of the magnetic field in the crystal. In the following sections, we shall calculate this angular dependence explicitly when the magnetic field is rotated around the a, b, and c' axes of the crystal.

6.3.4a Rotation of the Magnetic Field Around the a and c' Axes

As mentioned in Sect. 6.3.2 the coordinate system of the calculation (x,y,z) is chosen such that its z axis coincides with the direction of the magnetic field. The magnetic field is always rotated around the x axis, which is also the direction of the high-frequency magnetic field. If we wish to rotate the magnetic field around the a axis of the crystal, the crystal has to be oriented such that its a axis coincides with this x axis. By α we denote the angle between the magnetic field and the +b axis, which means that the magnetic field for $\alpha = 0^0$ is parallel to the b axis and for $\alpha = 90^0$ parallel to the c' axis.

For a rotation of the magnetic field around the c' axis of the crystal, this axis has to coincide with the x axis. α is now the angle between the direction of the magnetic field and the +a axis. Then for $\alpha = 0^0$ the magnetic field is parallel to the +a axis and for $\alpha = 90^0$ parallel to the +b axis of the crystal. In both cases the angular dependences of the positions and of the widths of the two ESR lines are given by

$$\mathrm{Im}\begin{Bmatrix} R^3 \\ R^1 \end{Bmatrix} = a_0^{\mp}(H_z,+1) + a_2^{\mp}(H_z,+1)\cos 2\alpha + a_4(H_z,+1)\cos 4\alpha \quad , \tag{6.99}$$

$$-\mathrm{Re}\{R^1\} = -\mathrm{Re}\{R^3\} = c_0(H_z) + c_2(H_z)\cos 2\alpha + c_4(H_z)\cos 4\alpha \tag{6.100}$$

for fixed value H_z of the field strength of the magnetic high field and

$$\left.\begin{matrix} H_z^3 \\ H_z^1 \end{matrix}\right\} = a_0^{\pm}(\tilde{R},-1) + a_2^{\pm}(\tilde{R},-1)\cos 2\alpha + a_4(\tilde{R},-1)\cos 4\alpha \quad , \tag{6.101}$$

$$-\mathrm{Re}\{R^1\} = -\mathrm{Re}\{R^3\} = c_0(\tilde{R}) + c_2(\tilde{R})\cos 2\alpha + c_4(\tilde{R})\cos 4\alpha \tag{6.102}$$

for fixed value $\tilde{R}$ of the frequency of the high-frequency magnetic field. The coefficients $a_i^{\pm}$ and $c_i^{\pm}$ are represented in Appendix C.1 for rotation of the magnetic field around the a axis and in Appendix C.3 for rotation around the c' axis.

6.3.4b Rotation of the Magnetic Field Around the b Axis

If we wish to rotate the magnetic field around the b axis, we have to orientate the crystal in such a way that its b axis is identical with the x axis of the coordinate system of the calculation. We have chosen the +b axis to show in the -x direction and α to be the angle between the +z direction, i.e. the direction of the static magnetic field, and the +a axis. Then for $\alpha = 0$ the magnetic field is parallel to the +a axis, and for $\alpha = 90^0$ the field is parallel to the +c' axis.

In this case we obtain for the line positions and line widths for a fixed value of the magnetic field strength

$$\begin{aligned} \mathrm{Im}\begin{Bmatrix} R^3 \\ R^1 \end{Bmatrix} &= a_0^{\mp}(H_z,+1) + a_2^{\mp}(H_z,+1)\cos 2\alpha + b_2^{\mp}(H_z,+1)\sin 2\alpha \\ &\quad + a_4(H_z,+1)\cos 4\alpha + b_4(H_z,+1)\sin 4\alpha \quad , \end{aligned} \tag{6.103}$$

$$-\mathrm{Re}\{R^1\} = -\mathrm{Re}\{R^3\} = c_0(H_z) + c_2(H_z)\cos 2\alpha + d_2(H_z)\sin 2\alpha \quad , \tag{6.104}$$

and in the case of fixed frequency $\tilde{R}$ of the high-frequency magnetic field we obtain

$$\begin{aligned} \left.\begin{matrix} H_z^3 \\ H_z^1 \end{matrix}\right\} &= a_0^{\pm}(\tilde{R},-1) + a_2^{\pm}(\tilde{R},-1)\cos 2\alpha + b_2^{\pm}(\tilde{R},-1)\sin 2\alpha \\ &\quad + a_4(\tilde{R},-1)\cos 4\alpha + b_4(\tilde{R},-1)\sin 4\alpha \quad , \end{aligned} \tag{6.105}$$

$$-\mathrm{Re}\{\tilde{R}_1\} = -\mathrm{Re}\{\tilde{R}_3\} = c_0(\tilde{R}) + c_2(\tilde{R})\cos 2\alpha + d_2(\tilde{R})\sin 2\alpha \quad . \tag{6.106}$$

The coefficients in these expressions are given in Appendix C.2.

The angular dependence of the ESR line positions and widths and especially their maxima and minima as a function of the parameters of the system have been discussed in /186/. There it was also shown that in the parameter range of interest the deviations between the exact eigenvalues, as represented in Sect. 6.3.3, and the approximate ones of this section are very small.

6.3.5 Comparison with Experimental Data

6.3.5a Line Positions

From (6.97) we immediately see that

$$H_z^1 - H_z^3 = -2\Delta M \quad ,$$

where ΔM has been defined in connection with (6.95) and contains the fine-structure parameters, which have been determined by evaluating /186/ experimental data of HAARER and WOLF /163,164/. The rotations of the magnetic field around the a, b, and c' axes yield by a least-square fit:

D	E	rotation around	
766 ± 15 G	- 99 ± 4 G	a axis	
734 ± 12 G	- 95.5 ± 8 G	b axis	(6.107)
796 ± 9 G	- 102 ± 4 G	c' axis ,	

the errors being purely statistical ones.

These values are of the same order of magnitude and also in reasonable agreement with the values

$$D = 743\,\mathrm{G} \quad \text{and} \quad E = -89.4\,\mathrm{G} \tag{6.108}$$

calculated by HAARER and WOLF. However, considering the errors determined from the mean-square deviations, we remark that the values are not completely consistent. A closer inspection of the deviations of the measured values $H_z^1 - H_z^3$ from those calculated from the theoretical curves, fitted by the method of least squares, reveals that the errors are not randomly distributed but that there are small systematic deviations.

For these deviations there may be two reasons. Firstly, there might be small errors in the orientation of the crystals. Further investigations have shown that the influence of such errors is much more severe on $H_z^1 - H_z^3$ than on $H_z^1 + H_z^3$. Secondly, a more physical reason might be that the g tensor has a small anisotropy and that this anisotropy should be taken into account in the theoretical investigations /185/.

6.3.5b Line Widths

In Fig. 6.23 the ESR line widths have been pictured for rotation of the magnetic high field around the a, b, and c' axes according to (6.102) and (6.106), respectively. The theoretical expressions have been evaluated for D = 750 G, E = - 100 G, which are very close to the mean values of Sect. 6.3.5a, and $\tilde{R}$ = 12 450 G. The small

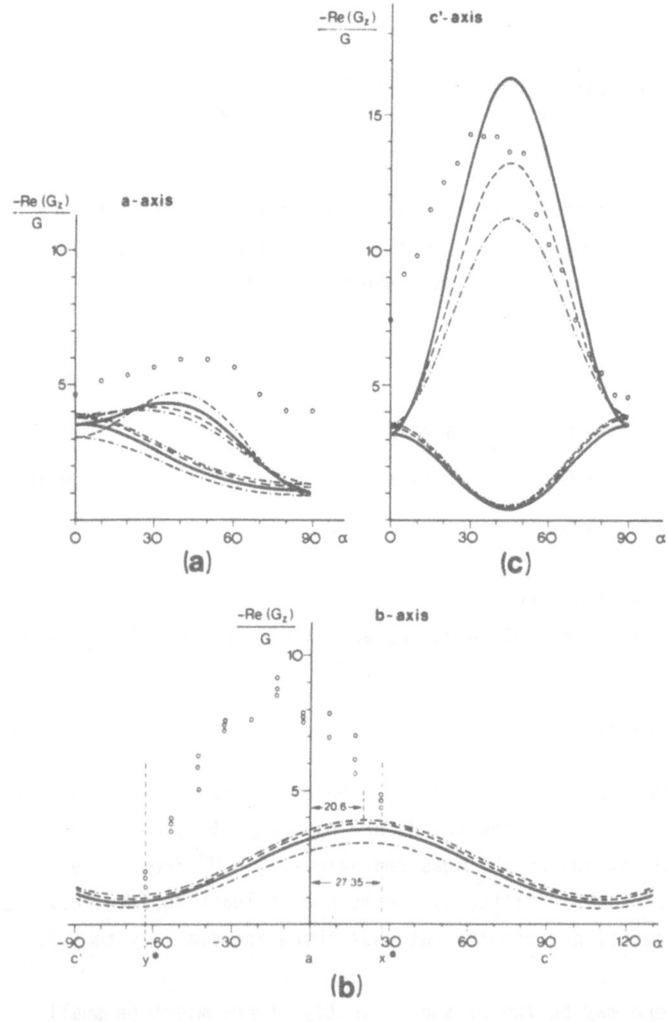

Fig. 6.23a-c. ESR line width for rotation of the magnetic field around the three crystals axes for D = 750 G, E = - 100 G, $\tilde{R}$ = 12 450 G and various values of Γ_1 /186/. In (a) and (c) for each value of Γ_1 the upper curve gives the total line width, the lower one the contribution of the non-secular terms of the Hamiltonian. The small circles correspond to measurements of HAARER and WOLF /163,164/. Γ_1 = 1.5 kG (--·---·), 2.0 kG (———), 2.5 kG (-----), 3.0 kG (-·-·-)

circles denote measured values of the ESR line width (half-width at half-maximum) of HAARER and WOLF /163,164/. In addition in Fig. 6.23b the angle $\alpha = 27.35^{o}$ of the extrema of the line positions and the angle $\alpha = 20.6^{o}$ of the maximum of the line width are shown for the values of D and E of the previous section.

Figure 6.23a shows that the quasi-incoherent motion of triplet excitons between the differently oriented molecules is able to account for a large part of the ESR line width. There is, however, a residual line width (RLW) of about 2 G on the average, which also seems to have a different angular dependence. This is still more obvious from Fig. 6.23b representing the ESR line width for rotation of the magnetic field around the b axis. Rotating the magnetic field around the c' axis, we have a large contribution to the ESR line width from the secular terms of the Hamiltonian which is given by the difference between the upper and the lower curves of Fig. 6.23c. Therefore, the relative contribution of the RLW mentioned above, which will be discussed below, is smaller and we have a reasonable agreement between theory and experiment. Assuming that as in the case of rotation around the a axis, the RLW gives an average contribution of about 2 G, we may estimate the effective hopping rate Γ_1 as

$$\Gamma_1 = (2500 \pm 500)\,\mathrm{G} \quad \text{or} \quad \Gamma_1 = (4.4 \pm 0.9) \times 10^{10}\ \mathrm{s}^{-1} \quad .$$

After the publication of the theoretical results, BIZZARO et al. re-examined the anthracene system /165/. Figure 6.24 shows that their data (circles) are fitted rather well by the theory. The triangles are the data of HAARER and WOLF.

One may imagine several reasons for the deviations between the theory and the experimental data of HAARER and WOLF. It is not possible in practice to avoid small errors in the orientation of the crystals, and therefore, when rotating the magnetic field around the b axis, a contribution to the ESR line width from the secular terms of the Hamiltonian also arises. Numerical calculations, however, show that in order to fit the experimental points, unreasonably high values (10° to 20°) for the deviations from the ideal orientation must be used.

Another contribution to the line width stems from the relaxation induced by the vibrations in the crystal. Intramolecular vibrations give rise to a time-dependent modulation of the fine-structure and spin-orbit interaction and thus to relaxation. On account of the librational motions of the molecules in an external magnetic field, the spin of the triplet exciton, which may be regarded as fixed in the molecular frame, sees a time-dependent magnetic field, giving rise to relaxation. Such a mechanism has also been discussed by KILPPER /267/ and KONZELMANN et al. /268/ for isolated molecules in an organic matrix. The relaxation, and thus the contribution to the ESR line width, should be largest where the slope of the curve representing the position of the ESR line versus the angular orientation of the magnetic field is largest. This slope is largest for $\alpha \approx -20^{o}$ and for $\alpha \approx 70^{o}$. Therefore, when rotating

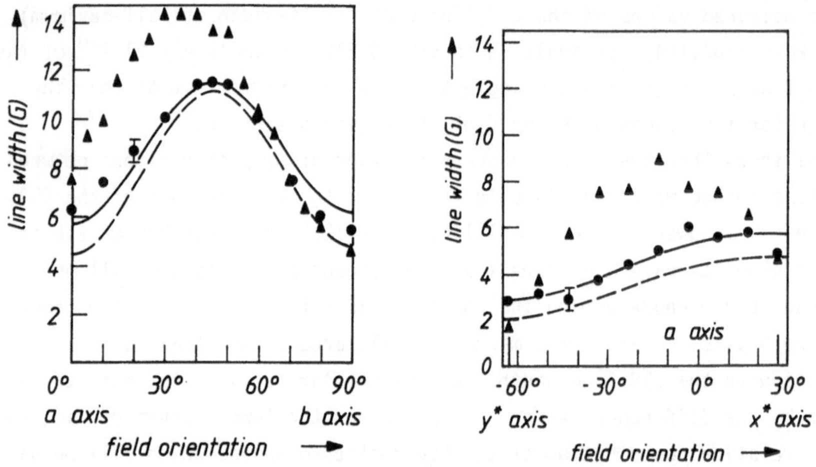

Fig. 6.24. (a) Angular dependence in the ab plane of the ESR line width of triplet excitons in single crystal anthracene at room temperature. The width refers to the half-width at half-maximum of a Lorentzian shape. The circles are experimental results /165/ which were measured at 24 GHz; the solid line is a least-squares fit to these data with $\Gamma_1 = 3285 \pm 75$ G and a residual line width (RLW) = 0.9 ± 0.2 G. The triangles are the data of /163,164/ which were measured at 35 GHz; the dashed line is the theoretical prediction for 35 GHz using the above parameters /165/.
(b) Similar to (a) but for the ac' plane and a fit with $\Gamma_1 = 3959 \pm 348$ G and RLW = 0.9 ± 0.2 G. The triangles are averages of the data points given in /163-165/

the magnetic field around the b axis, there should be a considerable contribution to the line width for $\alpha \approx -20^0$ and $\alpha \approx 70^0$. From Fig. 6.23b we see that this is indeed the case for $\alpha \approx -20^0$. In the theory these contributions have not been taken into account. However, if they would influence the line width in an essential manner, these contributions should also be present for $\alpha = 70^0$.

HAARER /269/ has suggested that the additional line width stems from an inhomogeneous orientation of small parts of the crystal. The contributions of small misorientations to an inhomogeneous broadening would again be largest for those orientations of the magnetic field with the largest slope of the $\underline{H}(\alpha)$ curve. Following this suggestion, ROSENTHAL et al. /255/ have subtracted from the observed line width of HAARER and WOLF a part which is proportional to the slope of the $\underline{H}(\alpha)$ curve. The results are represented in Figs. 6.25a,b for rotation of the magnetic field around the c' and b axes, respectively, and agree well with the theoretical line width curve calculated according to (6.102, 106).

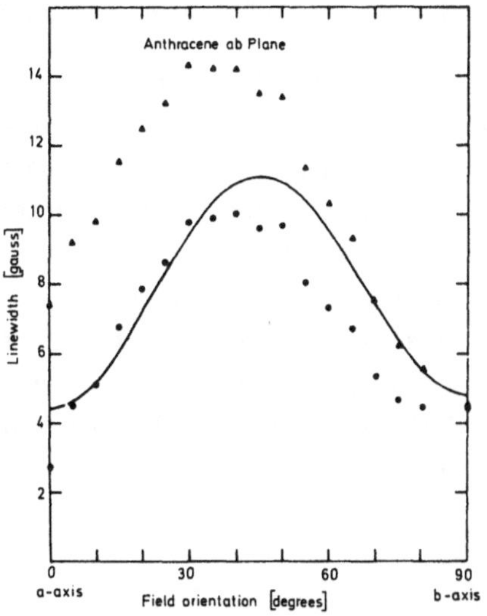

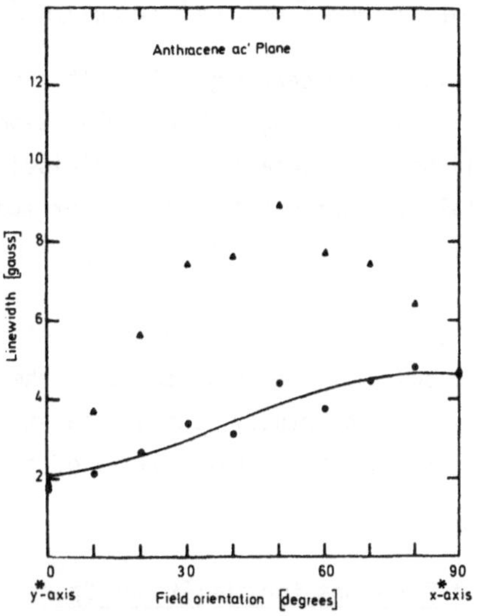

Fig. 6.25. (a) Angular dependence of the ESR line width of triplet excitons for the ab plane of a single crystal anthracene at room temperature. The width refers to the half-width at half-maximum of a Lorentzian shape. (b) The same as (a) but for the ac' plane. The solid curve is the theoretical prediction for 35 GHz deduced from the analysis in /165/. The triangles are the averages of the data given in /164/. The circles are the corrected data /255/

6.4 Investigation of the Coherent One-Dimensional Motion of Triplet Excitons by Spin Resonance

On account of its simple one-dimensional structure, dibromonaphthalene has been investigated by several methods, optical spectroscopy /117/, ESR /147-149/, ODMR /16/, and recently by Stark spectroscopy /119/. Calculations of ESR line shapes for this material have been carried out in /179/.

Another still more frequently investigated one-dimensional system is 1,2,4,5-tetrachlorobenzene. The relationship between the zero-field line shape and the band structure was investigated in /150,158/, transitions of excitons between localized and delocalized states were the subject of /151,159/, and the connection between dimer states and excitons was discussed in /132,133/. Results of ESR and ODMR measurements are represented in /154/ and optical data are reported in /160/. A calculation of the influence of the interchain interaction is carried out in /153/. Reviews are given by HARRIS /270/, HARRIS and ZWEMER /15/, and by BURLAND and ZEWAIL /82/. In very recent papers the problem of the connection between optical line width and spin dephasing times is investigated both experimentally and theoretically /155-157,271/. The theoretical description starts from the Bloch equations of two level systems for each of the k states of the exciton band and allows for scattering between different k states.

7. Influence of the Exciton Motion on Proton Relaxation in Organic Solids

Investigating anthracene crystals with nuclear-spin resonance (NSR), one observes a shortening of the longitudinal relaxation time /188,190/ if the crystal is irradiated by light. This behaviour is explained in the following way. The irradiation of the crystal by light creates triplet excitons. Their magnetic dipole moments interact with those of the protons via hyperfine-structure interaction. On account of the motion of the excitons between the sites of the crystal, this interaction is time dependent and thus gives rise to relaxation. The longitudinal relaxation time of the protons will not only depend on the exciton density, on whether they move coherently or incoherently, and on the strength of the magnetic field /37,181,190, 191/, but also on the dimensionality of the exciton motion /191/.

7.1 Bloch Equations for the Proton Spins

In order to calculate the longitudinal relaxation time of the protons, a simplified model will be used. We shall neglect the fine-structure interaction of the triplet spin of the exciton and also the dipole-dipole interaction between different proton spins. Furthermore, the interaction between the proton spins, mediated by the moving triplet spin of the exciton, will not be taken into account. In this way all many-body effects are neglected, and, therefore, we can consider the interaction of an exciton spin with a single proton spin, which may be assumed to sit at the origin of the coordinate system (on account of translational symmetry). The Hamiltonian of the system then reads /191/

$$H = H_{ex}(t) + H_S + H_p + H_W = H_0(t) + H_W \quad . \tag{7.1}$$

$H_{ex}(t)$ describes the coupled coherent and incoherent exciton motion and is given by (2.14) together with (2.5,10). H_S and H_p are the Zeeman energies of the exciton and proton spins in an external magnetic field

$$H_S = -\,\omega_e S_z \quad , \qquad \omega_e = \gamma_e H_z \quad , \tag{7.2}$$

$$H_p = -\,\omega_p I_z \quad , \qquad \omega_p = \gamma_p H_z \quad , \tag{7.3}$$

and H_W is the hyperfine interaction between exciton and proton spins. In a simplified description of this interaction we consider only its scalar part A. Furthermore, this interaction is non-zero only if the exciton and the proton are at the same site. Thus H_W explicitly reads

$$H_W = A\, b_0^+ b_0\, S \cdot I \quad . \tag{7.4}$$

In the interaction representation the equation of motion for the density operator reads

$$\dot{\tilde{\rho}} = -i\,[\tilde{H}_W(t),\tilde{\rho}] \tag{7.5}$$

with

$$\begin{aligned}\tilde{H}_W(t) &= A\,\vec{T}\exp\left(i\int_0^t H_{ex}(t')\,dt'\right) b_0^+b_0\,\overleftarrow{T}\exp\left(-i\int_0^t H_{ex}(t')\,dt'\right)\\ &\cdot\left(\frac{1}{2}\,S^+I^-\,e^{-i\Omega t} + \frac{1}{2}\,S^-I^+\,e^{i\Omega t} + S_zI_z\right)\quad.\end{aligned} \tag{7.6}$$

In (7.6) $\Omega = \omega_e - \omega_p$; on account of $\gamma_p \ll \gamma_e$ we have $\Omega \approx \omega_e$. Using a perturbation expansion up to second order, from (7.5) we arrive at

$$\begin{aligned}\tilde{\rho}(t) - \tilde{\rho}(0) &= -i\int_0^t dt_1\,\left(\tilde{H}_W(t_1)\right)^\times\,\tilde{\rho}(0)\\ &+ (-i)^2\int_0^t dt_1\int_0^{t_1} dt_2\,\left(\tilde{H}_W(t_1)\right)^\times\left(\tilde{H}_W(t_2)\right)^\times\,\tilde{\rho}(0)\quad.\end{aligned} \tag{7.7}$$

Multiplying this equation with the z component of the proton spin I_z and taking the trace we obtain

$$\begin{aligned}\langle I_z\rangle_t - \langle I_z\rangle_0 &= i\int_0^t dt_1\,\mathrm{Tr}\left\{\left(\tilde{H}_W(t_1)^\times\,I_z\right)\tilde{\rho}(0)\right\}\\ &+ (-i)^2\int_0^t dt_1\int_0^{t_1} dt_2\,\mathrm{Tr}\left\{\left(\tilde{H}_W(t_2)^\times\,\tilde{H}_W(t_1)^\times\,I_z\right)\tilde{\rho}(0)\right\}\end{aligned} \tag{7.8}$$

For the further evaluation of this expression we assume that at the initial time the density operator $\tilde{\rho}(0)$ factorizes /45/ into three parts,

$$\tilde{\rho}(0) = \tilde{\rho}_{ex}(0)\cdot\tilde{\rho}_S(0)\,\tilde{\rho}_I(0)\quad, \tag{7.9}$$

and that $\tilde{\rho}_{ex}(0)$ and $\tilde{\rho}_S(0)$ describe stationary states of the electronic and spin degrees of freedom of the exciton. With these assumptions the first term in (7.8) vanishes and from the second we have

$$\begin{aligned}\langle I_z\rangle_t - \langle I_z\rangle_0 &= -\frac{A^2}{2}\left(\langle S^-S^+\rangle\langle I^+I^-\rangle - \langle S^+S^-\rangle\langle I^-I^+\rangle\right)\\ &\cdot\int_0^t dt_1\int_0^{t_1} dt_2\,\langle\widetilde{b_0^+b_0(t_1-t_2)b_0^+b_0}\rangle\cos\Omega\,(t_1-t_2)\quad.\end{aligned} \tag{7.10}$$

The calculation of the spin expectation values gives

$$\begin{aligned}\langle S^-S^+\rangle\langle I^+I^-\rangle - \langle S^+S^-\rangle\langle I^-I^+\rangle &= 2(\langle \underline{S}^2\rangle - \langle S_z^2\rangle)\,\langle I_z\rangle - 2\langle S_z\rangle\,(\langle \underline{I}^2\rangle - \langle I_z^2\rangle)\\ &= 2\left(S(S+1) - \langle S_z^2\rangle\right)\langle I_z\rangle - \langle S_z\rangle\quad,\end{aligned}$$

where in the last equality we have made use of the fact that $\underline{I}$ describes the spin of protons. With the assumption that the correlation function decays rapidly as compared to the time interval t, which is of relevance in the perturbation theory, the double integral in (7.10) may be written as /191/ t $J(\Omega)$. Dividing by t we obtain in the limit of small t values

$$\frac{d}{dt} \langle I_z \rangle = - \frac{1}{T_{1,ex}} \langle I_z \rangle + C \quad . \tag{7.11}$$

In this expression, C /191/ describes the optical nuclear-spin polarization /272, 273/ and is given by

$$C = \frac{1}{2} A^2 \langle S_z \rangle J(\Omega) \quad . \tag{7.12}$$

$T_{1,ex}^{-1}$ is the longitudinal relaxation rate in the presence of the excitons and given by /190,191/

$$\frac{1}{T_{1,ex}} = A^2 [S(S+1) - \langle S_z \rangle] J(\Omega) \quad . \tag{7.13}$$

The spectral density $J(\Omega)$ is calculated from

$$J(\Omega) = \int_0^\infty d\tau \cos \Omega\tau \ \langle \widetilde{b_0^+ b_0(\tau)} \ b_0^+ b_0 \rangle \quad . \tag{7.14}$$

The expectation value is given by

$$\langle \widetilde{b_0^+ b_0(t)} \ b_0^+ b_0 \rangle$$
$$= \mathrm{Tr}\left\{ b_0^+ b_0 \ \overleftarrow{T} \exp\left(-i \int_0^t dt' \ H_{ex}(t')\right) b_0^+ b_0 \ \rho_{ex}(0) \ \vec{T} \exp\left(i \int_0^t dt' \ H_{ex}(t')\right)\right\} \quad . \tag{7.15}$$

The trace in this expression has to be calculated using electronic states only. The stationary density operator $\rho_{ex}(0)$ is given by

$$\rho_{ex}(0) = \frac{N_{ex}}{N} 1 \quad , \tag{7.16}$$

where 1 is the identity operator; N_{ex} is the number of molecules with excitons and N the total number of molecules. Then (7.15) becomes

$$\langle \widetilde{b_0^+ b_0(t)} \ b_0^+ b_0 \rangle$$
$$= \frac{N_{ex}}{N} \mathrm{Tr}\left\{ b_0^+ b_0 \ \overleftarrow{T} \exp\left(-i \int_0^t dt' \ H_{ex}(t')\right) \ b_0^+ b_0 \ \vec{T} \exp\left(i \int_0^t dt' \ H_{ex}(t')\right)\right\} \tag{7.17}$$
$$= \frac{N_{ex}}{N} \rho_{00}(t) \quad . \tag{7.18}$$

$\rho_{00}(t)$ is the conditional probability of finding the exciton at the origin at time t if it was there at the initial time. This quantity is obtained from the solution of the density matrix equation with the initial condition that the exciton is at the origin at the initial time. The spectral density, determining the relaxation rate $T_{1,ex}^{-1}$ of the protons, is then obtained as the Fourier-Laplace transform of this probability. In the following we shall consider explicitly the proton relaxation for the case of a coupled coherent and incoherent exciton motion in a dimer /37,142,181/. Furthermore, the spectral density will be represented for the purely coherent and the purely incoherent exciton motion on a linear chain /190/. The influence of the coupled coherent and incoherent motion and of its dimensionality on the proton relaxation has been investigated in detail by SCHWARZER and HAKEN /191/.

7.2 Calculation of the Spectral Density

7.2.1 Coupled Coherent and Incoherent Motion of Excitons in a Dimer

From (7.14,18) we have

$$J(\Omega) = \frac{N_{ex}}{N} \int_0^\infty d\tau \cos \Omega\tau \; \rho_{00}(\tau) \quad . \tag{7.19}$$

The conditional probability $\rho_{00}(\tau)$ of finding the exciton at a specific lattice site, e.g. at the origin of the coordinate system, if it was at that lattice site initially is obtained immediately from the solution of the density matrix equation satisfying that initial condition.

In Sect. 3.1 this probability was calculated for the coupled coherent and incoherent exciton motion in a dimer. From (3.9) and Table 3.2 we obtain ($H_1 = J$)

$$\begin{aligned} \rho_{00}(t) = \frac{1}{2} + \frac{1}{4}\left[1 + i\gamma_0(4J^2 - \gamma_0^2)^{-1/2}\right] e^{-\lambda_3 t} \\ + \frac{1}{4}\left[1 - i\gamma_0(4J^2 - \gamma_0^2)^{-1/2}\right] e^{-\lambda_4 t} \quad . \end{aligned} \tag{7.20}$$

With $\bar{\gamma}_1 = \gamma_1$ from (3.6c,d) the eigenvalues λ_3 and λ_4 are given by

$$\left.\begin{matrix}\lambda_3 \\ \lambda_4\end{matrix}\right\} = \gamma_0 + 4\gamma_1 \pm i(4J^2 - \gamma_0^2)^{1/2} \quad . \tag{7.21}$$

In the case of the incoherent exciton motion ($\gamma_0 > 2J$) we get

$$\begin{aligned} J(\Omega) = \frac{1}{2}\frac{N_{ex}}{N}\left\{\frac{1}{4}\left[1 - \gamma_0(\gamma_0^2 - 4J^2)^{-1/2}\right]\frac{2\tau_1}{1 + \Omega^2\tau_1^2}\right. \\ \left. + \frac{1}{4}\left[1 + \gamma_0(\gamma_0^2 - 4J^2)^{-1/2}\right]\frac{2\tau_2}{1 + \Omega^2\tau_2^2}\right\} \quad . \end{aligned} \tag{7.22}$$

The coherent exciton motion ($\gamma_0 < 2J$) gives the following expression for the spectral density

$$J(\Omega) = \frac{1}{4}\frac{N_{ex}}{N}\left\{\frac{\tau}{1+(\beta-\Omega)^2\tau^2}\left[1+\gamma_0(4J^2-\gamma_0^2)^{-1/2}\tau(\beta-\Omega)\right]\right.$$
$$\left. + \frac{\tau}{1+(\beta+\Omega)^2\tau^2}\left[1+\gamma_0(4J^2-\gamma_0^2)^{-1/2}\tau(\beta+\Omega)\right]\right\} \quad . \tag{7.23}$$

In (7.22,23) the following abbreviations have been used

$$\Omega = \omega_e - \omega_p \quad , \tag{7.24a}$$

$$\left.\begin{matrix}\tau_1^{-1}\\ \tau_2^{-1}\end{matrix}\right\} = \left\{\begin{matrix}\lambda_3\\ \lambda_4\end{matrix}\right\} = (\gamma_0+4\gamma_1) \pm (\gamma_0^2-4J^2)^{1/2} \quad , \tag{7.24b,c}$$

$$\tau^{-1} = \gamma_0 + 4\gamma_1 \quad , \tag{7.24d}$$

$$\beta = (4J^2 - \gamma_0^2)^{1/2} \quad . \tag{7.24e}$$

For high temperatures $\gamma_0 \gg J$ and (7.22) reduces to

$$J(\Omega) = \frac{1}{4}\frac{N_{ex}}{N}\frac{2\tau_2}{1+\Omega^2\tau_2^2} \quad , \tag{7.25}$$

with

$$\tau_2^{-1} \approx 2(2\gamma_1 + \frac{J^2}{\gamma_0}) \quad . \tag{7.26}$$

In this case of the quasi-incoherent exciton motion the spectral density is determined by a Debye relaxation term. The relaxation rate $T_{1,ex}^{-1}$ decreases for $\Omega\tau_2 > 1$. Here Ω is essentially the Larmor frequency of the exciton spin which gives rise to the field modulation at the site of the proton spin. τ_2^{-1} is the effective hopping rate of the exciton between the two molecules in the dimer and determines the time, for which the exciton stays at the site of the proton considered. Furthermore, the comparison with (4.17) shows that τ_2^{-1} is the quantity determining the diffusion constant in nearest-neighbour approximation.

7.2.2 Purely Coherent Exciton Motion on a Linear Chain

According to (3.58), the probability of finding an exciton at time t at the origin of a linear chain, if it was there initially is given by

$$\rho_{00}(t) = J_0^2(2H_1 t) \quad . \tag{7.27}$$

Inserting this expression into (7.19) results in

$$J(\Omega) = \frac{N_{ex}}{N} \frac{1}{2H_1} \int_0^\infty d\tau \cos \frac{\Omega}{2H_1} \tau \; J_0^2(\tau) \quad . \tag{7.28}$$

The integral can be evaluated /274/ with the following result /190/:

$$J(\Omega) = \begin{cases} \frac{N_{ex}}{N} \frac{1}{4H_1} \frac{2}{\pi} K\left[\sqrt{1 - \left(\frac{\Omega}{4H_1}\right)^2}\right] & 0 < \Omega < 4H_1 \; , \\ 0 & \text{else} \; . \end{cases} \tag{7.29}$$

Expansion of the complete elliptic integral of the first kind for $\Omega \approx 4H_1$ and $\Omega \ll H_1$ gives

$$K \approx \frac{\pi}{2} \left\{1 + \frac{1}{4}\left[1 - \left(\frac{\Omega}{4H_1}\right)^2\right]\right\} \quad , \qquad \Omega \approx 4H_1 \tag{7.30}$$

$$K \approx \ln \frac{16H_1}{\Omega} + \frac{1}{4}\left(\ln \frac{16H_1}{\Omega} - 1\right)\left(\frac{\Omega}{4H_1}\right)^2 , \qquad \Omega \ll H_1 \; . \tag{7.31}$$

From (7.30) we see that the spectral density has a finite minimum value for $\Omega = 4H_1$. If Ω is larger than the bandwidth of the one-dimensional coherent motion, the spectral density drops to zero. The second expression shows that $J(\Omega)$ has a logarithmic singularity for $\Omega \to 0$.

7.2.3 Purely Incoherent Exciton Motion on a Linear Chain

In the case of the purely hopping motion, the solution of the density matrix equation, satisfying the same initial condition as in the previous section, is given by (3.60)

$$\rho_{00}(t) = e^{-4\gamma_1 t} \, I_0(4\gamma_1 t) \quad . \tag{7.32}$$

The integral (7.19) can again be evaluated exactly /274/ with the result /190/

$$J(\Omega) = \frac{N_{ex}}{N} \frac{1}{8\sqrt{2}\,\gamma_1} \; \tilde{\Omega}^{-1/2} \left(1 + \tilde{\Omega}^2\right)^{-1/2} \left[\tilde{\Omega} + \left(1 + \tilde{\Omega}^2\right)^{-1/2}\right]^{-1/2} \tag{7.33}$$

with $\tilde{\Omega} = \Omega/8\gamma_1$. It should be mentioned that the same result is obtained if we replace the hopping rate $2\gamma_1$ of the purely hopping motion by the effective hopping rate W_1 of (4.29) for the quasi-incoherent motion.

We again consider limiting cases of (7.33) and obtain

$$J(\Omega) = \frac{N_{ex}}{N} \frac{1}{8\gamma_1} \begin{cases} \sqrt{\frac{4\gamma_1}{\Omega}} \; , & \Omega \ll 8\gamma_1 \quad (7.34) \\ 2\sqrt{2}\left(\frac{4\gamma_1}{\Omega}\right)^2 \; , & \Omega \gg 8\gamma_1 \; . \quad (7.35) \end{cases}$$

The comparison of this result with (7.25), representing the spectral density for the quasi-incoherent exciton motion in a dimer, shows that for $\Omega\tau_2 \gg 1$ both spectral densities decay quadratically with the frequency Ω. However, in the opposite limit of $\Omega\tau_2 \ll 1$ the spectral density in the dimer approaches a constant value, whereas in the case of the hopping motion on a linear chain we have a square root singularity.

These dimensionality effects as well as the spectral density in the case of the coupled coherent and incoherent motion on a linear chain have been discussed in great detail by SCHWARZER and HAKEN /191/.

8. Connection Between Stochastic Liouville Equation and Generalized Master Equation Treatments

In this final section - following closely the procedure of /81/ - we derive the connection between the stochastic Liouville equation treatment of the coupled coherent and incoherent exciton motion /34-36/ and the generalized master equation (GME) treatment proposed by KENKRE and KNOX /63,64/. To that end, starting from the stochastic Liouville equation of the full Haken-Strobl model for the exciton motion on a linear chain, the corresponding generalized master equation is obtained in an exact manner /80,81/ using the NAKAJIMA-ZWANZIG /65,66/ projection formalism. From this exact result the GMEs for the purely incoherent, the purely coherent, and the quasi-incoherent cases are rederived. Furthermore, the GME /77/ for a simplified version /196/ of the SLE mentioned above is derived and the consequences of the Born approximation are investigated. A more detailed discussion of the relation between SLEs and GMEs is given in this volume in the article by Kenkre. The relations between the various descriptions of exciton dynamics from the viewpoint of a microscopic treatment of the exciton-phonon interaction have been represented recently by SILBEY /57/.

8.1 Stochastic Liouville and Nakajima-Zwanzig Equations

The SLE of the Haken-Strobl model is given by (2.35) together with (2.11-13) and (2.32-34). We introduce a modified density operator $\bar{\rho}(t)$

$$\rho(t) = e^{-2\Gamma t}\, \bar{\rho}(t) \tag{8.1}$$

where Γ, the sum over all fluctuation strengths, is defined in (2.30). Using (8.1), the SLE from (2.35) transforms into

$$\dot{\bar{\rho}}(t) = \bar{L}\, \bar{\rho}(t) \quad . \tag{8.2}$$

The Liouville operator $\bar{L}$ is given by

$$\bar{L}\bar{\rho}(t) = -\sum_{k} \tilde{H}_k [b_k^+ b_k, \bar{\rho}(t)] + 2\sum_k \sum_{k'} \sum_{k''} (\tilde{\tilde{\gamma}}_{k+k''} + \tilde{\gamma}_{k'-k''}) b_k^+ b_{k'} \, \bar{\rho}(t) \, b_{k''}^+ b_{k+k''-k'} \tag{8.3}$$

and may be written in the form

$$\bar{L} = L_1 + L_2 + L_3 \quad , \tag{8.4}$$

where the three parts L_1, L_2, and L_3 contain the parameters $\tilde{H}_k$, $\tilde{\tilde{\gamma}}_k$, and $\tilde{\gamma}_k$, respectively. We are not interested in all of the information contained in the modified density operator $\bar{\rho}$ but only in that part describing the probability of finding the exciton at a lattice site n. This information is obtained from $\bar{\rho}$ by the application of a projection operator P /44,65,66,237/ projecting out the diagonal part $\bar{\rho}_{nn}$ of $\bar{\rho}$ and is given explicitly in (8.8). The equation of motion for $P\bar{\rho}$ is /44/

$$\frac{d}{dt} P \, \bar{\rho}(t) = P \bar{L} P \, \bar{\rho}(t) + \int_0^t dt' \; K(t') \, P \bar{\rho}(t-t') \quad . \tag{8.5}$$

The kernel K(t) in this expression is defined by

$$K(t) = P\bar{L} \; e^{(1-P)\bar{L}t} \; (1-P)\bar{L}P \quad . \tag{8.6}$$

In writing (8.5) we have used the fact that an inhomogeneous term vanishes in this equation /275/ (for details see Appendix C of /276/).

8.2 Calculation of the Kernel K(t)

8.2.1 Definition of the Projection Operator

In the following we shall define a projection operator P_κ which projects out the diagonal part of the density operator (more exactly its Fourier transform). The diagonal matrix elements of the density operator in the site representation are expressed in the following way by its Fourier transform:

$$\rho_{nn} = \sum_{\kappa} e^{i\kappa n} \; \tilde{\rho}_\kappa \quad , \tag{8.7}$$

where $\tilde{\rho}_\kappa = \sum_k \rho_{k,k-\kappa}$. The projection operator projecting out this component of the density operator is given by

$$P_\kappa \Omega = \frac{1}{N} n_\kappa \, (n_\kappa, \Omega) = \frac{1}{N} n_\kappa \, \mathrm{Tr}\{n_\kappa^+ \Omega\} \quad , \tag{8.8}$$

the scalar product having been defined by the second equality. The operators n_κ are defined by

$$n_\kappa = \sum_k n_{k\kappa} \quad , \tag{8.9}$$

where $n_{k\kappa} = |k\rangle\langle k-\kappa|$ and $\langle|$ and $|\rangle$ denote Dirac's bra and ket vectors. From these definitions we immediately derive

$$(n_{k_1\kappa}, n_{k_2\kappa'}) = \delta_{k_1 k_2}\, \delta_{\kappa\kappa'} \; . \tag{8.10}$$

With this result we easily prove the fundamental property of projection operators $P_\kappa^2 \Omega = P_\kappa \Omega$.

The projection operator (8.8) has the form

$$P\Omega = A(B,\Omega) \equiv A\, \mathrm{Tr}\{B^+ \Omega\} \quad , \tag{8.11}$$

which is considered in /277/ with $A = N^{-1} n_\kappa$, $B = n_\kappa$. On account of $P^2 = P$ we must have $(B,A) = 1$.

8.2.2 Transformation of the Kernel

We first express the Kernel K(t) by its Laplace transform. After a little algebra it may be written in the following way:

$$K(t) = \frac{1}{2\pi i} \int_{c-i\infty}^{c+i\infty} dz\, e^{zt}\, P\bar{L} \left\{ \frac{z}{z - (1-P)\,\bar{L}} - 1 \right\} P \quad . \tag{8.12}$$

The path of integration has to be chosen parallel to the imaginary axis of the z plane in such a way that all singularities are situated to its left. For $z \to \infty$ the Laplace transform disappears, and, therefore, the path of integration may be *closed* by a semicircle in the left half plane. Denoting this closed path by C, from (8.12) we arrive at

$$K(t) = \frac{1}{2\pi i} \int_C dz\, e^{zt}\, P\bar{L} \left(\frac{z}{z - (1-P)\,\bar{L}} - 1 \right) P \quad . \tag{8.13}$$

Using the identity

$$\frac{1}{z - \bar{L} + P\bar{L}} = \frac{1}{z - \bar{L}} - \frac{1}{z - \bar{L}}\, P\bar{L}\, \frac{1}{z - \bar{L} + P\bar{L}} \quad , \tag{8.14}$$

we obtain

$$\left(P \frac{1}{z - \bar{L}} P\right) \left(P\bar{L} \frac{z}{z - (1-P)\bar{L}} P\right) = P \left(\frac{1}{z - \bar{L}} - 1\right) P \quad . \tag{8.15}$$

Inserting the projection operator form (8.11), we have

$$A \left(B, \frac{1}{z - \bar{L}} A\right) \left(B, \bar{L} \frac{1}{z - (1-P)\bar{L}} A\right) \left(B, \ldots\right) = A \left(B, \left[\frac{z}{z - \bar{L}} - 1\right] A\right) \left(B, \ldots\right) \quad . \tag{8.16}$$

Dividing by $\left(B, \frac{1}{z-\bar{L}} A\right)$, we get

$$A\left(B, \bar{L} \frac{z}{z-(1-P)\bar{L}} A\right)\left(B, \ldots\right) = z\,A\left(B, \ldots\right) - \left(B, \frac{1}{z-\bar{L}} A\right)^{-1} A\left(B, \ldots\right) \quad . \tag{8.17}$$

The left-hand side of (8.17) is just the first term in the integrand of (8.12). Therefore K(t) reads

$$K(t) = \frac{1}{2\pi i} \int_C dz\, e^{zt} \left\{ z - \left(B, \frac{1}{z-\bar{L}} A\right)^{-1} - (B, \bar{L}A) \right\} P \quad . \tag{8.18}$$

On account of the closed path of integration, the first and third terms in (8.18) disappear and we arrive at

$$K(t) = -\frac{1}{2\pi i} \int_C dz\, e^{zt} \left(B, \frac{1}{z-\bar{L}} A\right)^{-1} A(B, \ldots) \quad . \tag{8.19}$$

Inserting A and B from (8.8,11), we have

$$\left(B, \frac{1}{z-\bar{L}} A\right) = \frac{1}{N} \left(n_\kappa, \frac{1}{z-\bar{L}} n_\kappa\right) \quad . \tag{8.20}$$

8.2.3 Calculation of Matrix Elements

The kernel (8.19) is determined mainly by the matrix element in (8.20). Using the definition (8.9), we have

$$\left(n_\kappa, \frac{1}{z-\bar{L}} \frac{1}{N} n_\kappa\right) = \frac{1}{N} \sum_{k_1} \sum_{k_2} \left(n_{k_1\kappa}, \frac{1}{z-\bar{L}} n_{k_2\kappa}\right) \quad . \tag{8.21}$$

The decomposition of the Liouville operator $\bar{L} = L_1 + L_2 + L_3$ introduced in (8.4) allows one to write the following identity:

$$\frac{1}{z-\bar{L}} = \frac{1}{z-L_1} + \frac{1}{z-\bar{L}} (L_2 + L_3) \frac{1}{z-L_1} \quad . \tag{8.22}$$

Using (8.22) and

$$L_1 n_{k_0,\kappa} = -i(\tilde{H}_{k_0} - \tilde{H}_{k_0-\kappa})\, n_{k_0,\kappa} \quad , \tag{8.23}$$

$$L_2 n_{k_0,\kappa} = 2 \sum_k \tilde{\tilde{\gamma}}_{k+k_0-\kappa} n_{k,\kappa} \quad , \tag{8.24}$$

$$L_3 n_{k_0,\kappa} = 2\tilde{\gamma}_\kappa n_\kappa \quad , \tag{8.25}$$

we obtain the following equation for the matrix elements in (8.21):

$$\sum_{k''}\left\{\delta_{k''k}-\frac{2\tilde{\gamma}_{\kappa}}{z+i(\tilde{H}_k-\tilde{H}_{k-\kappa})}-\frac{2\tilde{\tilde{\gamma}}_{k''+k-\kappa}}{z+i(\tilde{H}_k-\tilde{H}_{k-\kappa})}\right\}\left(n_{\kappa},\frac{1}{z-\mathcal{L}}n_{k''\kappa}\right)$$

$$=\frac{1}{z+i(\tilde{H}_k-\tilde{H}_{k-\kappa})} \quad . \tag{8.26}$$

Equation (8.26) represents a set of inhomogeneous algebraic equations for the matrix elements of interest. The equations for different matrix elements are coupled via $\tilde{\tilde{\gamma}}_{k''+k-\kappa}$. Inserting its Fourier transform (2.34), using the transformation $k \to k+\frac{\kappa}{2}$, $k'' \to k''+\frac{\kappa}{2}$, and applying $\frac{1}{N}\sum_k e^{ikn}$, we have from (8.26)

$$f_n - 2\tilde{\gamma}_{\kappa}NI_n\, f_0 - \sum_{n'} 2\bar{\gamma}_{n'}(1-\delta_{n'0})\, I_{n-n'}\, f_{-n'} = -I_n \quad . \tag{8.27}$$

Here the abbreviations

$$f_n = -\frac{1}{N}\sum_k e^{ikn}\left(n_{\kappa},\frac{1}{z-\mathcal{L}}\,|k+\tfrac{\kappa}{2}\rangle\langle k-\tfrac{\kappa}{2}|\right) \tag{8.28}$$

and

$$I_n = \frac{1}{N}\sum_k e^{ikn}\frac{1}{z+i(\tilde{H}_{k+\kappa/2}-\tilde{H}_{k-\kappa/2})} \tag{8.29}$$

have been introduced. For the calculation of the kernel of the GME we only need

$$-f_0 = \left(n_{\kappa},\frac{1}{z-\mathcal{L}}\frac{1}{N}n_{\kappa}\right) \quad . \tag{8.30}$$

In order to obtain explicit results, we consider nearest-neighbour interaction. In this case (8.27) gives

$$(1-2\bar{\gamma}_1 I_{-2})\, f_{-1} - 2\tilde{\gamma}_{\kappa}NI_{-1}\, f_0 - 2\bar{\gamma}_{-1}I_0\, f_1 = -I_{-1} \quad , \tag{8.31a}$$

$$-2\bar{\gamma}_1 I_{-1}\, f_{-1} + (1-2\tilde{\gamma}_{\kappa}NI_0)\, f_0 - 2\bar{\gamma}_{-1}I_1\, f_1 = -I_0 \quad , \tag{8.31b}$$

$$-2\bar{\gamma}_1 I_0\, f_{-1} - 2\tilde{\gamma}_{\kappa}NI_1\, f_0 + (1-2\bar{\gamma}_{-1}I_2)\, f_1 = -I_1 \quad . \tag{8.31c}$$

For an infinite chain of molecules and nearest-neighbour interaction (8.29) transfroms to

$$I_n = \frac{1}{2\pi}\int_{-\pi}^{\pi} dk\,\frac{e^{ikn}}{z-i4H_1\sin k\,\sin\frac{\kappa}{2}} \quad . \tag{8.32}$$

Evaluation of this integral in the complex plane results in

$$I_n = \frac{1}{a_{\kappa}^n}\frac{\left(z-(z^2+a_{\kappa}^2)^{1/2}\right)^n}{(z^2+a_{\kappa}^2)^{1/2}} \quad , \qquad n \geqslant 0 \tag{8.33}$$

with $I_{-n} = (-1)^n I_n$ and $a_{\kappa} = 4H_1\sin\kappa/2$. Assuming still $\bar{\gamma}_1 = \bar{\gamma}_{-1}$, from the solution of (8.31) we obtain

$$- \frac{1}{f_0(\kappa,z)} = (n_\kappa , \frac{1}{z - \mathcal{L}} \frac{1}{N} n_\kappa)^{-1}$$
$$= - 2\tilde{\gamma}_\kappa N + z + \frac{(z^2 + a_\kappa^2)^{1/2} - z}{1 + (4\bar{\gamma}_1/a_\kappa^2)[(z^2 + a_\kappa^2)^{1/2} - z]} \quad . \tag{8.34}$$

With (8.19,20,34) the kernel of the GME may be written in the following way:

$$K(t)\Omega = \frac{1}{2\pi i} \int_C dz \; e^{zt} \frac{1}{f_0(\kappa,z)} \frac{1}{N} n_\kappa \; (n_\kappa,\Omega) \tag{8.35}$$

(Ω again being an arbitrary operator), and the retarded term in (8.5) becomes

$$\int_0^t dt' \; K(t') \; P_\kappa \bar{\rho}(t - t') = \frac{1}{2\pi i} \int_0^t dt' \int_C dz \; e^{zt'} \frac{1}{f_0(\kappa,z)} \frac{1}{N} n_\kappa \; \tilde{\bar{\rho}}_\kappa(t - t') \quad , \tag{8.36}$$

where $\tilde{\bar{\rho}}_\kappa(t)$ is the Fourier transform of $\bar{\rho}_{nn}(t)$.

8.3 Generalized Master Equation for the Coupled Coherent and Incoherent Motion of Excitons

In order to obtain explicit results for the memory function of the GME, in this section its kernel is represented in a form that easily allows computer evaluation. The first (non-retarded) term of (8.5) may be immediately evaluated:

$$P_\kappa \; \mathcal{L} \; P_\kappa \; \bar{\rho}(t) = n_\kappa \; 2\tilde{\gamma}_\kappa \tilde{\bar{\rho}}_\kappa = 2N\tilde{\gamma}_\kappa \, P_\kappa \, \bar{\rho}(t) \quad . \tag{8.37}$$

In arriving at (8.37) use has been made of (8.23-25) and of $\sum_k H_k = \sum_k H_{k-\kappa}$. With (8.36,37) and dropping now the operator n_κ occuring in each term, from (8.5) we arrive at

$$\frac{d}{dt} \tilde{\bar{\rho}}_\kappa(t) = 2N\tilde{\gamma}_\kappa \; \tilde{\bar{\rho}}_\kappa(t) + \frac{1}{2\pi i} \int_0^t dt' \int_C dz \; e^{zt'} \frac{1}{f_0(\kappa,z)} \tilde{\bar{\rho}}_\kappa(t - t') \quad . \tag{8.38}$$

Using (8.1), this equation assumes the following form

$$\frac{d}{dt} \tilde{\rho}_\kappa(t) = -2(\Gamma - N\tilde{\gamma}_\kappa) \; \tilde{\rho}_\kappa(t) + \frac{1}{2\pi i} \int_0^t dt' \int_C dz \; e^{(z-2\Gamma)t'} \frac{1}{f_0(\kappa,z)} \tilde{\rho}_\kappa(t - t') \; . \tag{8.39}$$

Transforming now to the site representation, (8.39) reads

$$\frac{d}{dt} \rho_n(t) = - 2\Gamma \; \rho_n(t) + \sum_{n'} 2\gamma_{n-n'} \; \rho_{n'}(t)$$
$$+ \sum_{n'} \int_0^t dt' \int_C \frac{dz}{2\pi i} \; e^{(z-2\Gamma)t'} \frac{1}{N} \sum_\kappa e^{i\kappa(n-n')} \tag{8.40}$$
$$\left\{ 2N\tilde{\gamma}_\kappa - z - \frac{(z^2 + a_\kappa^2)^{1/2} - z}{1 + \frac{4\bar{\gamma}_1}{a_\kappa^2} \left((z^2 + a_\kappa^2)^{1/2} - z\right)} \right\} \rho_{n'}(t - t') \quad ,$$

where (2.34) and (8.7) with $\rho_{nn}(t) \equiv \rho_n(t)$ have been used. As described in Sect. 8.2.2, the path of integration C in the z plane runs parallel to the imaginary axis and is closed in the left half plane including all singularities of the integrand. On account of the analyticity with respect to z, the contributions of the first and second terms in the integrand disappear and we obtain

$$\frac{d}{dt}\rho_n(t) = -2\Gamma\rho_n(t) + \sum_{n'} 2\gamma_{n-n'}\rho_{n'}(t) + \sum_{n'}\int_0^t dt'\, e^{-2\Gamma t'} K_{n-n'}(t')\,\rho_{n'}(t-t') \quad . \tag{8.41}$$

For a linear chain of molecules of infinite length we have

$$K_{n-n'}(t) = \int_{-\pi}^{\pi}\frac{d\kappa}{2\pi}\, e^{i\kappa(n-n')}\,\tilde{K}_\kappa(t) \quad , \tag{8.42}$$

where

$$\tilde{K}_\kappa(t) = -\int_C \frac{dz}{2\pi i}\, e^{zt}\,\frac{a_\kappa^2}{(z^2+a_\kappa^2)^{1/2} + z + 4\bar{\gamma}_1} \quad . \tag{8.43}$$

The evaluation of the integrals has been carried out in some detail in /81,276/. With

$$K'_m(\tau) = K_m\left(\frac{\tau}{2H_1}\right) \quad , \qquad \tau = 2H_1 t \quad , \qquad \bar{\gamma} = \bar{\gamma}_1/H_1 \quad , \tag{8.44}$$

we finally obtain the previously reported result /80/

$$K'_m(\tau) = 4H_1^2\int_0^1 dx\; x[1-\bar{\gamma}\tau(1-x^2)]\; e^{-\bar{\gamma}\tau(1-x^2)}\left[J_{m+1}^2(\tau x) + J_{m-1}^2(\tau x) - 2J_m^2(\tau x)\right] . \tag{8.45}$$

In Fig. 8.1a we have pictured

$$\bar{K}_m(\tau) = K'_m(\tau)/4H_1^2 \quad , \qquad m = 1,\ldots,10 \tag{8.46}$$

as a function of τ for $\bar{\gamma} = \bar{\gamma}_1/H_1 = 0$, i.e. for the completely coherent case. For $\tau = 0$ only $\bar{K}_1(\tau)$, describing transitions between nearest neighbours, is different from zero. The other $\bar{K}_m$ s, describing transitions between neighbours a larger distance apart, are zero at the initial time. With increasing time all transition rates show oscillations with decreasing amplitude. This case of $\bar{\gamma}_1 = 0$ has also been discussed by KENKRE /75,76/, who especially stresses that long-distance generalized transfer rates evolve despite the fact that the Liouville equation contains only nearest-neighbour matrix elements. These non-local generalized transfer rates obviously originate from the elimination of the non-diagonal part of the density operator, that describes phase relations between different lattice sites. These phase relations are naturally most important in the case of the coherent, wave-like motion of the exciton. In Figs. 8.1b,c the same quantities are represented for $\bar{\gamma} = 0.5$ and

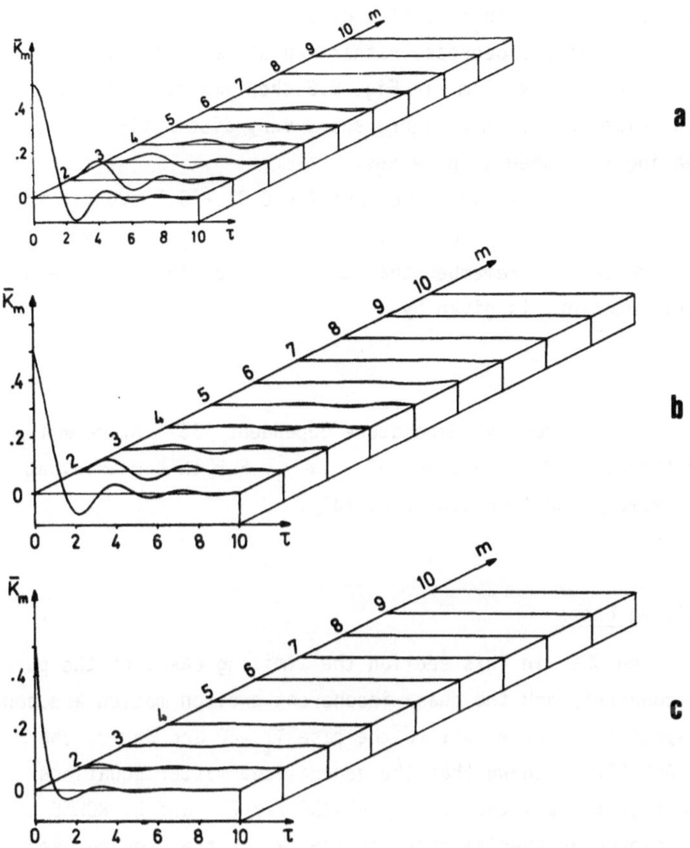

Fig. 8.1a-c. Generalized transition rates $\bar{K}_m(\tau)$ according to (8.46) for m = 1,...,10 and $\bar{\gamma}$ = 0.0 (a), $\bar{\gamma}$ = 0.5 (b), and $\bar{\gamma}$ = 2.0 (c) /81/

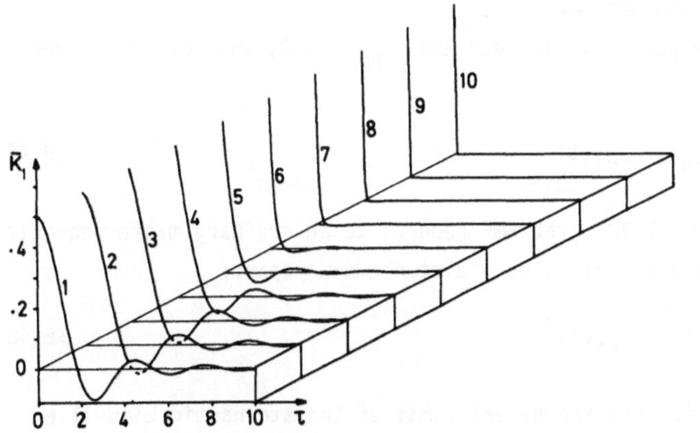

Fig. 8.2. Generalized transition rates $\bar{K}_1(\tau)$ according to (8.46). The numbering 1,...,10 of the curves corresponds to $\bar{\gamma}$ = 0 , 0.1 , 0.5 , 1 , 2 , 5 , 10 , 20 , 50 and 100 /81/

$\bar{\gamma} = 2.0$. The essential points are that the initial amplitude of $\bar{K}_1(\tau)$ now drops more rapidly, and that the oscillations decrease rather rapidly as well with increasing time as with increasing values of m. In Fig. 8.2 the time behaviour of $\bar{K}_1(\tau)$ is shown for various values of $\bar{\gamma}$. This figure shows very distinctly the rapid decay of $\bar{K}_1(\tau)$ with increasing time when $\bar{\gamma}$ increases.

In the case of single crystals of anthracene we have $\bar{\gamma} \approx 0.01 - 0.1$ even at room temperature /35/. Therefore the real situation is described approximately by Fig. 8.1a. However, it is very important to remember that on account of (8.1) the kernel of the GME, i.e. the memory function, is given by

$$e^{-2\Gamma t} K_m(t) \quad .$$

In this expression $\Gamma = \gamma_0 + 2\gamma_1$ is strongly temperature dependent /35/. At room temperature Γ is much larger than $\gamma_1 \approx \bar{\gamma}_1$ on account of $\gamma_0 \gg \gamma_1$. For this reason, at room temperature the total memory function decays rapidly.

8.4 Investigation of Limiting Cases

Starting from (8.41,45) for the GME, in this section the limiting cases of the purely incoherent, the purely coherent, and the quasi-incoherent exciton motion are considered. Furthermore, we specialize our result to the case $\bar{\gamma}_1 = 0$ and derive the Born approximation of the GME. It is shown that the generalized master equations obtained recently for a part of these situations by KENKRE /75-77/ and by KOHNE and REINEKER /78,79,93/ are recovered as special cases of the GME of the previous section.

8.4.1 Purely Incoherent Exciton Motion

In this case $H_{n-n'} = 0$. From (8.45) it is seen that $K'_m(\tau) = 0$, and (8.41) is then given by

$$\dot{\rho}_n(t) = -2\Gamma \rho_n(t) + \sum_{n'} 2\gamma_{n-n'} \rho_{n'}(t) \quad . \tag{8.47}$$

With the definition of Γ from (2.30), the GME reduces to an ordinary master equation with hopping rates $2\gamma_{n-n'}$ between molecules a distance n-n' apart:

$$\dot{\rho}_n(t) = \sum_{n'(\neq n)} 2\gamma_{n-n'} [\rho_{n'}(t) - \rho_n(t)] \quad . \tag{8.48}$$

This equation is identical with the incoherent limit of the stochastic Liouville equation of HAKEN, REINEKER, and STROBL /34-36,93/.

8.4.2 Purely Coherent Exciton Motion

In the limit of $\gamma_{n-n'} = \bar{\gamma}_{n-n'} = \Gamma = 0$ the kernel (8.45) of the GME becomes ($\tau x = x'$)

$$K'_m(\tau) = \frac{4H_1^2}{\tau^2}\int_0^\tau dx'\; x'[J_{m+1}^2(x') + J_{m-1}^2(x') - 2J_m^2(x')] \quad . \qquad (8.49)$$

Using a series representation for products of Bessel functions /242/, from (8.49) we arrive at

$$K'_m(\tau) = \frac{4H_1^2}{\tau}\frac{d}{d\tau}J_m^2(\tau) \quad . \qquad (8.50)$$

In complete coincidence with /78/ we obtain for the GME

$$\frac{d}{dt}\rho_n(t) = \sum_{n'}\int_0^t dt'\left(\frac{1}{t'}\frac{d}{dt'}J_{n-n'}^2(2H_1t')\right)\rho_{n'}(t-t') \quad . \qquad (8.51)$$

It may be shown that the memory function of KENKRE /76/ may also be written in this form.

8.4.3 Quasi-Incoherent Exciton Motion

Using (8.45), the last term of the GME (8.41) is written

$$\begin{aligned} A_{nn'}(t) &= \int_0^t dt'\; e^{-2\Gamma t'}\; K_{n-n'}(t')\;\rho_{n'}(t-t') \\ &= \int_0^t dt'\; e^{-2\Gamma t'}\; 4H_1^2\int_0^1 dx\; x[1-2\bar{\gamma}_1 t'(1-x^2)]\; e^{-2\bar{\gamma}_1 t'(1-x^2)} \qquad (8.52) \\ &\cdot [J_{n-n'+1}^2(2H_1t'x) + J_{n-n'-1}^2(2H_1t'x) - 2J_{n-n'}^2(2H_1t'x)]\;\rho_{n'}(t-t') \;. \end{aligned}$$

In the quasi-incoherent case $\Gamma \gg H_1$, and $\exp(-2\Gamma t')$ decays rapidly with time as compared to the changes of the Bessel functions. Therefore, the main contribution of the t' integral stems from $t' \approx 0$. For this reason we evaluate the Bessel functions and $\rho_{n'}$ at $t' = 0$ and let the upper limit of the t' integral approach infinity. The evaluation of the integral and of the Bessel function results in

$$A_{nn'}(t) = [\delta_{n-n'+1,0} + \delta_{n-n'-1,0} - 2\delta_{n-n',0}]\;\rho_{n'}(t)\;\frac{H_1^2}{\Gamma+\bar{\gamma}_1} \quad . \qquad (8.53)$$

Taking into account nearest-neighbour interaction also in the incoherent terms, the GME (8.41) becomes

$$\frac{d}{dt}\rho_n(t) = \left(2\gamma_1 + \frac{H_1^2}{\Gamma+\bar{\gamma}_1}\right)(\rho_{n+1} + \rho_{n-1} - 2\rho_n) \quad . \qquad (8.54)$$

The comparison with (8.48) shows that in this case the influence of the coherent interaction merely leads to a modification of the incoherent hopping rate /93/.

8.4.4 The Case $\bar{\gamma} = 0$

Using $\bar{\gamma} = 0$ in (8.45), we see that $K'_m(\tau)$ reduces to the expression (8.49) of the purely coherent exciton motion. (However, now $\Gamma \neq 0$.) Using (8.50,44), from (8.41) we obtain

$$\frac{d}{dt} \rho_n(t) = -2\Gamma\rho_n(t) + \sum_{n'} 2\gamma_{n-n'} \rho_{n'}(t) + \sum_{n'} \int_0^t dt' \, e^{-2\Gamma t'} \left(\frac{1}{t'} \frac{d}{dt'} J^2_{n-n'}(2H_1 t')\right) \rho_{n'}(t-t') \quad . \tag{8.55}$$

This result has been achieved recently by KENKRE /77/. As compared to the completely coherent case of Sect. 8.4.2, all memory functions are multiplied by the same damping factor $\exp(-2\Gamma t')$.

8.4.5 Derivation of the Generalized Master Equation in the Born Approximation

The Born approximation has been used by KENKRE and KNOX /64/ in order to determine the kernel of the generalized master equation from optical spectra, generalizing in this way the Markovian result of FÖRSTER /32/. SILBEY /57/ and KENKRE /75,76/ have commented recently on its inapplicability for the case of highly coherent motion. Therefore, it will be instructive to compare the exactly calculated kernel with the approximate one.

For the application of this approximation it is convenient to start from (8.5,6) /79/. In the following calculations we again use (8.4) and $P = P_K$ from (8.8). The first (non-retarded) term is given in (8.37). After some algebra we obtain

$$\frac{d}{dt} P_K \bar{\rho}(t) = 2N \tilde{\gamma}_K P_K \bar{\rho}(t) + \int_0^t dt' \, P_K L_1 \exp\left\{[(1-P_K) L_1 + L_2 + L_3] t'\right\} L_1 P_K \bar{\rho}(t-t') \quad . \tag{8.56}$$

Up to this point the calculations are exact. The Born approximation now consists in neglecting the term $(1-P_K)L_1$ in the exponent, stemming from the coherent interaction.

We now use the explicit form (8.8) of the projection operator and drop the factor n_K, common to all terms:

$$\frac{d}{dt} \tilde{\bar{\rho}}_K = 2N\tilde{\gamma}_K \tilde{\bar{\rho}}_K + N^{-1} \int_0^t dt' \, (n_K, L_1 \, e^{(L_2+L_3)t'} \, L_1 n_K) \, \tilde{\bar{\rho}}_K(t-t') \quad . \tag{8.57}$$

The scalar product in (8.57) may be written as follows:

$$\begin{aligned} S &= (n_\kappa\, , L_1\, e^{(L_2+L_3)t'}\, L_1 n_\kappa) \\ &= -\sum_k \sum_{k'} (\widetilde{H}_k - \widetilde{H}_{k-\kappa})\, (\widetilde{H}_{k'} - \widetilde{H}_{k'-\kappa})\, (n_{k'\kappa}\, , e^{L_2 t'}\, n_{k\kappa}) \quad . \end{aligned} \tag{8.58}$$

Here we have used $\sum_k (\widetilde{H}_k - \widetilde{H}_{k-\kappa}) = 0$.

Transforming to the site representation, with the help of (2.33,34) and (8.7) we have

$$\begin{aligned} \frac{d}{dt}\, \bar{\rho}_n(t) &= \sum_{n'} 2\gamma_{n-n'}\, \bar{\rho}_{n'}(t) \\ &+ 4 \sum_\kappa \sum_m \sum_{m'} \sum_{m''} N^{-2} \int_0^t dt'\, e^{i\kappa(n-m'')}\, H_m H_{m'}\, \sin\frac{\kappa}{2}\, m'\, \sin\frac{\kappa}{2}\, m \\ &\times \sum_k \sum_{k'} e^{i\kappa m/2}\, e^{i\kappa m'/2}\, e^{-ik'm'}\, e^{-ikm}\, (n_{k'\kappa}\, , e^{L_2 t}\, n_{k\kappa})\, \bar{\rho}_{m''}(t - t') \quad . \end{aligned} \tag{8.59}$$

We denote the prefactor of $\bar{\rho}_{m''}(t - t')$ in the last line by $A_{m',-m}$, and from the differential equation for this expression we get

$$A_{m',-m}(t) = \alpha\, e^{2\bar{\gamma}_m t} - \beta\, e^{-2\bar{\gamma}_m t} \tag{8.60}$$

with $\alpha = \frac{N}{2}\, (\delta_{m',m} + \delta_{m',-m})$ and $\beta = \frac{N}{2}\, (\delta_{m',m} - \delta_{m',-m})$.

Inserting now (8.60) into (8.59), evaluating the Kronecker symbols, and using once more the inversion symmetry, i.e. $H_m = H_{-m}$, with the help of (8.1) we finally obtain

$$\begin{aligned} \frac{d}{dt}\, \rho_n(t) &= \sum_m 2\gamma_m \big(\rho_{n+m}(t) - \rho_n(t)\big) \\ &+ \sum_m \int_0^t dt'\, 2H_m^2\, e^{-2(\Gamma+\bar{\gamma}_m)t'} \big(\rho_{n+m}(t - t') - \rho_n(t - t')\big) \quad . \end{aligned} \tag{8.61}$$

The comparison of this result with the exact evaluation in Sect. 8.3 and with the purely coherent motion discussed in Sect. 8.4.2 and in /75,78/ reveals differences in two points. Firstly, in the case of nearest-neighbour interaction, $H_m = H_1(\delta_{m,1} + \delta_{m,-1})$, the kernel of (8.61) gives interaction only between nearest neighbours, whereas in the two other cases mentioned above also an interaction between non-nearest neighbours exists. This point has also been stressed by KENKRE. Secondly, in the exact evaluation and in the limit of the purely coherent motion the kernels show an oscillatory behaviour in time, whereas the kernel in (8.61) decays exponentially. The Markovian limit, i.e. the quasi-incoherent motion, is obtained from (8.61) if the exponential decays so rapidly that the essential contribution to the integral in (8.61) is obtained for $t' \approx 0$. Then $\rho_{n+m}(t - t') - \rho_n(t - t') \approx$

$\rho_{n+m}(t) - \rho_n(t)$, and the upper limit in the integral may be replaced by infinity. In this way one arrives at the ordinary master equation (8.54).

Appendix A. Equation of Motion for the Density Operator

In this section we shall show that the stochastic density operator equation /41/, which in Sect. 2.3 has been derived in a perturbative manner, is exact if the stochastic variable represents a δ correlated Gaussian process. As an intermediate step in this derivation we shall arrive at an equation of motion /37,38/ that is valid also for random variables that are not δ correlated. In order to arrive at this equation, we use generalized cumulants /234,251/.

The Hamiltonian of the system is given by (2.5,10) and consists of a time-independent part H_0 and a stochastically time-dependent part $H_1(t)$

$$H = H_0 + H_1(t) \quad . \tag{A.1}$$

The time-dependent part is given by

$$H_1(t) = \sum_n \sum_{n'} h_{nn'}(t)\, b_n^+ b_{n'} = h_{nn'}(t)\, b_n^+ b_{n'} \quad , \tag{A.2}$$

where $h_{nn'}(t)$ is a c number and represents the stochastic process, the operator part of $H_1(t)$ is given by $b_n^+ b_{n'}$, and in the last equality Einstein's summation convention is used.

The equation of motion for the density operator is

$$\dot{W} = -i\,[H,W] \equiv -i\,H^\times W = -i\,L\,W \quad , \tag{A.3}$$

where the operators $H^\times = L$ are defined by this equation. In the same way as the Hamiltonian H, these operators split into a time-independent and a stochastically time-dependent part:

$$H^\times = H_0^\times + H_1^\times(t) \quad , \tag{A.4}$$

$$L = L_0 + L_1(t) \quad . \tag{A.5}$$

Introducing the density operator in the interaction picture $\tilde{W}(t)$ by

$$W(t) = e^{-iL_0 t}\, \tilde{W}(t) \quad , \tag{A.6}$$

the equation of motion for $\tilde{W}(t)$ becomes

$$\dot{\tilde{W}}(t) = -i\, \tilde{L}_1(t)\, \tilde{W}(t) \quad , \tag{A.7}$$

where

$$\tilde{L}_1(t) = e^{iL_0t}\, L_1(t)\, e^{-iL_0t} \quad . \qquad \text{(A.8)}$$

The solution of (A.7) is given by

$$\tilde{W}(t) = \overleftarrow{T} \exp\left[-i \int_0^t d\tau\, \tilde{L}_1(\tau)\, \rho(0)\right] \quad , \qquad \text{(A.9)}$$

where $\overleftarrow{T}$ is a time ordering operator which orders operators with increasing time from right to left.

We are not interested in the operator $W(t)$ still containing the fluctuations but in the quantity $\rho(t) = \langle W(t)\rangle$ averaged over the fluctuations:

$$\rho(t) = \langle W(t)\rangle = e^{-iL_0t} \langle\tilde{W}(t)\rangle = e^{-iL_0t}\, \tilde{\rho}(t) \quad . \qquad \text{(A.10)}$$

The angular brackets in (A.10) indicate the ensemble average, and with the help of generalized cumulants /234/ we arrive at

$$\tilde{\rho}(t) = \langle \overleftarrow{T} \exp\left(-i \int_0^t d\tau\, \tilde{L}_1(\tau)\right) \rho(0)\rangle = \overleftarrow{T}\, e^{K(t)}\, \rho(0) \quad . \qquad \text{(A.11)}$$

The time ordering operator $\overleftarrow{T}$ in the last expression is again a total time ordering operator, ordering *all* operators from right to left with increasing time. Assuming now that the stochastic process is Gaussian with disappearing mean value, the cumulant function $K(t)$ is given by

$$K(t) = -\frac{1}{2} \int_0^t d\tau_1 \int_0^t d\tau_2\, k_2(\tau_1,\tau_2) = -\int_0^t d\tau_1 \int_0^{\tau_1} d\tau_2\, k_2(\tau_1,\tau_2) \quad , \qquad \text{(A.12)}$$

and the second cumulant $k_2(\tau_1,\tau_2)$ is

$$k_2(\tau_1,\tau_2) = \langle\tilde{L}_1(\tau_1)\, \tilde{L}_1(\tau_2)\rangle \qquad \text{(A.13)}$$

$$= \langle h_{n_1n_1'}(\tau_1)\, h_{n_2n_2'}(\tau_2)\rangle\, \tilde{\ell}_{n_1n_1'}(\tau_1)\, \tilde{\ell}_{n_2n_2'}(\tau_2) \quad . \qquad \text{(A.14)}$$

Here again the summation convention has been used and the operators $\tilde{\ell}_{nn'}(\tau)$ are defined by

$$\tilde{\ell}_{nn'}(\tau) = \left(\tilde{b}_n^+(\tau)\, \tilde{b}_{n'}(\tau)\right)^x \quad . \qquad \text{(A.15)}$$

In order to arrive at an equation of motion for the averaged density operator $\rho(t)$, we differentiate (A.11) with respect to time and get

$$\dot{\tilde{\rho}}(t) = \overleftarrow{T} \left[\dot{K}(t)\, e^{K(t)}\right] \rho(0) \quad , \qquad \text{(A.16)}$$

and from (A.12) we have

$$\dot{K}(t) = - \int_0^t d\tau \; k_2(t,\tau) \quad . \tag{A.17}$$

With the help of (A.14) and splitting the double integral in K(t), the equation of motion (A.16) may be written in the following way:

$$\begin{aligned}\dot{\tilde{\rho}}(t) = -\overleftarrow{T} \int_0^t d\tau \, \langle h_{n_1n_1'}(t) \, h_{n_2n_2'}(\tau)\rangle \, \tilde{\ell}_{n_1n_1'}(t) \, \exp\Big[- \int_\tau^t d\tau_1 \int_\tau^{\tau_1} d\tau_2 \; k_2(\tau_1,\tau_2)\Big] \\ \cdot \tilde{\ell}_{n_2n_2'}(\tau) \, \exp\Big[- \int_\tau^t d\tau_1 \int_0^\tau d\tau_2 \, \langle h_{n_3n_3'}(\tau_1) \, h_{n_4n_4'}(\tau_2)\rangle \, \tilde{\ell}_{n_3n_3'}(\tau_1) \, \tilde{\ell}_{n_4n_4'}(\tau_2)\Big] \\ \cdot \exp\Big[- \int_0^\tau d\tau_1 \int_0^{\tau_1} d\tau_2 \; k_2(\tau_1,\tau_2)\Big] \, \rho(0) \quad .\end{aligned} \tag{A.18}$$

In this expression a partial time ordering is executed. The first exponential, taken alone, would describe the time development of a density operator from the initial time τ to the final time t. The last exponential, also taken alone, would represent the corresponding operator in the interval $0...\tau$. The second exponential, however, is more complicated. The time ordering requires that the exponential has to be represented by its power series and all operators operating in the time interval $\tau...t$ have to be placed on the left of $\tilde{\ell}_{n_2n_2'}(\tau)$. Furthermore these operators have to be ordered within the terms stemming from the power series expansion of the first exponential. In an analogous manner operators operating in the interval $0...\tau$ have to stand on the right of $\tilde{\ell}_{n_2n_2'}(\tau)$ and should be ordered within the terms stemming from the third exponential.

If this complicated time ordering is neglected and only the ordering within $\dot{K}(t)$ and within exp[K(t)] is taken into account, and if, furthermore, only local fluctuations are considered, we arrive at the density operator equation of BLUMEN and SILBEY /125/. As will be seen in the discussion of the equation of motion for the density operator in the limit of a δ function correlation of the stochastic process, with respect to the ordering of operators the procedure of BLUMEN and SILBEY is closely related to this δ correlation limit. Nevertheless, in this paper as well as that of SUMI /124/, important aspects of the influence of a non-δ correlated Gaussian stochastic process on the properties of excitons in molecular crystals are discussed.

In this limit, according to HAKEN, REINEKER, and STROBL /34-36/, the correlation function is written as

$$<h_{n_1n_1'}(t)\; h_{n_2n_2'}(\tau)> = 2\Lambda(n_1,n_1',n_2,n_2')\;\delta(t-\tau) \quad . \tag{A.19}$$

On account of the δ function we have $t = \tau$ and the integrals in the first and second exponents of (A.18) disappear. The exponents may be replaced by 1 and the τ integration may be executed. We thus arrive at

$$\dot{\tilde{\rho}}(t) = -\Lambda(n_1,n_1',n_2,n_2')\;\tilde{\ell}_{n_1n_1'}(t)\;\tilde{\ell}_{n_2n_2'}(t)\;\tilde{\rho}(t) \quad . \tag{A.20}$$

Of course, this result may also be obtained directly from (A.16) without the intermediate steps. Transforming back to the Schrödinger picture, we have

$$\dot{\rho}(t) = -i\,L_0\,\rho(t) - \Lambda(n_1,n_1',n_2,n_2')\;\ell_{n_1n_1'}\;\ell_{n_2n_2'}\;\rho(t) \quad . \tag{A.21}$$

This equation consists of a commutator describing the coherent part of the motion of the excitons and a double commutator stemming from the fluctuating part of the Hamiltonian. According to HAKEN, REINEKER and STROBL /34-36/, $\Lambda(n_1,n_1',n_2,n_2')$ is given by

$$\Lambda(n_1,n_1',n_2,n_2') = \gamma_{|n_1-n_1'|}\;\delta_{n_1n_2'}\;\delta_{n_1'n_2} + \bar{\gamma}_{n_1-n_1'}\;\delta_{n_1n_2}\;\delta_{n_1'n_2'}\;(1-\delta_{n_1n_1'}) \tag{A.22}$$

and (A.21) may be written in the following matrix notation, where n denotes the localization of the electron-hole pair:

$$\begin{aligned}\dot{\rho}_{nn'}(t) &= -i\,[H_0,\rho]_{nn'}\\ &\quad + \delta_{nn'}\sum_{n_1} 2\gamma_{|n_1-n|}(\rho_{n_1n_1}-\rho_{nn}) - (1-\delta_{nn'})(2\Gamma\rho_{nn'} - 2\bar{\gamma}_{n-n'}\rho_{n'n}) \quad .\end{aligned} \tag{A.23}$$

In this equation

$$\Gamma = \sum_{n_1}\gamma_{|n-n_1|} \quad . \tag{A.24}$$

The two-molecule model is obtained for $n,n',n_1 \in \{1,2\}$ and $\Gamma = \gamma_0 + \gamma_1$.

Appendix B. Green's Function of the Exciton Motion on a Linear Chain

The inhomogeneous difference equation for the Green's function of the exciton motion on a linear chain is given by (3.23):

$$G_{m+1}(z) + G_{m-1}(z) - 2\cos z\;G_m(z) = \frac{2}{ic_k}\,\delta_{m0} \quad . \tag{B.1}$$

With the ansatz

$$G_m(z) = a^m \tag{B.2}$$

from the homogeneous part of (B.1) we get

$$a = e^{\pm iz} \quad . \tag{B.3}$$

The Green's function is determined by a superposition of these solutions, and taking into account that $G_m(z)$ has to be invariant with respect to inversion, we get

$$G_m(z) = A\, e^{iz|m|} + B\, e^{-iz|m|} \quad . \tag{B.4}$$

The arbitrary constants A and B have to be determined from the inhomogeneous equation and the periodic boundary condition, i.e. from the equations for m = 0 and m = N/2:

$$G_1(z) + G_{-1}(z) - 2\cos z\, G_0(z) = \frac{2}{ic_k} \quad , \tag{B.5}$$

$$G_{N/2+1}(z) + G_{N/2-1}(z) - 2\cos z\, G_{N/2}(z) = 0 \quad . \tag{B.6}$$

With $G_{N/2+1}(z) = G_{-N/2+1}(z) = G_{N/2-1}(z)$ we get

$$\left.\begin{matrix} A \\ B \end{matrix}\right\} = (2ic_k \sin z \;\sin Nz/2)^{-1}\, e^{\mp izN/2} \quad . \tag{B.7}$$

Inserting A and B into (B.4), we get

$$G_m(z) = (ic_k \sin z \;\sin Nz/2)^{-1} \cos[(N/2 - |m|)z] \quad . \tag{B.8}$$

Appendix C. Angular Dependence of ESR Line Position and Width

By definition the fine-structure tensor $\underline{\underline{F}}_1$ of molecule 1 in the unit cell of anthracene crystals is diagonal in the principal axes system (ξ,η,ζ) of this molecule. (The same is naturally true for the second molecule of the unit cell in its principal axes system.) Thus the tensor has only diagonal elements in this coordinate system:

$$F_{\xi\xi} = A \quad , \qquad F_{\eta\eta} = B \quad , \qquad F_{\zeta\zeta} = C \quad . \tag{C.1}$$

It is well known /278/ that the trace of the fine-structure tensor disappears. Therefore, the three diagonal elements may be expressed by only two parameters D and E

$$A = E - \frac{1}{3}D \quad , \qquad B = -E - \frac{1}{3}D \quad , \qquad C = \frac{2}{3}D \quad . \tag{C.2}$$

With the knowledge of the direction cosines between the molecular principal axes system and the crystal axes system, the tensor components in the crystal axes system are given by

$$
\begin{aligned}
F_{11} &= \cos^2\chi A + \cos^2\chi' B + \cos^2\chi'' C \quad , \\
F_{12} &= \cos\chi \cos\psi A + \cos\chi' \cos\psi' B + \cos\chi'' \cos\psi'' C \quad , \\
F_{13} &= \cos\chi \cos\omega A + \cos\chi' \cos\omega' B + \cos\chi'' \cos\omega'' C \quad , \\
F_{22} &= \cos^2\psi A + \cos^2\psi' B + \cos^2\psi'' C \quad , \\
F_{23} &= \cos\psi \cos\omega A + \cos\psi' \cos\omega' B + \cos\psi'' \cos\omega'' C \quad , \\
F_{33} &= \cos^2\omega A + \cos^2\omega' B + \cos^2\omega'' C \quad .
\end{aligned}
\qquad (C.3)
$$

C.1 Rotation Around the a Axis

The components M_{ij} and D_{ij} of the fine-structure tensors in the system (x,y,z) are expressed in the following way by the components in the crystal axes system (C.3) of the tensor $\underline{\underline{F}}_1$:

$$
\begin{aligned}
M_{xx} &= F_{11} \quad , & D_{xx} &= 0 \quad , \\
M_{xy} &= -\cos\alpha\, F_{13} \quad , & D_{xy} &= \sin\alpha\, F_{12} \quad , \\
M_{xz} &= \sin\alpha\, F_{13} \quad , & D_{xz} &= \cos\alpha\, F_{12} \quad , \\
M_{yy} &= \sin^2\alpha\, F_{22} + \cos^2\alpha\, F_{33} \quad , & D_{yy} &= -2\sin\alpha \cos\alpha\, F_{23} \quad , \\
M_{yz} &= \sin\alpha \cos\alpha\, (F_{22} - F_{33}) \quad , & D_{yz} &= (\sin^2\alpha - \cos^2\alpha)\, F_{23} \quad , \\
M_{zz} &= \cos^2\alpha\, F_{22} + \sin^2\alpha\, F_{33} \quad , & D_{zz} &= 2\cos\alpha \sin\alpha\, F_{23} \quad ;
\end{aligned}
\qquad (C.4)
$$

for $\alpha = 0$ $H_z \parallel +b$ and for $\alpha = \pi/2$ $H_z \parallel +c'$.

The coefficients in (6.99-102) are given by

$$
\begin{aligned}
a_0^{\mp}(x,s) &= x \mp \frac{3}{4} F_{11} + \frac{s}{8x} \left(2F_{11}^2 + \frac{7}{8} F_{22}^2 + \frac{7}{8} F_{33}^2 - \frac{3}{4} F_{22}F_{33} + 4F_{13}^2\right) \\
&\quad + \frac{s}{2} \frac{x}{16\Gamma_1^2 + 4x^2} \left(2F_{12}^2 + \frac{1}{2} F_{23}^2\right) + \frac{s}{4} \frac{x}{16\Gamma_1^2 + x^2} \left(F_{23}^2 + F_{12}^2\right) \quad , \\
a_2^{\mp}(x,s) &= \mp \frac{3}{4} (F_{33} - F_{22}) + \frac{3s}{16x} F_{11}(F_{22} - F_{33}) \\
&\quad - s\left(\frac{x}{16\Gamma_1^2 + 4x^2} - \frac{1}{4} \frac{x}{16\Gamma_1^2 + x^2}\right) F_{12}^2 \quad ,
\end{aligned}
\qquad (C.5)
$$

$$a_4(x,s) = -\frac{3s}{64x} - \frac{s}{4}\left(\frac{x}{16\Gamma_1^2 + 4x^2} - \frac{x}{16\Gamma_1^2 + x^2}\right)F_{23}^2 ,$$

$$c_0(x) = \frac{9}{32}\frac{F_{23}^2}{\Gamma_1} + \frac{\Gamma_1}{16\Gamma_1^2 + 4x^2}\left(2F_{12}^2 + \frac{1}{2}F_{23}^2\right) + 3\frac{\Gamma_1}{16\Gamma_1^2 + x^2}\left(F_{23}^2 + F_{12}^2\right) ,$$

$$c_2(x) = \left(-2\frac{\Gamma_1}{16\Gamma_1^2 + 4x^2} + 3\frac{\Gamma_1}{16\Gamma_1^2 + x^2}\right)F_{12}^2 , \tag{C.5}$$

$$c_4(x) = \left(-\frac{9}{32\Gamma_1} - \frac{1}{2}\frac{\Gamma_1}{16\Gamma_1^2 + 4x^2} + 3\frac{\Gamma_1}{16\Gamma_1^2 + x^2}\right)F_{23}^2 .$$

C.2 Rotation Around the b Axis

In this case M_{ij} and D_{ij} are given by

$$\begin{aligned}
M_{xx} &= F_{22} , & D_{xx} &= 0 ,\\
M_{xy} &= 0 , & D_{xy} &= -\sin\alpha\, F_{12} + \cos\alpha\, F_{23} ,\\
M_{xz} &= 0 , & D_{xz} &= -\cos\alpha\, F_{12} - \sin\alpha\, F_{23} ,\\
M_{yy} &= \sin^2\alpha\, F_{11} - 2\sin\alpha\cos\alpha\, F_{13} + \cos^2\alpha\, F_{33} , & D_{yy} &= 0 ,\\
M_{yz} &= \sin\alpha\cos\alpha\,(F_{11} - F_{33}) - (\cos^2\alpha - \sin^2\alpha)\, F_{13} , & D_{yz} &= 0 ,\\
M_{zz} &= \cos^2\alpha\, F_{11} + 2\sin\alpha\cos\alpha\, F_{13} + \sin^2\alpha\, F_{33} , & D_{zz} &= 0 ;
\end{aligned} \tag{C.6}$$

for $\alpha = 0$ $H_z \parallel +a$ and for $\alpha = \pi/2$ $H_z \parallel +c'$.

The coefficients in (6.103-106) are calculated as

$$a_0^{\mp}(x,s) = x \mp \frac{3}{4}F_{22} + \frac{s}{8x}\left(2F_{22}^2 + \frac{7}{8}F_{11}^2 + \frac{7}{8}F_{33}^2 + \frac{5}{2}F_{13}^2 - \frac{3}{4}F_{11}F_{33}\right)$$
$$+ s\left(\frac{x}{16\Gamma_1^2 + 4x^2} + \frac{1}{4}\frac{x}{16\Gamma_1^2 + x^2}\right)\left(F_{12}^2 + F_{23}^2\right) ,$$

$$a_2^{\mp}(x,s) = \mp\frac{3}{4}(F_{33} - F_{11}) + \frac{3s}{16x}F_{22}(F_{11} - F_{33})$$
$$- s\left(\frac{x}{16\Gamma_1^2 + 4x^2} - \frac{1}{4}\frac{x}{16\Gamma_1^2 + x^2}\right)\left(F_{12}^2 - F_{23}^2\right) , \tag{C.7}$$

$$b_2^{\mp}(x,s) = \pm\frac{3}{2}F_{13} + \frac{3s}{8x}F_{22}F_{13} - s\left(\frac{x}{16\Gamma_1^2 + 4x^2} - \frac{1}{4}\frac{x}{16\Gamma_1^2 + x^2}\right)2F_{12}F_{23} ,$$

$$a_4(x,s) = \frac{s}{8x}\left(-\frac{3}{8}F_{11}^2 + \frac{3}{2}F_{13}^2 - \frac{3}{8}F_{33}^2 + \frac{3}{4}F_{11}F_{33}\right) ,$$

$$b_4(x,s) = -\frac{3s}{16x} F_{13}(F_{11} - F_{33}) \quad ,$$

$$c_0(x) = \left(2\frac{\Gamma_1}{16\Gamma_1^2 + 4x^2} + 3\frac{\Gamma_1}{16\Gamma_1^2 + x^2}\right)(F_{12}^2 + F_{23}^2) \quad ,$$

$$c_2(x) = \left(-2\frac{\Gamma_1}{16\Gamma_1^2 + 4x^2} + 3\frac{\Gamma_1}{16\Gamma_1^2 + x^2}\right)(F_{12}^2 - F_{23}^2) \quad , \qquad (C.7)$$

$$d_2(x) = \left(-2\frac{\Gamma_1}{16\Gamma_1^2 + 4x^2} + 3\frac{\Gamma_1}{16\Gamma_1^2 + x^2}\right) 2F_{12}F_{23} \quad .$$

C.3 Rotation Around the c' Axis

M_{ij} and D_{ij} are determined as

$$\begin{aligned}
M_{xx} &= F_{33} \ , & D_{xx} &= 0 \ , \\
M_{xy} &= \sin\alpha\, F_{13} \ , & D_{xy} &= -\cos\alpha\, F_{23} \ , \\
M_{xz} &= \cos\alpha\, F_{13} \ , & D_{xz} &= \sin\alpha\, F_{23} \ , \\
M_{yy} &= \sin^2\alpha\, F_{11} + \cos^2\alpha\, F_{22} \ , & D_{yy} &= -2\sin\alpha\cos\alpha\, F_{12} \ , \\
M_{yz} &= \sin\alpha\cos\alpha\,(F_{11} - F_{22}) \ , & D_{yz} &= -(\cos^2\alpha - \sin^2\alpha)\, F_{12} \ , \\
M_{zz} &= \cos^2\alpha\, F_{11} + \sin^2\alpha\, F_{22} \ , & D_{zz} &= 2\sin\alpha\cos\alpha\, F_{12} \ ;
\end{aligned} \qquad (C.8)$$

for $\alpha = 0$ $H_z \parallel a$ and for $\alpha = \pi/2$ $H_z \parallel b$.

The coefficients in (6.99-102) are given by

$$a_0^{\mp}(x,s) = x \mp \frac{3}{4} F_{33} + \frac{s}{8x}\left(2F_{33}^2 + \frac{7}{8}F_{11}^2 + \frac{7}{8}F_{22}^2 - \frac{3}{4}F_{11}F_{22} + 4F_{13}^2\right)$$
$$+ \frac{s}{2}\frac{x}{16\Gamma_1^2 + 4x^2}\left(2F_{23}^2 + \frac{1}{2}F_{12}^2\right) + \frac{s}{4}\frac{x}{16\Gamma_1^2 + x^2}(F_{12}^2 + F_{23}^2) \quad ,$$

$$a_2^{\mp}(x,s) = \mp\frac{3}{4}(F_{22} - F_{11}) + \frac{3s}{16x}F_{33}(F_{11} - F_{22}) \qquad (C.9)$$
$$+ s\left(\frac{x}{16\Gamma_1^2 + 4x^2} - \frac{1}{4}\frac{x}{16\Gamma_1^2 + x^2}\right)F_{23}^2 \quad ,$$

$$a_4(x,s) = -\frac{3s}{64x}(F_{11} - F_{22})^2 - \frac{s}{4}\left(\frac{x}{16\Gamma_1^2 + 4x^2} - \frac{x}{16\Gamma_1^2 + x^2}\right)F_{12}^2 \quad ,$$

$$c_0(x) = \frac{9}{32}\frac{F_{12}^2}{\Gamma_1} + \frac{\Gamma_1}{16\Gamma_1^2 + 4x^2}\left(2F_{23}^2 + \frac{1}{2}F_{12}^2\right) + 3\frac{\Gamma_1}{16\Gamma_1^2 + x^2}\left(F_{12}^2 + F_{23}^2\right) \quad ,$$

$$c_2(x) = \left(2\frac{\Gamma_1}{16\Gamma_1^2 + 4x^2} - 3\frac{\Gamma_1}{16\Gamma_1 + x^2}\right)F_{23}^2 \quad , \tag{C.9}$$

$$c_4(x) = \left(-\frac{9}{32\Gamma_1} - \frac{1}{2}\frac{\Gamma_1}{16\Gamma_1^2 + 4x^2} + 3\frac{\Gamma_1}{16\Gamma_1^2 + x^2}\right)F_{12}^2 \quad .$$

Acknowledgments. The author is greatly indebted to Prof. H. Haken, who stimulated this review article and in whose institute part of the work presented here has been carried out. It is a pleasure to acknowledge the interest of Prof. H. Risken. The author likes to thank K. Kassner, A. Scheuing, B. Schmid, and U. Schmid for the help in proofreading the manuscript and for several suggestions for its improvement. This article was completed during the author's sabbatical leave at the IBM Research Laboratory in San Jose, U.S.A. Several discussions with Dr. H. Morawitz are gratefully acknowledged. Last but not least the author is deeply grateful to Mrs. I. Gruhler and especially to Mrs. H. Wenning, who did the typewriting and performed the final version of the manuscript with infinite skill.

References

1 J. Frenkel: Phys. Rev. *37*, 17 (1931); *37*, 1276 (1931)
2 H. Haken: Fortschr. Phys. *6*, 271 (1958)
3 R.S. Knox: "Theory of Excitons", in *Solid State Physics Supplement*, Vol.5, ed. by H. Ehrenreich, F. Seitz, D. Turnbull, 2nd print. (Academic Press, New York 197
4 H. Haken, S. Nikitine (eds.): *Springer Tracts in Modern Physics*, Vol. 73 Excitons at High Density (Springer, Berlin, Heidelberg, New York 1975)
5 E. Hanamura, H. Haug: Phys. Rep. *33C*, 209 (1977)
6 A.S. Davydov: *Theory of Molecular Excitons* (McGraw-Hill, New York 1962)
7 A.S. Davydov: *Theory of Molecular Excitons* (Plenum Press, New York, London 1971)
8 D.P. Craig, S.H. Walmsley: *Excitons in Molecular Crystals* (W.A. Benjamin, New York 1968)
9 E.A. Silinsh: *Organic Molecular Crystals, Their Electronic States*, Springer Series in Solid State Sciences, Vol.16 (Springer, Berlin, Heidelberg, New York 1980)
10 Z.G. Soos: Ann. Rev. Phys. Chem. *25*, 121 (1974)
11 H.C. Wolf: In *Solid State Physics*, Vol.9, ed. by H. Ehrenreich, F. Seitz, D. Turnbull (Academic Press, New York 1959) pp. 1-81
12 H.C. Wolf: Adv. At. Mol. Phys. *3*, 119 (1967)
13 P. Avakian, R.E. Merrifield: Mol. Cryst. *5*, 37 (1968)
14 D.M. Hanson: CRC Crit. Rev. Solid State Sci. *3*, 243 (1973)
15 C.B. Harris, D.A. Zwemer: Ann. Rev. Phys. Chem. *29*, 473 (1978)
16 R. Schmidberger, H.C. Wolf: Chem. Phys. Lett. *32*, 21 (1975)
17 H. Möhwald, E. Sackmann: Chem. Phys. Lett. *21*, 43 (1973)
18 H. Möhwald, A. Böhm: Chem. Phys. Lett. *43*, 49 (1976)
19 H. Möhwald, E. Erdle, A. Thaer: Chem. Phys. *27*, 79 (1978)

20 W. Steudle, J.U. von Schütz, H. Möhwald: Chem. Phys. Lett. *54*, 461 (1978); J.U. von Schütz, W. Steudle, H.C. Wolf, P. Reineker, U. Schmid: Chem. Phys. Lett. *79*, 1 (1981)
21 D. Haarer, N. Karl: Chem. Phys. Lett. *21*, 49 (1973)
22 D. Haarer: Chem. Phys. Lett. *27*, 91 (1974)
23 D. Haarer: J. Chem. Phys. *67*, 4076 (1977)
24 H. Möhwald: Habilitationsschrift, Universität Ulm (1978)
25 D. Haarer: "Optical and Photoelectric Properties of Organic Charge Transfer Crystals", in *Festkörperprobleme*, Vol. XX, ed. by J. Treusch (Vieweg, Braunschweig 1980) p. 341
26 W. Goldacker, K.H. Hausser, D. Schweitzer, H.A. Staab: J. Lumin.*18/19*, 415 (1979)
27 D. Schweitzer: Habilitationsschrift, Universität Heidelberg (1980)
28 N. Karl: "Organic Semiconductors", in *Festkörperprobleme*, Vol. XIV, ed. by J. Treusch (Vieweg, Braunschweig 1974) p. 261
29 J.S. Miller, A.J. Epstein (eds.): "Synthesis and Properties of Low-Dimensional Materials", Ann. N.Y. Acad. Sciences *313* (1978)
30 T.D. Schultz: "Open Questions in Physics of Quasi One-Dimensional Metals", in *Festkörperprobleme*, Vol. XX, ed. by J. Treusch (Vieweg, Braunschweig 1980) p. 463
31 R.E. Merrifield: J. Chem. Phys. *28*, 647 (1958)
32 T. Förster: Ann. Phys. (Leipzig) *2*, 55 (1948)
33 M. Trlifaj: Czech. J. Phys. *8*, 510 (1958)
34 H. Haken, G. Strobl: "Exact Treatment of Coherent and Incoherent Triplet Exciton Migration", in *The Triplet State*, ed. by A. Zahlan (Cambridge Univ. Press, London 1967) pp. 311-314
35 H. Haken, F. Reineker: Z. Phys. *249*, 253 (1972)
36 H. Haken, G. Strobl: Z. Phys. *262*, 135 (1973)
37 P. Reineker: Dissertation, Universität Stuttgart (1971)
38 P. Reineker: Phys. Rev. B*19*, 1999 (1979)
39 C. Aslangul, Ph. Kottis: Phys. Rev. B*10*, 4364 (1974)
40 C. Aslangul, Ph. Kottis: Adv. Chem. Phys. *41*, 321 (1980)
41 R. Kubo: J. Math. Phys. *4*, 174 (1963)
42 H. Haken, P. Reineker: Acta Univ. Carol.-Math. Phys. *14*, 23 (1973)
43 P. Reineker, H. Haken: "The Coupled Coherent and Incoherent Motion of Frenkel Excitons in Molecular Crystals", in *Localization and Delocalization in Quantum Chemistry*, Vol. II, ed. by O. Chalvet, R. Daudel, R. Diner, P. Malrieu (Reidel, Dortrecht, Boston 1976) pp. 285-300
44 H. Haken, P. Reineker: "Comments on the Interaction of Excitons and Phonons", in *Excitons, Magnons and Phonons in Molecular Crystals*, ed. by A. Zahlan (Cambridge Univ. Press, London 1968) pp. 185-194
45 W. Weidlich, F. Haake: Z. Phys. *185*, 30 (1965); *186*, 203 (1965); F. Haake: "Statistical Treatment of Open Systems by Generalized Master Equations", in *Springer Tracts in Modern Physics*, Vol. 66 (Springer, Berlin, Heidelberg, New York 1973) pp. 98-168
46 A. Scheuing: Diplomarbeit, Universität Ulm (1979); P. Reineker, A. Scheuing: To be published
47 M. Grover, R. Silbey: J. Chem. Phys. *52*, 2099 (1970)
48 M. Grover, R. Silbey: J. Chem. Phys. *54*, 4843 (1971)
49 T. Holstein: Ann. Phys. (N.Y.) *8*, 325 (1959)
50 D. Emin: Ann. Phys. (N.Y.) *64*, 336 (1971)
51 S. Rackovsky, R. Silbey: Mol. Phys. *25*, 61 (1972)
52 I.I. Abram, R. Silbey: J. Chem. Phys. *63*, 2317 (1975)
53 D.R. Yarkony, R. Silbey: J. Chem. Phys. *67*, 5818 (1978)
54 J.W. Allen, R. Silbey: Chem. Phys. *43*, 341 (1979)
55 R.W. Munn, W. Siebrand: J. Chem. Phys. *52*, 47 (1970)
56 R.W. Munn, R. Silbey: J. Chem. Phys. *68*, 2439 (1978)
57 R. Silbey: Ann. Rev. Phys. Chem. *27*, 203 (1976)
58 R. Silbey, R.W. Munn: J. Chem. Phys. *72*, 2763 (1980)
59 R.W. Munn, R. Silbey: Mol. Cryst. Liq. Cryst. *57*, 131 (1980)
60 D.L. Dexter: J. Chem. Phys. *21*, 836 (1953)
61 L. van Hove: Physica *21*, 517 (1955); *23*, 441 (1957)

62 L. Prigogine, P. Resibois: Physica *27*, 629 (1961)
63 V.M. Kenkre: Phys. Lett. *47A*, 119 (1974)
64 V.M. Kenkre, R.S. Knox: Phys. Rev. B *9*, 5279 (1974); Phys. Rev. Lett. *33*, 803 (1974)
65 S. Nakajima: Prog. Theor. Phys. *20*, 948 (1958)
66 R. Zwanzig: J. Chem. Phys. *33*, 1338 (1960)
67 V.M. Kenkre, R.S. Knox: J. Lumin. *12/13*, 187 (1976)
68 V.M. Kenkre, T.S. Rahman: Phys. Lett. *50A*, 170 (1974)
69 D.L. Smith: Phys. Lett. *53A*, 271 (1975)
70 F.F. Sokolov, V.V. Hizhnyakov: Phys. Status Solidi (b) *75*, 669 (1976)
71 F.F. Sokolov: Phys. Status Solidi (b) *76*, K131 (1976)
72 V. Câpek, I. Rips: Phys. Status Solidi (b) *97*, K93 (1980)
73 V.M. Kenkre: Phys. Rev. B *11*, 1741 (1975)
74 V.M. Kenkre: Phys. Rev. B *12*, 2150 (1975)
75 V.M. Kenkre: Phys. Lett. *63A*, 367 (1977)
76 V.M. Kenkre: Phys. Rev. B *18*, 4064 (1978)
77 V.M. Kenkre: Phys. Lett. *65A*, 391 (1978)
78 R. Kühne, P. Reineker: Phys. Status Solidi (b) *89*, 131 (1978)
79 P. Reineker, R. Kühne: Phys. Status Solidi (b) *92*, 123 (1979)
80 R. Kühne, P. Reineker: Solid State Commun. *29*, 279 (1979)
81 P. Reineker, R. Kühne: Phys. Rev. B *21*, 2448 (1980)
82 D.M. Burland, A.H. Zewail: Adv. Chem. Phys. *40*, 369 (1979)
83 P. Avakian, R.E. Merrifield: Phys. Rev. Lett. *13*, 541 (1964)
84 V. Ern, P. Avakian, R.E. Merrifield: Phys. Rev. *148*, 862 (1966)
85 M. Levine, J. Jortner, A. Szöke: J. Chem. Phys. *45*, 1591 (1966)
86 D.F. Williams, J. Adolph: J. Chem. Phys. *46*, 4252 (1967)
87 G. Durocher, D.F. Williams: J. Chem. Phys. *51*, 1675 (1969)
88 V. Ern: Phys. Rev. Lett. *22*, 343 (1969)
89 V. Ern, A. Suna, Y. Tomkiewicz, P. Avakian, R.P. Groff: Phys. Rev. B *5*, 3222 (1972)
90 P. Avakian, V. Ern, R.E. Merrifield, A. Suna: Phys. Rev. *165*, 974 (1968)
91 E. Schwarzer, H. Haken: Phys. Lett. *42A*, 317 (1972)
92 P. Reineker: Phys. Lett. *42A*, 389 (1973); Z. Physik *261*, 187 (1973)
93 P. Reineker, R. Kühne: Z. Phys. B *22*, 193 (1975)
94 R. Kühne, P. Reineker: Z. Phys. B *22*, 201 (1975)
95 M.A. Davidovich, R.S. Knox: Chem. Phys. Lett. *68*, 391 (1979)
96 C. Aslangul, Ph. Kottis: Phys. Rev. B *18*, 4462 (1978)
97 R.C. Powell, Z.G. Soos: J. Lumin. *11*, 1 (1975)
98 D.M. Hanson: J. Chem. Phys. *52*, 3409 (1970)
99 H. Port, D. Vogel, H.C. Wolf: Chem. Phys. Lett. *34*, 23 (1975)
100 R.H. Clarke, R.M. Hochstrasser: J. Chem. Phys. *49*, 3313 (1968)
101 G. Castro, G.W. Robinson: J. Chem. Phys. *50*, 1159 (1969)
102 H. Port, H.C. Wolf: Z. Naturforsch. *30a*, 1290 (1975)
103 S.L. Robinette, G.J. Small, S.H. Stevenson: J. Chem. Phys. *68*, 4790 (1978)
104 R.H. Clarke, R.M. Hochstrasser: J. Chem. Phys. *46*, 4532 (1967)
105 H. Port, D. Rund: Chem. Phys. Lett. *54*, 474 (1978)
106 H. Port, D. Rund, G.J. Small, V. Yakhot: Chem. Phys. *39*, 175 (1979)
107 H. Port, D. Rund: Chem. Phys. Lett. *69*, 406 (1980)
108 H. Port, D. Rund: J. Mol. Struct. *45*, 455 (1978)
109 G.C. Morris, M.G. Sceats: Chem. Phys. *3*, 332 (1974)
110 G.C. Morris, M.G. Sceats: Chem. Phys. *3*, 342 (1974)
111 M.R. Philpott, J.M. Turlet: J. Chem. Phys. *64*, 3852 (1976)
112 K. Syassen, M.R. Philpott: J. Chem. Phys. *68*, 4870 (1978)
113 H. Port, K. Mistelberger: J. Lumin. *12/13*, 351 (1976)
114 K. Mistelberger, H. Port: Mol. Cryst. Liq. Cryst. *57*, 203 (1980)
115 H. Port, K. Mistelberger, D. Rund: Mol. Cryst. Liq. Cryst. *50*, 11 (1979)
116 D.M. Burland, D.E. Cooper, M.D. Fayer, C.R. Gochanour: Chem. Phys. *52*, 279 (1977)
117 R.M. Hochstrasser, J.D. Whiteman: J. Chem. Phys. *56*, 5949 (1972)
118 D.M. Burland, U. Konzelmann, R.M. Macfarlane: J. Chem. Phys. *67*, 1926 (1977)
119 R.M. Hochstrasser, L.W. Johnson, C.M. Klimcak: J. Chem. Phys. *73*, 156 (1980)
120 C. Aslangul, Ph. Kottis: Phys. Rev. B *13*, 5544 (1976)

121 E. Schwarzer, H. Haken: Opt. Commun. *9*, 64 (1973)
122 P. Reineker: Z. Naturforsch. *29a*, 282 (1974)
123 R. Kubo: "A Stochastic Theory of Line Shape"; in *Advances in Chemical Physics* XV, ed. by K.E. Shuler (John Wiley & Sons, New York 1969) pp. 23–68
124 H. Sumi: J. Chem. Phys. *67*, 2943 (1977)
125 A. Blumen, R. Silbey: J. Chem. Phys. *69*, 3589 (1978)
126 I.B. Rips, V. Cápek: Phys. Status Solidi (b) *100*, 451 (1980)
127 L.A. Dissado: Chem. Phys. Lett. *33*, 57 (1975)
128 L.A. Dissado: Chem. Phys. *8*, 289 (1975)
129 D.P. Craig, L.A. Dissado: Chem. Phys. *14*, 89 (1976)
130 T.S. Rahman, R.S. Knox, V.M. Kenkre: Chem Phys. *44*, 197 (1979)
131 R.M. Hochstrasser, A.H. Zewail: Chem. Phys. *4*, 142 (1974)
132 A.H. Zewail, C.B. Harris: Chem. Phys. Lett. *28*, 8 (1974)
133 A.H. Zewail, C.B. Harris: Phys. Rev. B *11*, 935 (1975); 952 (1975)
134 B.J. Botter, A.J. van Strien, J. Schmidt: Chem. Phys. Lett. *49*, 39 (1978)
135 M. Schwoerer, H.C. Wolf: Proc. Colloq. AMPERE *14*, 544 (1966)
136 M. Schwoerer, H.C. Wolf: Mol. Cryst. *3*, 177 (1967)
137 C.A. Hutchison, B.W. Mangum: J. Chem. Phys. *34*, 908 (1961)
138 M. Schwoerer: Dissertation, Universität Stuttgart (1967)
139 C.L. Braun, H.C. Wolf: Chem. Phys. Lett. *9*, 260 (1971)
140 Ph. Kottis: Chem. Phys. Lett. *6*, 133 (1970)
141 D.M. Hanson: Chem. Phys. Lett. *11*, 175 (1971)
142 B.J. Botter, C.J. Nonhof, J. Schmidt, J.H. van der Waals: Chem. Phys. Lett. *43*, 210 (1976)
143 C.A. van't Hof, J. Schmidt: Chem. Phys. Lett. *36*, 460 (1975)
144 H. Hinkel: Dissertation, Universität Stuttgart (1977)
145 H. Hinkel, H. Port, H. Sixl, M. Schwoerer, P. Reineker, D. Richart: Chem. Phys. *31*, 101 (1978)
146 P. Reineker, H. Sixl: Solid State Commun. *27*, 389 (1978)
147 R. Schmidberger, H.C. Wolf: Chem. Phys. Lett. *16*, 402 (1972)
148 R. Schmidberger, H.C. Wolf: Chem. Phys. Lett. *25*, 185 (1974)
149 R. Schmidberger, H.C. Wolf: Chem. Phys. Lett. *32*, 18 (1975)
150 A.H. Francis, C.B. Harris: Chem. Phys. Lett. *9*, 181 (1971); 188 (1971)
151 M.D. Fayer, C.B. Harris: Phys. Rev. B *9*, 748 (1974)
152 C.B. Harris, M.D. Fayer: Phys. Rev. B *10*, 1784 (1974)
153 S.J. Sheng, D.M. Hanson: Chem. Phys. Lett. *33*, 451 (1975)
154 J. Zieger, H.C. Wolf: Chem. Phys. *29*, 209 (1978)
155 R.D. Wieting, M. D. Fayer: J. Chem. Phys. *73*, 744 (1980)
156 W.G. Breiland, M.C. Saylor: J. Chem. Phys. *72*, 6485 (1980)
157 (a) B.J. Botter, A.I.M. Dicker, J. Schmidt: Mol. Phys. *36*, 129 (1978)
(b) A.J. van Strien, J.F.C. van Kooten, J. Schmidt: Chem. Phys. Lett. *76*, 7 (1980)
158 A.H. Francis, C.B. Harris: J. Chem. Phys. *57*, 1050 (1972)
159 M.D. Fayer, C.B. Harris: Chem. Phys. Lett. *25*, 149 (1974)
160 W. Güttler, J.U. von Schütz, H.C. Wolf: Chem. Phys. *24*, 159 (1977)
161 D.A. Zweemer, C.B. Harris: Chem. Phys. *38*, 139 (1979)
162 D. Haarer, H.C. Wolf: Phys. Status Solidi (b) *33*, K117 (1969)
163 D. Haarer: Dissertation, Universität Stuttgart (1969)
164 D. Haarer, H.C. Wolf: Mol. Cryst. Liq. Cryst. *10*, 359 (1970)
165 W. Bizzaro, J. Rosenthal, N.F. Berk, L. Yarmus: Phys. Status Solidi (b) *84*, 27 (1977)
166 L. Yarmus, J. Rosenthal, M. Chopp: Chem. Phys. Lett. *16*, 477 (1972)
167 W. Bizzaro, L. Yarmus, J. Rosenthal, N.F. Berk: Phys. Rev. B*23*, 5673 (1981)
168 W. Bizzaro, L. Yarmus, J. Rosenthal, N.F. Berk: Chem. Phys. Lett. *53*, 49 (1978)
169 N.F. Berk, W. Bizzaro, J. Rosenthal, L. Yarmus: Phys. Rev. B*23*, 5661 (1981)
170 E.B. Iturbe, M. Sharnoff: Mol. Cryst. Liq. Cryst. *57*, 227 (1980)
171 P.W. Anderson: J. Phys. Soc. Jap. *9*, 316 (1954)
172 R. Kubo, K. Tomita: J. Phys. Soc. Jap. *9*, 888 (1954)
173 R. Kubo: "A Stochastic Theory of Line Shape and Relaxation" in *Fluctuation, Relaxation and Resonance in Magnetic Systems*, ed. by D. ter Haar (Oliver & Boyd, Edinburgh 1962) p. 23
174 A. Hudson, A.D. McLachlan: J. Chem. Phys. *43*, 1518 (1965)

175 M. Blume: Phys. Rev. *174*, 351 (1968)
176 L.T. Muus, P.W. Atkins: *Electron Spin Relaxation in Liquids* (Plenum, New York, London 1972)
177 H. Sternlicht, M. McConnel: J. Chem. Phys. *35*, 1793 (1961)
178 G.O. Berim, A.R. Kessel: Phys. Status Solidi (b) *76*, 827 (1976); *79*, 489 (1977)
179 J.P. Lemaistre, Ph. Kottis: J. Chem. Phys. *68*, 2730 (1977)
180 P. Reineker: Phys. Status Solidi (b) *52*, 439 (1972)
181 P. Reineker, H. Haken: Z. Physik *250*, 300 (1972)
182 P. Reineker: Solid State Commun. *14*, 153 (1974)
183 P. Reineker: Phys. Status Solidi (b) *70*, 189 (1975)
184 P. Reineker: Phys. Status Solidi (b) *70*, 471 (1975)
185 P. Reineker: Chem. Phys. *16*, 425 (1976)
186 P. Reineker: Phys. Status Solidi (b) *74*, 121 (1976)
187 P. Reineker: Solid State Commun. *25*, 859 (1978)
188 G. Maier, H.C. Wolf: Z. Naturforsch. *23a*, 1068 (1968)
189 H. Kolb, H.C. Wolf: Z. Naturforsch. *27a*, 51 (1972)
190 H. Haken, E. Schwarzer: Chem. Phys. Lett. *27*, 41 (1974)
191 E. Schwarzer, H. Haken: Phys. Status Solidi (b) *84*, 253 (1977)
192 A. Hammer, H.C. Wolf: Mol. Cryst. *4*, 191 (1968)
193 H. Auweter, U. Mayer, D. Schmid: Z. Naturforsch. *33a*, 651 (1978)
194 H. Auweter, A. Braun, U. Mayer, D. Schmid: Z. Naturforsch. *34a*, 761 (1979)
195 D. Schmid: Abstracts of the 9th Molecular Crystal Symposium, Mittelberg, Kleinwalsertal (1980) pp. 234-239
196 R.P. Hemenger, K. Lakatos-Lindenberg, R.M. Pearlstein: J. Chem. Phys. *60*, 3271 (1974)
197 T. Markvart: Phys. Status Solidi (b) *73*, 689 (1976); *74*, 135 (1976)
198 V.M. Kenkre: Chem. Phys. *36*, 377 (1979)
199 Y.M. Wong, V.M. Kenkre: Phys. Rev. B *20*, 2438 (1979)
200 R.C. Johnson, R.E. Merrifield, P. Avakian, R.B. Flippen: Phys. Rev. Lett. *19*, 285 (1967)
201 R.C. Johnson, R.E. Merrifield: Phys. Rev. B *1*, 896 (1970)
202 L. Altwegg, M. Chabr, I. Zschokke-Gränacher: Phys. Rev. B *14*, 1963 (1979)
203 S. Fujiwara, T. Nakayama, N. Itoh: Phys. Status Solidi (b) *78*, 519 (1976)
204 J. Fünfschilling, K. von Burg, I. Zschokke-Gränacher: Chem. Phys. Lett. *55*, 344 (1978)
205 L. Altwegg, M.A. Davidovich, J. Fünfschilling, I. Zschokke-Gränacher: Phys. Rev. B *18*, 4444 (1978)
206 L. Altwegg: Chem. Phys. Lett. *63*, 97 (1979)
207 L. Altwegg, I. Zschokke-Gränacher: Phys. Rev. B *20*, 4326 (1979)
208 W. Chabr, D.F. Williams: Phys. Rev. B *19*, 5206 (1979)
209 V.M. Kenkre: Phys. Rev. B *22*, 2089 (1980)
210 R.E. Merrifield: J. Chem. Phys. *48*, 4318 (1968)
211 A. Suna: Phys. Rev. B *1*, 1716 (1970)
212 C.E. Swenberg, N.F. Geacintov: "Exciton Interaction in Organic Solids", in *Organic Molecular Photophysics*, ed. by J.B. Birks (Wiley, New York 1973) p.489
213 R. Kopelman: J. Chem. Phys. *80*, 2191 (1976)
214 J. Hoshen, R. Kopelman: Phys. Status Solidi (b) *81*, 479 (1977)
215 J. Klafter, J. Jortner: Solid State Commun. *28*, 663 (1978)
216 K. Godzik, J. Jortner: J. Chem. Phys. *72*, 4471 (1980)
217 A. Blumen, R. Silbey: J. Chem. Phys. *70*, 3707 (1979)
218 J. Klafter, R. Silbey: Phys. Lett. *76*A, 143 (1980)
219 J. Klafter, R. Silbey: J. Chem. Phys. *72*, 843 (1980); 72, 849 (1980)
220 H. Metiu, K. Kitahara, R. Silbey, J. Ross: Chem. Phys. Lett. *43*, 189 (1976)
221 A. Madhukar, W. Post: Phys. Rev. Lett. *39*, 1424 (1977); *40*, 70E (1978)
222 V.M. Kenkre, R. Kühne, P. Reineker: Z. Phys. B *41*, 177 (1981)
223 R. Kühne, P. Reineker, V.M. Kenkre: Z. Phys. B *41*, 181 (1981)
224 P. Reineker, V.M. Kenkre, R. Kühne: Phys. Lett. *84*A, 294 (1981)
225 L.B. Schein, A.R. McGhie: Phys. Rev. B *20*, 1631 (1979)
226 N. Karl: Abstracts of the 9th Molecular Crystal Symposium, Mittelberg, Kleinwalsertal (1980) pp. 149-152

227 H. Haken: *Quantenfeldtheorie des Festkörpers* (B.G. Teubner, Stuttgart 1973)
228 R.F. Fox: J. Math. Phys. *13*, 1196 (1972); Phys. Rep. *48*, 179 (1978)
229 A. Schenzle, H. Brand: Phys. Rev. B *20*, 1628 (1979)
230 G.L. Sewell: Phys. Rev. *129*, 597 (1963)
231 R.L. Stratonovich: *Topics in the Theory of Random Noise*, Vol. 1 (Gordon and Breach, New York 1963)
232 W. Feller: *An Introduction to Probability Theory and its Applications* (John Wiley, New York 1968)
233 K. Kitahara, J.W. Haus: Z. Phys. B *32*, 419 (1979)
234 R. Kubo: J. Phys. Soc. Jap. *17*, 1100 (1962)
235 N.G. van Kampen: Phys. Rep. C *24*, 171 (1976)
236 W.H. Louisell: *Quantum Statistical Properties of Radiation* (John Wiley & Sons, New York 1973) pp. 1-76
237 G.S. Agarwal: "Master Equation Methods in Quantum Optics", in *Progress in Optics XI*, ed. by E. Wolf (North Holland, Amsterdam 1973)
238 G. Strobl: Diplomarbeit, Universität Stuttgart (1966)
239 P. Reineker: Phys. Lett. *44A*, 429 (1973)
240 E. Schwarzer: Dissertation, Universität Stuttgart (1974)
241 E. Schwarzer: Z. Phys. B *20*, 185 (1975)
242 M. Abramowitz, I. Stegun: *Handbook of Mathematical Functions* (Dover Publications, New York 1965)
243 M. Lax: Phys. Rev. *129*, 2342 (1963); *145*, 110 (1966); *157*, 213 (1967); *172*, 350 (1968)
244 H. Haken, W. Weidlich: Z. Phys. *205*, 96 (1967)
245 R. Kubo: J. Phys. Soc. Jap. *12*, 570 (1957)
246 P. Reineker, R. Kühne: Phys. Lett. *70A*, 187 (1979)
247 Y. Toyozawa: "Exciton Lattice Interaction - Fluctuation, Relaxation and Defects Formation", in *Proc. 4th Int. Conf.* on Vacuum Ultraviolet Radiation Physics, ed. by E. Koch, R. Haensel, C. Kunz (Pergamon, New York 1974) pp. 317-330; J. Lumin. *12/13*, 13 (1976)
248 Y. Toyozawa: "Electrons, Holes and Excitons in Deformable Lattice", in *Relaxation of Elementary Excitations*, ed. by R. Kubo, E. Hanamura, Springer Series in Solid State Sciences 18 (Springer, Berlin, Heidelberg, New York 1980) pp. 3-18
249 J.H. Freed: J. Chem. Phys. *49*, 376 (1968)
250 J.H. Freed: in *ESR Relaxation in Liquids*, ed. by L.T. Muus, P.W. Atkins (Plenum, New York 1972) pp. 165-212
251 B. Yoon, J.M. Deutch, J. Freed: J. Chem. Phys. *62*, 4687 (1975)
252 K. Kassner, P. Reineker: To be published
253 W.G. Breiland, H.C. Brenner, C.B. Harris: J. Chem. Phys. *62*, 3458 (1975)
254 N.F. Berk, L. Yarmus, W. Bizzaro, J. Rosenthal: Phys. Status Solidi (b) *89*, K167 (1978)
255 J. Rosenthal, L. Yarmus, N.F. Berk, W. Bizzaro: Chem. Phys. Lett. *56*, 214 (1978)
256 G.A. Baker, Jr.: *Essentials of Padé Approximants* (Academic Press, New York 1975)
257 D.W.J. Cruickshank: Acta Crystallogr. *10*, 504 (1957)
258 P. Reineker, D. Richardt, U. Schmid: To be published
259 A. Abragam: *The principles of Nuclear Magnetism* (Clarendon, Oxford 1970)
260 J. Grad, M.A. Brebner: in *Collected Algorithms from CACM*, Algorithm 343 (Service of the ACM, New York)
261 J.H. Wilkinson: *The Algebraic Eigenvalue Problem* (Clarendon, Oxford 1965)
262 R. Wertheimer, R. Silbey: Chem. Phys. Lett. *75*, 243 (1980); J. Chem. Phys. *74*, 686 (1981)
263 M. Schwoerer: Habilitationsschrift, Universität Stuttgart (1972)
264 R.A. Sack: Mol. Phys. *1*, 163 (1958)
265 P. Reineker: Z. Phys. B *21*, 409 (1975)
266 H. Haken: Rev. Mod. Phys. *47*, 67 (1975); *Synergetics, An Introduction*, 2nd ed., Springer Series in Synergetics, Vol. 1 (Springer, Berlin, Heidelberg, New York 1978)
267 D. Kilpper: Dissertation, Universität Stuttgart (1974)
268 U. Konzelmann, D. Kilpper, M. Schwoerer: Z. Naturforsch. *30a*, 754 (1975); F. Dietz, U. Konzelmann, H. Port, M. Schwoerer: Chem. Phys. Lett. *58*, 565 (1978)
269 D. Haarer: private communication

270 C.B. Harris: Pure Appl. Chem. *37*, 73 (1974)
271 A.H. Zewail: J. Chem. Phys. *70*, 5759 (1979)
272 G. Maier, U. Haeberlen, H.C. Wolf, K.H. Hausser: Phys. Lett. *25A*, 384 (1967)
273 H. Schuch, D. Stehlik, K.H. Hausser: Z. Naturforsch. *26a*, 1944 (1971)
274 I.S. Gradshteyn, I.M. Ryzhik: *Table of Integrals, Series and Products* (Academic Press, New York 1980)
275 V.M. Kenkre: J. Stat. Phys. *19*, 333 (1978)
276 P. Reineker, R. Kühne: "Dynamics of the Coupled Coherent and Incoherent Exciton Motion: Exact Derivation and Solution of the Nakajima-Zwanzig generalized Master Equation and Discussion of Approximate Treatments for the Full Haken-Strobl Model", PABS PRBMDO-72-2448-76, (American Institute of Physics, New York, 1979)
277 R. Kühne, P. Reineker: Z. Phys. B *31*, 105 (1978)
278 A. Carrington, A.D. McLachlan: *Introduction to Magnetic Resonance* (Harper & Row, New York; Evanston, London; John Weatherhill, Tokyo 1967)